Rhizosphere Biology

Series Editor

Anil K. Sharma
Biological Sciences, CBSH, G.B.Pant University of Agriculture & Technology, Pantnagar, Uttarakhand, India

The Series Rhizosphere Biology, emphasizes on the different aspects of Rhizosphere. Major increase in agricultural productivity, to meet growing food demands of human population is imperative, to survive in the future. Along with methods of crop improvement, an understanding of the rhizosphere biology, and the ways to manipulate it, could be an innovative strategy to deal with this demand of increasing productivity. This Series would provide comprehensive information for researchers, and encompass all aspects in field of rhizosphere biology. It would comprise of topics ranging from the classical studies to the most advanced application being done in the field. Rhizoshpere is a dynamic environment, and a series of processes take place to create a congenial environment for plant to grow and survive. There are factors which might hamper the growth of plants, resulting in productivity loss, but, the mechanisms are not very clear. Understanding the rhizosphere is needed, in order to create opportunities for researchers to come up with robust strategies to exploit the rhizosphere for sustainable agriculture.

There are titles already available in the market in the broad area of rhizosphere biology, but there is a major lack of information as to the functions and future applications of this field. These titles have not given all the up-to-date information required by the today's researchers and therefore, this Series aims to fill out those gaps.

More information about this series at http://www.springer.com/series/15861

Anil K. Sharma • Pratibha Sharma
Editors

Trichoderma

Host Pathogen Interactions and Applications

Springer

Editors
Anil K. Sharma
Biological Sciences, CBSH
G.B.Pant University of Agriculture & Technology
Pantnagar, Uttarakhand, India

Pratibha Sharma
Division of Plant Pathology
ICAR-Indian Agricultural Research Institute
New Delhi, India

ISSN 2523-8442 ISSN 2523-8450 (electronic)
Rhizosphere Biology
ISBN 978-981-15-3320-4 ISBN 978-981-15-3321-1 (eBook)
https://doi.org/10.1007/978-981-15-3321-1

This Springer imprint is published by the registered company Springer Nature Singapore Pte Ltd.
The registered company address is: 152 Beach Road, #21-01/04 Gateway East, Singapore 189721, Singapore

Dedicated to Parents, Children and God

Foreword

Trichoderma spp. (perfect stage: *Hypocrea*) are a widely studied filamentous ascomycetous fungus used as a biocontrol agent against major phytopathogens. They comprise different economically important species, *viz. T. harzianum, T. asperellum, T. viride, T. atroviride*, *T. virens*, and *T. reesei.* They are an omnipresent mycoflora and major colonizers of numerous soils in all the ecosystems and different agroclimatic zones. These species are known for their ability to colonize cellulosic materials and, in the rhizosphere of plants, can induce systemic resistance against plant pathogens. These fungi are characterized by hasty growth, intense green conidia, and repeatedly branched conidiophores. By virtue of their high antagonistic and mycoparasitic potential, some *Trichoderma* strains encompass the ability to decrease the severity of plant diseases by inhibiting plant pathogens, mainly in the soil or on plant roots. They do control a wide range of plant pathogens, including fungi, bacteria, oomycetes, and even viral diseases, through elicitation of induced systemic resistance or localized resistance. They have antagonistic effects on soil-borne fungal plant pathogens belonging to different groups (ascomycetes, deuteromycetes, and basidiomycetes) and also foliar fungal plant pathogens. They are well known as a proficient biocontrol agent because of their high reproductive capability, ability to stay alive under very adverse conditions, effectiveness in nutrient utilization, ability to alter the rhizosphere, sturdy aggressiveness against plant pathogens, and efficiency in promoting plant growth and defense mechanisms. *Trichoderma* establishes well in the rhizospheric region and assists the progress of the positive interaction with plants because of the availability of fungal prey and root-derived nutrients. Also some rhizosphere-competent *Trichoderma* strains have been shown to have direct effects on plants, increasing their growth potential and nutrient uptake, fertilizer use efficiency, percentage and rate of seed germination, and stimulation of plant defenses against biotic and abiotic damages.

The present volume on *Trichoderma: Host Pathogen Interactions and Applications* provides detailed information on rhizospheric interaction of *Trichoderma* spp. in response to host and pathogen systems. This volume consists of 15 chapters. Chapter 1 by Lavy and Horwitz describes the experimental framework to approach the genetic basis of root colonization, and a more systematic genetic analysis might finally identify central regulators of colonization in the fungal genome. Chapter 2 by Monti et al. provides the vocabulary of *Trichoderma*-plant interactions. The chapter provides the known "*words*" that *Trichoderma* spp. use to chat with plants. Chapter 3 by Poveda et al. presents the process of root colonization by *Trichoderma*, the indirect benefits obtained by plants through their symbiotic relationship with *Trichoderma*, and the thin line that separates mutualism from parasitism in this interaction. Chapter 4 by Nakkeeran et al. is about harnessing the perception of *Trichoderma* signal molecules in rhizosphere to improve soil health and plant health. This chapter deals with molecular communication that occurs between *Trichoderma* and the rhizosphere to improve plant health and soil health. Chapter 5 by Gupta and Maya Bar elaborates plant immunity, priming, and systemic resistance as mechanisms for *Trichoderma* spp. biocontrol and also provides an overview of the current knowledge of plant immunity and mechanisms underlying induced systemic resistance triggered by *Trichoderma* spp. Chapter 6 by Chen et al. presents the compilation of the induced resistance of *Trichoderma* species on maize plants against different plant pathogens and the mechanisms. Induced immunity developed by *Trichoderma* species in plants has been described by Chakraborty et al. in Chap. 7. Chapter 8 by Swain and Mukherjee presents the effect of host-pathogen-*Trichoderma* interaction during abiotic and biotic stresses. Trichoderma-based biological control strategies of Fusarium wilt and its interactions during pathogen interactions were highlighted by Sharma and Sharma in Chap. 9. Chapter 10 by Navi and Yang explains the use of *Trichoderma* in the management of diseases in North American row crops. In this chapter, they added the *Trichoderma*-based products that are labeled for major row crops like corn, soybean, cotton, and wheat. Chapter 11 is about the potential of *Trichoderma* spp. for pest management and plant growth promotion in North East India. This chapter by Bora and his team summarizes the comprehensive information on research findings on *Trichoderma* including plant disease and insect pest management, plant growth promotion, *Trichoderma*-mediated host defense response, and role in organic agriculture in NE India. Chapter 12 by Dutta and his colleagues provides detailed information on the potential use of *Trichoderma* spp. for the management of tea diseases, mode of action of *Trichoderma* against the pathogens causing tea diseases, and the role of different biotic and abiotic factors on *Trichoderma* spp. Anandaraj and Umadevi discuss in Chap. 13 the multipartite interactions of *Trichoderma harzianum* (MTCC 5179), beneficial interactions, increase in rhizospheric efficiency, and gene editing techniques. In Chap. 14, Sharma and her team present the outcomes of *Trichoderma*-based consortia which have shown effective results in protecting the crops from various diseases, and also an explanation on the evaluation or assessment of microbial consortium on host and plant pathogens through mathematical models has been added in this chapter. Chapter 15 authored by William Rivera-Méndez details the different aspects related to research with native isolates with the aim of improving the inoculation techniques of the fungus and its establishment in the rhizosphere of onion, garlic, and sweet pepper crops, under tropical environmental conditions in Central America.

Overall, great efforts have been carried out by Dr. Anil Sharma and Prof. Pratibha Sharma and scientists from different countries to compile this book as a highly unique, up-to-date source on *Trichoderma: Host Pathogen Interactions and Applications* for students, researchers, scientists, and academics. I hope that readers will find this book highly useful and interesting during their pursuit on detailed aspects of *Trichoderma*.

GBPAUT, Pantnagar, India Tej Partap

Preface

Presently, agriculture is passing through a transitional phase of change over from chemicals to non-chemical practices by switching over to microbial technologies in the form of biopesticides and biofertilizers. There is a growing concern for organic food and agriculture amongst the agriculture clientele. With the climate change, there are large problems related to agriculture mainly drought, salinity, decreased soil fertility, and pests. At this juncture, we really need an eco-friendly biocontrol agent that is a multitasker in terms of resolving the above-mentioned problems. Trichoderma is one such genus that is ubiquitously present in the environment which is a very important component and resident of the rhizosphere. Trichoderma spp. are common soil and root inhabitants, which have been widely studied due to their multiple mechanisms including antagonism, plant growth promotion, production of different kinds of enzymes, and induction of defense responses in plants which have a significant role in biotic and abiotic stress tolerance. Furthermore, Trichoderma, in association with plant roots, can trigger systemic resistance and improve plant nutrient uptake.

Trichoderma-based products have been particularly noted as successful biological control agents for contrasting plant pathogens, mainly soil-borne pathogens, as well as inducing resistance to biotic stresses. However, in addition to their success as a biocontrol agent, some Trichoderma strains have been proven to have a biostimulant activity, plant growth promotion, improved yield, and nutritional quality, as well as mitigating the detrimental effect of abiotic stresses. Trichoderma have gained importance as microbial plant biostimulants also. Plant biostimulants, which include organic and inorganic natural substances, are used to enhance nutrient uptake, and crop production could be considered as a sustainable and environmentally friendly approach to secure yield stability under low-input conditions. There are reports on the mechanism of stimulation by Trichoderma which involves a multilevel root–shoot communication.

Therefore, it is not surprising that Trichoderma now is known as a successful beneficial microbial biological agent, which is used in agriculture as biopesticide, biofertilizer, bio-growth enhancer, and biostimulant marketed worldwide for conventional and organic agricultural production.

The success of *Trichoderma* in the rhizosphere is due to saprophytic nature, endophytism, rhizosphere competency, efficiency in the utilization of nutrients, capacity to modify the rhizosphere, and strong aggressiveness against plant pathogenic fungi responses of *Trichoderma* to fluctuations in environmental conditions. The literature compiled on these versatile topics will reveal innovative ways plants adopt in response to fungal communities to improve survival in highly dynamic environments and agroecosystem. Therefore, rhizosphere biology of Trichoderma is a subject of great importance to students and teachers. Lots of efforts are being made to explore the rhizosphere to understand the tripartite interactions between plant pathogens, host, and biocontrol agents. However, there is no compilation of such work in the form of book, which will be useful for researchers working in the area of microbiology, soil sciences, plant pathology, botany, and also biotechnology.

This book will be the compilation of work done in the area of Trichoderma Rhizosphere Biology. It will cover microbial interaction, cross talk between plants and microbes, interactions with abiotic and biotic factors, advances in biocontrol agents, biofertilizers, and biostimulants. This volume will cover all the aspects of *Trichoderma* dealing mainly with microbes, plant microbes, and environment interactions.

Pantnagar, India Anil K. Sharma
New Delhi, India Pratibha Sharma

Contents

Editors and Contributors

About the Editors

Anil K. Sharma is a Professor in the Department of Biological Sciences at CBSH G.B. Pant University of Agriculture and Technology, Pantnagar. After completing his postdoctoral studies at GSU, Louisiana, USA, he served as a Visiting Scientist at the University of Basel, Switzerland, and at the University of Helsinki, Lahti, Finland. Prof. Sharma has extensive research and teaching experience, holds two patents (in plant biology and microbiology, respectively), and has received a number of prestigious grants. He has published more than 60 research articles and 30 review articles.

Pratibha Sharma is an ICAR Emeritus Scientist and former Principal Scientist and Professor in the Division of Plant Pathology at ICAR Indian Agricultural Research Institute, New Delhi. Her major area of research concerns biological control agents, especially *Trichoderma*, *Metarhizium*, *Beauveria,* and *Pseudomonas* and their use in IPM and IDM for various crop systems. Prof. Sharma has more than 100 peer-reviewed research papers in national and international journals, as well as a patent on *Trichoderma*, to her credit. Having received several awards from professional societies, Prof. Sharma is currently working on the application of *Trichoderma harzianum* and *Metarhizium anisopliae* to combat major pests impacting groundnut and cumin.

Contributors

Patricia Abril-Urias Institute of Environmental Sciences of Castilla-La Mancha (ICAM), University of Castilla-La Mancha, Toledo, Spain

Ariella Alperovitch-Lavy Faculty of Biology, Technion – Israel Institute of Technology, Haifa, Israel

M. Anandaraj ICAR - Indian Institute of Spices Research, Kozhikode, Kerala, India

Maya Bar Department of Plant Pathology and Weed Research, ARO, The Volcani Center, Rishon LeZion, Israel

R. P. Bhuyan Department of Tea Husbandry and Technology, Assam Agricultural University, Jorhat, Assam, India

L. C. Bora Department of Plant Pathology, Assam Agricultural University, Jorhat, Assam, India

Popy Bora Department of Plant Pathology, Assam Agricultural University, Jorhat, Assam, India

B. N. Chakraborty Department of Biological Sciences, Aliah University, Kolkata, India

U. Chakraborty Department of Botany, University of North Bengal, Siliguri, India

Jie Chen School of Agriculture and Biology, Shanghai Jiao Tong University, Shanghai, P. R. China

Pranab Dutta Department of Plant Pathology, Assam Agricultural University, Jorhat, Assam, India

Daniel Eugui Blue Agro Bioscience, Noain, Navarra, Spain

Monoj Gogoi Department of Plant Pathology, Assam Agricultural University, Jorhat, Assam, India

L. Gualtieri Institute for Sustainable Plant Protection (IPSP-CNR), Portici, NA, Italy

Rupali Gupta Department of Plant Pathology and Weed Research, ARO, The Volcani Center, Rishon LeZion, Israel

Benjamin A. Horwitz Faculty of Biology, Technion – Israel Institute of Technology, Haifa, Israel

Prashant Prakash Jambhulkar Department of Plant Pathology, Rani Lakshmi Bai Central Agricultural University, Jhansi, Uttar Pradesh, India

Shaily Javeria Division of Seed Science and Technology, ICAR- Indian Agricultural Research Institute, New Delhi, India

Valliappan Karuppiah School of Agriculture and Biology, Shanghai Jiao Tong University, Shanghai, P. R. China

M. M. Monti Institute for Sustainable Plant Protection (IPSP-CNR), Portici, NA, Italy

Arup K. Mukherjee Molecular Plant Pathology Laboratory, Division of Crop Protection, ICAR-National Rice Research Institute, Cuttack, Odisha, India

Malaiyandi Muthamilan Department of Plant Pathology, Tamil Nadu Agricultural University, Coimbatore, India

Sevugapperumal Nakkeeran Department of Plant Pathology, Tamil Nadu Agricultural University, Coimbatore, India

Shrishail S. Navi Department of Plant Pathology & Microbiology, Iowa State University, Ames, IA, USA

P. A. Pedata Institute for Sustainable Plant Protection (IPSP-CNR), Portici, Italy

Jorge Poveda Spanish-Portuguese Institute for Agricultural Research (CIALE), University of Salamanca, Salamanca, Spain

M. Raja Department of Plant Pathology, SKN Agricultural University, Jaipur, Rajasthan, India

Suppaiah Rajamanickam Department of Plant Pathology, Tamil Nadu Agricultural University, Coimbatore, India

Perumal Renukadevi Department of Plant Pathology, Tamil Nadu Agricultural University, Coimbatore, India

William Rivera-Méndez InstitutoTecnológico de Costa Rica, Centro de Investigación en Biotecnología, Cartago, Costa Rica

M. Ruocco Institute for Sustainable Plant Protection (IPSP-CNR), Portici, Italy

Satish Kumar Sain ICAR-Central Institute for Cotton - Research Regional Station, Sirsa, Haryana, India

Anil K. Sharma Department of Biological Sciences, College of Basic Sciences and Humanities, GBPUAT, Pantnagar, India

Ishwar Prakash Sharma Department of Biological Sciences, College of Basic Sciences and Humanities, GBPUAT, Pantnagar, India

Pratibha Sharma Department of Plant Pathology, ICAR-Indian Agricultural Research Institute, New Delhi, India

K. Sunar Department of Botany, Balurghat Mahila Mahavidyalaya, Balurghat, India

Harekrushna Swain Molecular Plant Pathology Laboratory, Division of Crop Protection, ICAR-National Rice Research Institute, Cuttack, Odisha, India

P. Umadevi ICAR – Indian Institute of Spices Research, Kozhikode, Kerala, India

Murugappan Vallikkannu School of Agriculture and Biology, Shanghai Jiao Tong University, Shanghai, P. R. China

Murugavel Vanthana Department of Plant Pathology, Tamil Nadu Agricultural University, Coimbatore, India

X. B. Yang Department of Plant Pathology & Microbiology, Iowa State University, Ames, IA, USA

Abbreviations

6PP	6-n-Pentyl-6H-pyran-2-one
ABA	Abscisic acid
ACC	1-Aminocyclopropane-1-carboxylic acid
Aib	α-Aminoisobutyric acid
ALA	Alamethicin
AMF	Arbuscular mycorrhiza fungus
APEDA	Agricultural and Processed Food Products Export Development Authority
ATCC	American Type Culture Collection
BCAs	Biological control agents
BLB	Bacterial leaf blight
BR	Brassinolide
cAMP	Cyclic adenosine monophosphate
CAT	Catalase
CF	Culture filtrate
CFU	Colony forming unit
CHS	Chalcone synthase
CIB	Central Insecticides Board
CMV	Cucumber mosaic virus
CP	Cerato-platanin
CPP	Cerato-platanin protein
CWDEs	Cell wall degrading enzymes
D3G	Deoxynivalenol-3-glucoside
DAMPs	Damage-associated molecular patterns
DAT	Days After Treatment
EIX	Ethylene biosynthesis-inducing xylanase
ESTs	Expressed sequence tags
ET	Ethylene
ETI	Effector-triggered immunity
FYM	Farm yard manure

GC-MS	Gas chromatography-mass spectrometry
GFP	Green fluorescent protein
GPCR	G protein-coupled receptor
GUS	β-Glucuronidase
HA	Harzianic acid
HA	Harzianum A
HFB7	Class II hydrophobin
HR	Hypersensitive response
HYTLO1	7.2 kDa class II hydrophobin
IAA	Indole-3-acetic acid
IAAld	Indole-3-acetaldehyde
IDM	Integrated Disease Management
IPM	Integrated Pest Management
Iso-HA	Isoharzianic acid
ISR	Induced systemic resistance
ITCC	Indian Type Culture Collection
JA	Jasmonic acid
LAMP	Loop-mediated isothermal amplification
LRR	Leu-rich repeat
MAMP	Microbe-associated molecular patterns
MAP Kinases	Mitogen-activated protein kinases
MBCAs	Microbial biological control agents
MP	Microbial phytohormones
MTCC	Microbial Type Culture Collection
MTI	MAMP-triggered immunity
MVOCs	Microbial volatile organic compounds
NBS	Nucleotide-binding site
NGS	Next-generation sequencing
NRPSs	Nonribosomal peptide synthestases
OHC	Oxygen heterocyclic compounds
PAL	Phenylalanine ammonia lyase
PAMPs	Pathogen-associated molecular patterns
PER	Peroxidase
PGP	Plant growth promotion
PGPM	Plant growth promoting microorganism
PGPR	Plant growth promoting rhizobacteria
PI	Propidium iodide
PKS	Polyketide synthase
PPTases	4-Phosphopantetheinyl transferases
PR Protein	Pathogenesis-related protein
PRRs	Pattern recognition receptors
PSB	Phosphorus solubilizing bacteria
PTI	PAMP-triggered immunity
QTL	Quantitative trait loci

REMI	Restriction enzyme-mediated integration
ROS	Reactive oxygen species
RSI	Reduced systemic immunity
SA	Salicylic acid
SAR	Systemic acquired resistance
SEM	Scanning electron microscope
SF	Synergy factor
SM	Secondary metabolites
SOD	Superoxide dismutase
SSCP	Small secreted cysteine-rich proteins
SSPs	Small secreted proteins
Tas-ACDS	1-Aminocyclopropane-1-carboxylate deaminase
TASHYD1	*T. asperellum* class I hydrophobin
TD	Trichodiene
TEM	Transmission electron microscope
T-ISR	Trichoderma-induced systemic resistance
TK VI	Trichokonin VI
T-RFLPs	Terminal restriction fragment length polymorphisms
VOCs	Volatile organic compounds
WGS	Whole-genome shotgun

Chapter 1
Can We Define an Experimental Framework to Approach the Genetic Basis of Root Colonization?

Ariella Alperovitch-Lavy and Benjamin A. Horwitz

Abstract The contributions of Trichoderma to the biocontrol of soilborne disease and plant growth promotion depend on the interaction between the biocontrol fungus, the plant, fungal pathogens, and other organisms in the rhizosphere. In some interactions, Trichoderma colonizes the root surface, epidermis, and outer cortex layers. In contrast to arbuscular or even ectomycorrhizae, the fungal hyphae are not precisely localized to a structure, making it somewhat of a challenge to define and quantify colonization. Following ingress into the root, hyphae grow in the apoplast in a parallel pattern defined by the root tissue, much like pathogens such as *Fusarium oxysporum*. Trichoderma, however, does not penetrate to inner cortex layers or the vascular tissues. To elucidate the genetic basis of root colonization, one would ideally isolate mutants defective in the process, and this needs a good definition of colonization phenotypes. Here, we compare different methods to follow and quantify Trichoderma entry and proliferation in roots and discuss data obtained for mutants.

Keywords Colonization · Root cortex · Rhizosphere · Fungal-plant interaction · Symbiont · Apoplast

1.1 Introduction

1.1.1 Trichoderma as an Opportunistic Symbiont of Plant Roots

Species belonging to the ascomycete genus Trichoderma grow in the soil. In the rhizosphere, defined as the soil layer contacting plant roots, these fungi grow as saprophytes and interact both with other fungi and roots. Biocontrol of soilborne

A. Alperovitch-Lavy · B. A. Horwitz (✉)
Faculty of Biology, Technion – Israel Institute of Technology, Haifa, Israel
e-mail: lariella@campus.technion.ac.il; horwitz@technion.ac.il

A. K. Sharma, P. Sharma (eds.), *Trichoderma*, Rhizosphere Biology,
https://doi.org/10.1007/978-981-15-3321-1_1

pathogens is a result of competition as well as direct attack and breakdown of other fungi, known as mycoparasitism or mycotrophy (Druzhinina et al. 2011; Kubicek et al. 2011). Many isolates with potential for biocontrol of soilborne fungal pathogens thus act as necrotrophic mycoparasites, destroying the host fungus (Karlsson et al. 2017a). The interaction with roots is not an essential part of the fungal life cycle. Likewise, plants do not depend on Trichoderma in a structurally defined, tightly linked mutualism of the kind seen in AMF mycorrhizae (Bonfante and Genre 2010). Root colonization and coexistence has, therefore, been defined as an opportunistic symbiosis (Harman et al. 2004; Shoresh et al. 2010). Trichoderma normally grows in the soil, but can be intentionally applied to leaves and stems, providing antagonism against foliar pathogens (Elad 2003). Interaction with plant roots primes the plant immune response, promoting systemic defense against pathogens, as discussed further elsewhere in this volume. Rhizosphere-competent strains, therefore, have a whole dimension of biocontrol capability that depends on the Trichoderma-plant interaction. The interaction, when optimal, provides many benefits, promoting the plant's response to stresses of biotic and abiotic origin and increasing growth rate and yield (Lorito et al. 2010). Proteins and metabolites released by Trichoderma into the rhizosphere can reach the roots even without contact. An example is release by the fungus of compounds that are detected by the plant as auxins (Contreras-Cornejo et al. 2009). Even the growth medium used for the fungal inoculum influences its ability to promote growth (González-Pérez et al. 2018) so that standardization is important in setting up these experiments. Growth on and in the root, however, allows much greater opportunities for direct surface interactions and for secretion directly into the plant cell, as known for some pathogen effectors (Białas et al. 2018; Franceschetti et al. 2017). Colonization of the root surface and tissues, therefore, has become a focus for studies of the Trichoderma-root interaction. In contrast to better knowledge about mycorrhizae (Plett et al. 2014, 2011; Tisserant et al. 2013; Oldroyd et al. 2009), less is known about the genetic basis for colonization of plant roots by Trichoderma. A genetic approach is often the best to define mechanism, and several of the biocontrol-competent Trichoderma species are amenable to gene knockout and overexpression. To screen for phenotypes related to root colonization, good assay methods are essential. Here, we evaluate the strategies used and the success so far in identification of genes with roles in root colonization.

Following inoculation with conidia or mycelia, hyphae proliferate on the root surface and can be visualized inside the cortex, mainly growing between cells. In the *T. harzianum*-tomato interaction, fungal cells with a morphology reminiscent of yeast were observed inside plant cells (Chacón et al. 2007). Hyphae inside root cells were observed more recently by confocal imaging in the *T. virens*-maize interaction (Nogueira-Lopez et al. 2018). Although inoculation under laboratory conditions is more massive than likely to be found in nature, it is useful in experiments designed to study the course of colonization and may also reflect conditions found in agricultural biocontrol settings. In the laboratory, conidia, mycelia, or chlamydospores are inoculated at a determined time. In the field or greenhouse, the fungus is present or inoculated in the soil as mycelia or spores. In either case, actively growing hyphae

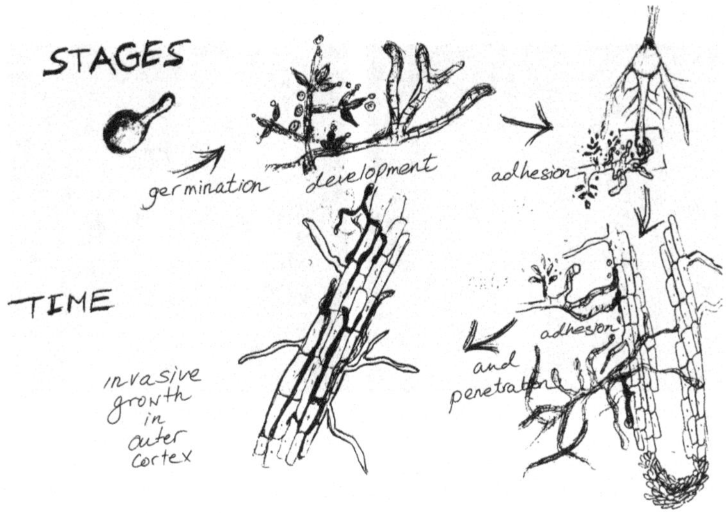

Fig. 1.1 A schematic time course of colonization. The stages shown are completed in 3–4 days in hydroponic culture; however, what actually happens in a natural soil environment is likely to be very different

reach plant roots, proliferate, and penetrate the outer root layers (Fig. 1.1). To do so, the fungus must be able to tolerate, or actively suppress, the immune response of the plant and survive the effects of preexisting plant defense compounds. Nevertheless, colonization is limited to the outer root cortex and hyphae do not invade the vascular cylinder. It is not completely clear why Trichoderma stops short of reaching the vascular cylinder, but the plant immune response is a likely player. Indeed, the salicylic acid (SA) pathway, a well-known induced defense response, is needed to limit the invasive growth of Trichoderma (Alonso-Ramírez et al. 2014; Martínez-Medina et al. 2017).

At this stage, there is a fundamental difference between mutualistic Trichoderma and pathogens like *Fusarium oxysporum*. Fusarium infection, initially, is remarkably similar to Trichoderma (Fig. 1.2), but invasive growth of the pathogen proceeds to the xylem elements of the root (Fig. 1.2a) where it causes blockage and wilting. Although the distinction between pathogen and mutualist seems very sharp, it may be more blurred than thought in the past (Plett and Martin 2011). At the very least, some of the molecular mechanisms for the early stages of colonization could be shared, helping to guide studies on the agriculturally important Trichoderma-root interaction about which less is known than for pathogens. Likewise, molecular work done in the past decade on the ecto- and arbuscular mycorrhizae can guide our insight for Trichoderma. Another example is the basidiomycete *P. indica* (Lahrmann et al. 2013; Thürich et al. 2018). If the same principles are at work in all these root-fungal interactions, we expect that both Trichoderma and plant release, and respond to, signals from the other partner.

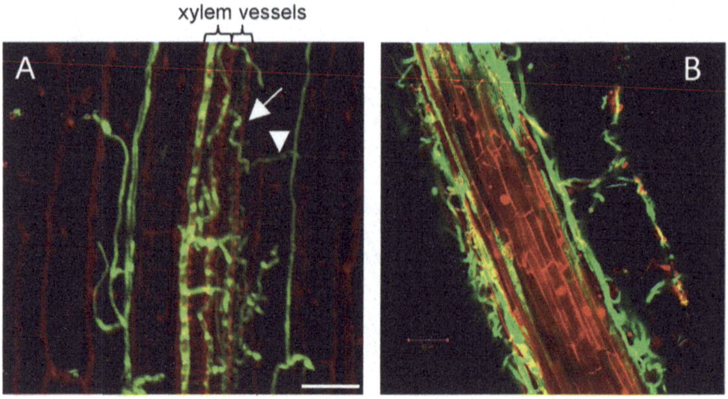

Fig. 1.2 Ingress into tomato roots by *Trichoderma virens* or *Fusarium oxysporum* f. sp. *lycopersici* (Fol). (**a**) Fol expressing a reporter construct, Psix1::GFP, which labels hyphae inside the root tissue (see text). The red channel shows plant autofluorescence; the scale bar indicates 25 μm. As indicated by the arrows, hyphae are not restricted to a single vascular bundle (Reproduced from (van der Does et al. 2008) with permission from the publisher). (**b**) *Trichoderma virens* proliferating on the surface and in the outer cell layers of a tomato root. Root section was stained with WGA-Alexa Fluor™ 488 (green channel) and propidium iodide (red channel) and imaged in a Zeiss confocal 510 microscope. The scale bar indicates 50 μm. (A. Alperovitch-Lavy, unpubl)

1.1.2 Root Colonization and the Plant Immune Response

As noted above, there is a spectrum from pathogens through endophytes and mycorrhizae (Rodriguez et al. 2009; Plett and Martin 2011), and the place of Trichoderma in this spectrum is not completely clear yet. There is no doubt though that the plant immune system responds to interaction of Trichoderma with the roots. Genes involved in host defense are upregulated, for example, within the first 24 h in maize inoculated with *T. virens* (Djonovic et al. 2007) and over a few days' time course in *Arabidopsis* interacting with *T. hamatum* (Mathys et al. 2012). Several classes of compounds secreted by *Trichoderma* spp. initiate resistance responses in plants by the activation of ISR mediated by jasmonic acid and ethylene and/or systemic acquired resistance (SAR), which involves salicylic acid (Djonovic et al. 2007; Shoresh et al. 2010; Martinez et al. 2001; Salas-Marina et al. 2015). The salicylic acid pathway is a major factor in halting ingress into the root, as shown by studies with a SA-deficient mutant of Arabidopsis (Alonso-Ramírez et al. 2014), explaining, in large part, the dramatic distinction between mutualistic Trichoderma and pathogenic *Fusarium oxysporum* illustrated in Fig. 1.2. The first effector of plant responses discovered from *Trichoderma* was a 22-kDa xylanase (Rotblat et al. 2002). It was later shown that the xylanase, produced by several species, does not need to be enzymatically active for the plant to recognize it as an effector. *T. virens* produces proteins and peptides that induce terpenoid phytoalexins in cotton; these promote disease resistance. One of these proteins was SM1, which has already been described as being primarily responsible for ISR in maize (Djonovic et al. 2007). Predictions from the genome (Schmoll et al. 2016) suggest that Trichoderma could secrete tens, or even of the order of a hundred, distinct small-protein inducer/effector-like molecules, making it a formidable challenge to identify the active ones. In the apoplast secretome of the *T. virens*-maize root interaction, 95 Trichoderma and 43 maize secreted proteins were identified (Nogueira-Lopez et al. 2018). The abundance of maize secreted proteins, in particular glycosyl hydrolases and peroxidases, was reduced by Trichoderma. Among the *T. virens* secreted proteins are cell wall hydrolyzing enzymes, proteins providing tolerance toward reactive oxygen species, and predicted effectors. The fungal partner thus seems able to manipulate the plant immune response, and this is evident at the level of the apoplast secretome (Nogueira-Lopez et al. 2018). Another group of different, lower-molecular-weight compounds known to elicit plant defense reactions are peptaibols. These are antimicrobial peptides produced by several *Trichoderma* spp. including *T. virens*, which produces an 18-mer form that elicits plant defense reactions (Viterbo et al. 2007). Alamethicin, another peptaibol, also induces resistance responses (Engelberth et al. 2001; Dotson et al. 2018). In *Arabidopsis*, alamethicin triggered deposition of callose, production of phenolic compounds, and induction of defense gene expression (Rippa et al. 2010). Trichokonins are a class of antimicrobial peptaibols first isolated from *T. pseudokoningii*. Luo et al. (2010) demonstrated that trichokonins induced defense responses and provided systemic resistance in tobacco against tobacco mosaic virus infection. Thus, there is evidence that many compounds produced by *Trichoderma* spp. can contribute to induced host resistance against plant pathogens. The compounds described above as inducers of plant defense responses, however, do not have any obvious role in colonization of plant roots by Trichoderma. In principle, any Trichoderma MAMP could act to prime the plant immune system; different protein or small-molecule MAMPs are presented to the plant by different Trichoderma species and strains. Another important point is that ISR and colonization are not necessarily linked. To give one example, Sm2, a paralog of Sm1, was needed for full colonization of maize roots by *T. virens*. Nevertheless, there was no significant effect on ISR in this interaction (Crutcher et al. 2015). In experiments with maize, *T. virens* was a stronger inducer of ISR than *T. atroviride*; however, comparing the same strains, we found preliminary evidence (Fig. 1.3) that *T. atroviride* is a better colonizer of roots. This comparison needs to be strengthened and extended. In general, the competitive ability of different natural and engineered Trichoderma strains as soil and rhizosphere saprophytes (Ryder et al. 2012; Weaver and Kenerley 2008) may be a factor in subsequent colonization of roots.

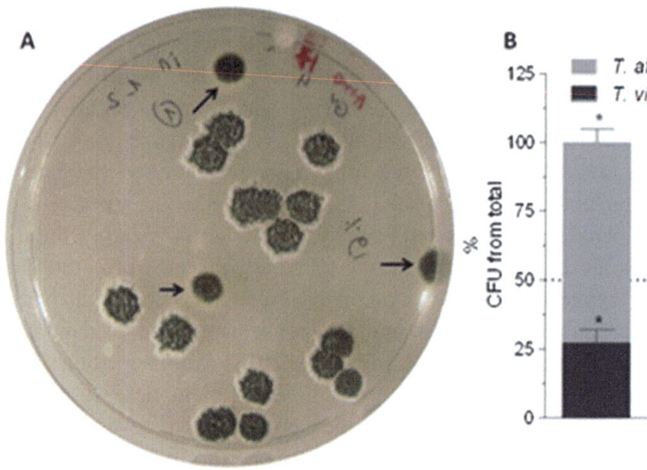

Fig. 1.3 Competition assay for colonization of maize roots. Equal numbers of *T. atroviride* and *T. virens* conidia were inoculated with maize roots in a hydroponic system and assayed after 3 days for colony-forming units. (**a**) Fewer *T. virens* colonies grow (arrows). (**b**) Average colony-forming units (CFU). (N. Lamdan, PhD thesis, Technion)

1.2 Mapping and Quantifying the Trichoderma-Root Interaction

1.2.1 Imaging

Electron microscopy (Yedidia et al. 1999) provided the first, and remains the most precise, localization. A dynamic and more quickly accessible visualization is provided by Gfp expression (Lu et al. 2004) or staining (Nogueira-Lopez et al. 2018) that have been used to map the location of Trichoderma hyphae on and in the root. Hyphae of several Trichoderma species have been visualized growing on the roots of several plant hosts (Arabidopsis, maize, tomato, cucumber), on the root surface, and inside the cortex, mainly growing between cells using either transmission electron micrographs or confocal laser microscopy. A fluorescent wheat germ agglutinin (WGA-Alexa Fluor™ 488) probe specifically stains fungal hyphae. This probe is being used together with propidium iodide (PI) or the plasma membrane dye FM 4-64 for staining the root cells. Using this method, a time course of root colonization by *T. virens* was visualized. Twenty-four hours after inoculation (hpi) of maize roots with spores, there was limited colonization of the surface, which increased dramatically between 48 and 72 hpi; by 96 hpi, WGA fluorescence from hyphae growing parallel to the long axis of the cells was visualized inside the root cortex in a confocal plane located two cell layers from the root surface (Fig. 1.4). *T. virens* Gv29-8 hyphae were also observed colonizing maize roots between the plasma membrane and the cell wall and in the intracellular spaces (Nogueira-Lopez et al. 2018).

In 1991, J. L. Parke (1991) defined root colonization "as the proliferation of microorganisms in, on and around the growing root." Most studies attempt to focus

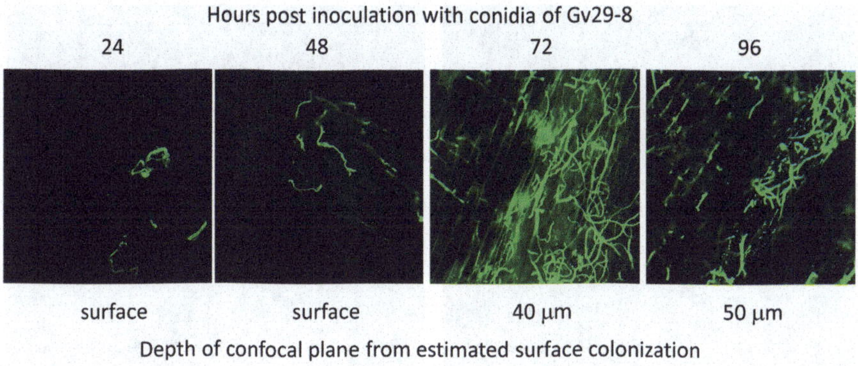

Fig. 1.4 Images of the time course of colonization of maize roots by *T. virens*. At the indicated times following inoculation with conidia of the sequenced reference *T. virens* strain Gv29-8, root sections were stained with WGA-Alexa Fluor™ 488 and imaged in a Zeiss confocal 510 microscope. (A. Alperovitch-Lavy and Nitsan Dahan, unpubl.)

on the "in" part of this picture. The reason is the assumption that hyphae internal to the root provide the close association needed to deliver effectors. New microscopic technologies are being developed to visualize plant-pathogen interactions. Using confocal microscopy enables to identify phenotype variation in fungi penetration styles into the host plant. Minker et al. (2018) developed a platform using 3D confocal microscopy to characterize and quantify leaf infection by three different fungal pathogens of maize. Application of 3D imaging in combination with computational analysis is a powerful tool in modeling fungi plant colonization. The authors emphasize the importance of the clearing procedure, and differences were demonstrated between the different protocols, as well as choosing the right microscopy. Application of this method to the roots of plants, while calibrating the system for testing the roots colonized by fungi, can expand and establish the information about the patterns and quantity of fungus inhabiting the roots.

The most effective way to follow the fungus inside the root would be to visualize a fluorescent protein that is expressed only inside living root tissues. A promoter that is uniquely expressed when the hyphae are in contact with live cells, or in internal layers, would be ideal. Such a reporter gene construct could supply extensive information about plant-fungus interaction. This strategy was applied to *Fusarium oxysporum* f. sp. *lycopersici* (van der Does et al. 2008, see Fig. 1.2a), where expression of the coding sequence for the green fluorescent protein (Gfp) was driven by the Six1 (*Secreted In Xylem* 1) promoter. Six1 confers expression only within living root tissues. The signal was observed immediately following hyphal penetration into the layers of the root, or in interaction with plant cells in culture. This was an indication of the fungus ability to specifically recognize live host plant root tissues (van der Does et al. 2008). To apply this principle to Trichoderma, Gfp was expressed under the upstream control of the promoter of a glycosyl hydrolase gene of *T. virens* whose expression was found to strongly increase when the fungus encounters roots, when grown together in hydroponic culture (Rubio et al. 2012;

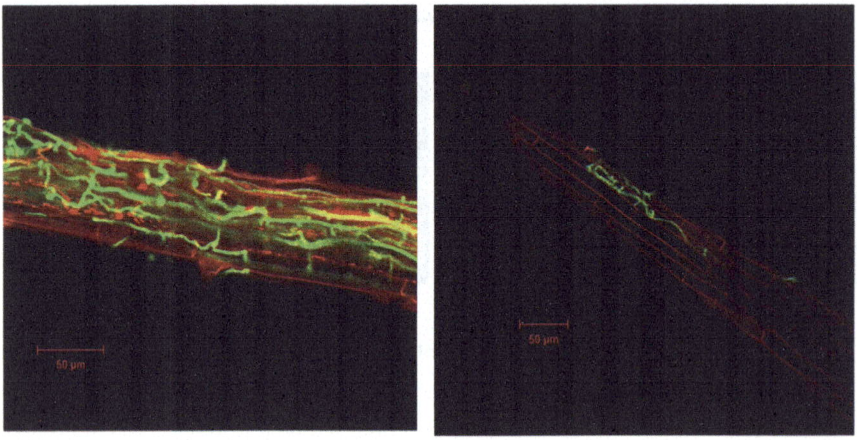

Fig. 1.5 Reporter gene strategy to follow the presence of Trichoderma on or inside Arabidopsis roots. Confocal images of *T. virens* expressing Gfp under control of the promoter of a cellobiohydrolase gene (*CBH1*), JGI ID number 90504, construct from Morán-Diez et al. (2015). Scale bars indicate 50 µm. Green channel, Gfp fluorescence; red channel, propidium iodide. (Images by Rinat Zaid and Ariella Alperovitch-Lavy, unpubl)

Morán-Diez et al. 2015). A strong GFP signal was induced by the plant root-fungus interaction. Unlike the Six1-GFP model (Fig. 1.2), the transgenic line p90504-GFP *T. virens* exhibited signal also on the external surface of the root (Fig. 1.5). This strategy could be developed and refined by screening promoter constructs from a larger sampling of different genes induced by coculture with plant roots.

1.2.2 Quantification of Trichoderma *Root Colonization*

As mentioned above, our ability to differentiate between colonization of the root surface and the cortex still needs to be greatly improved. Quantification protocols include either surface sterilization, sonication (Martínez-Medina et al. 2017), or brushing, or sometimes merely a gentle washing step, to detach fungal hyphae, conidia, and germinlings from the root surface. Can "in" be separated from "on"? The result will obviously depend on what technique is used to "clean" the root surface. A colonization phenotype defined by a single value, whether cfu or qPCR, could be made up of a massive hyphal growth on the root surface with limited penetration and vice versa. Because of this, the full picture should be evaluated. Bleach might enter the roots and destroy the hyphae inside, particularly for the delicate roots of Arabidopsis, which contain only one cortex layer, as noted by (Martínez-Medina et al. 2017). In this method, plants are cocultivated with Trichoderma spores or hyphae, grown in soil, agar, or hydroponic systems. Roots are detached at different time points postinoculation and washed in water or PBS. Surface sterilization of the roots is usually done in 1% NaOCl (bleach) for a 1–2-min exposure, or by sonication according to (Martínez-Medina et al. 2017). Regardless of the quantitation method, the separation of Trichoderma growing on the root surface from colonization inside the root tissues is difficult. Once this problem has been overcome (partially) or at least clearly defined, the actual quantitation is more straightforward. The main methods used are colony-forming units (cfu) or qPCR amplification from genomic DNA isolated from the combined plant and fungal tissue sample. Colony-forming units provide a semiquantitative measure of fungal biomass; the number of colonies per biomass depends on how the fungus is dispersed, for example, by homogenization. More drastic homogenization will increase the number of colonies, as long as fungal cells are not irreversibly damaged. Roots are washed, weighed (this step can be done before surface sterilization), and homogenized. Serial dilutions for cfu counting are plated on growth medium; 0.1% of Triton X-100 can be added to restrict colony expansion (Lamdan et al. 2015; Brotman et al. 2008; Crutcher et al. 2015; Martínez-Medina et al. 2017). For qPCR amplification, primer pairs are designed to selectively amplify from plant or fungal DNA, both of which are present in the combined template genomic DNA obtained from the Trichoderma-root interaction. After choosing the primers to optimize specificity for plant or fungal DNA amplification and adjusting the template concentration to ensure linearity, qPCR provides a signal that is directly proportional to the ratio of fungal to plant nuclei present. This a good measure of fungal biomass relative to plant biomass, to the extent that the fungal hyphal compartments contain, on average, the same number of nuclei throughout the stages of colonization.

1.3 Physiological and Environmental Factors

The definition of "root colonization" should ideally refer to the soil, the natural plant root-microorganism interaction zone. In colonization assays, plants are sometimes grown in soil or sand, but also often in agar-solidified media or hydroponic systems, with different artificial medium compositions. Do the environmental parameters influence Trichoderma strategies of penetration and invasive growth? Is the fungus beneficial or harmful to the plant under these conditions? Many laboratory assays might not reflect the behavior of Trichoderma with the host plant, in the natural habitat, so the results always need to be considered in the context of the microenvironment and parameters affecting the physiology of both the plant and fungus.

A reduction in the fresh weight of *A. thaliana* cocultivated with *T. asperellum LU1370* in sterile soil was observed, while on agar plates a beneficial effect of *T. asperellum on A. thaliana* was shown. In the same study, a strong expression of GUS was seen in the leaves and primary roots of DR5::GUS in an *A. thaliana* expression line cocultured with Trichoderma strains, grown on PDA medium containing tryptophan. In contrast, only *T. reesei* QM6a strain grown on MS medium gave a positive signal in both roots and leaves (Nieto-Jacobo et al. 2017).

The effect of medium composition on the interaction between Trichoderma and *A. thaliana* was also studied in a split system (González-Pérez et al. 2018).

T. atroviride IMI 206040 and *T. virens* Gv29-8 were grown on either MS plates or PDA, with or without direct contact with the plants. The beneficial effect on the plant's growth was seen under split conditions, mainly on PDA media. *T. virens* and *T. atroviride* were able to rescue *A. thaliana* mutant *rhd6* to form root hairs, in a split system both on MS and PDA medium, or with direct interaction. The strong effect was under the growth of *T. virens* on PDA plates in a split system, expressed in a higher number of root hairs. The authors referred to the ability of Trichoderma strains to revert (recover) root hair deficiency as the outcome of volatile organic compound (VOC) generation, such as precursors of the auxin biosynthesis or auxin signaling pathway. In contrast, after 5 dpi, a negative effect of inhibition on the elongation of the primary root was observed under split conditions with *T. atroviride* grown on PDA plates. This was explained due to the higher presence (abundance) of 6-pentyl-2H-pyran-2-one (6-PP) in PDA compared to MS medium. VOC production was shown to be influenced by the medium composition and strain-specific, which might explain these effects of Trichoderma on the plant (González-Pérez et al. 2018).

1.4 Trichoderma Genetic Determinants of Colonization

Despite the uncertainties and the need to better define colonization, progress has indeed been made in defining some of the genes needed for Trichoderma to proliferate on and in the root. From the genome projects, we have the genomic sequences of *T. reesei*, noted for its ability to degrade cellulose, and of two species that interact with plants and are also mycotrophic: *T. atroviride* and *T. virens* (Kubicek et al. 2011). The sequence is a central resource facilitating identification of relevant genes, construction of mutants, and transcriptome studies. Several reverse genetic approaches prior to the Trichoderma genome projects have recognized a few genes involved in root colonization. For instance, a gene (*TasSwo*) encoding an expansin-like protein (swollenin) from *T. asperellum* was identified from culture medium with the fungus and cucumber seedlings by LC-MS. Root colonization was reduced in mutants that were silenced for swollenin gene expression (Brotman et al. 2008). Colonization of tomato roots decreased upon silencing of the *ThPG1* gene (which encodes an endopolygalacturonase) from *T. harzianum* (Morán-Diez et al. 2009). This protein was identified from proteins whose expression was dependent on the three-way interaction of *T. harzianum* with tomato roots and a plant pathogen. *T. virens* PG2, in contrast, is needed for ISR induction but is dispensable for colonization (Sarrocco et al. 2017). A cellulase from *T. harzianum* is involved in ISR of maize (Saravanakumar et al. 2016). Enzymes that begin to degrade the plant cell walls, thus, can induce ISR directly or via the products released and may also facilitate ingress by the fungus. Hydrophobins, which are surface-active proteins unique to fungi, could modulate attachment or penetration of the Trichoderma hyphae. If attachment and surface colonization are rate-limiting for later entry into the root, this stage may be a critical one. Viterbo and Chet (Viterbo and Chet 2006), using a differential mRNA display analysis of *T. asperellum* mycelia in the presence of plant roots, identified a gene (*TasHyd1*) encoding a hydrophobin-like protein. Deletion mutants had altered surface properties and were impaired in their ability to attach to and colonize roots. Several hydrophobin genes were shown to be important for ISR induction and colonization (Guzmán-Guzmán et al. 2017; Karlsson et al. 2017b). Class II hydrophobin ThHyd1 of *T. harzianum* elicits ISR in maize, and a plant-membrane localized target protein was identified (Yu et al. 2019). Swollenin and pectinases could act to increase the flexibility of the cell walls; partial degradation of the pectin middle lamella would permit fungal growth between cells. Thus, secreted expansin-like proteins and pectinases would seem to favor growth of Trichoderma in the apoplast. Many CWDEs are expressed in interaction with roots (Rubio et al. 2012; Morán-Diez et al. 2015) and their function could be to soften the cell wall. We note that gene knockout experiments have not been done yet for the vast majority of these candidates. One difficulty is that the large number of regulated CWDEs poses a formidable problem of functional redundancy. Two other, interrelated, factors in cell wall extensiblity are pH (acid growth) and auxin. Thus, in addition to expansin-like proteins (swollenin being the best) characterized (Saloheimo et al. 2002), fungal signaling pathways that detect or alter pH (Peñalva et al. 2008; Masachis et al. 2016; Bi et al. 2016), or the ion pumps that support alkalinization or acidification, would be good candidates to test for involvement in colonization. *Trichoderma atroviride* was shown to acidify the medium (Pelagio-Flores et al. 2017); however, alkalinization was observed, for *T. virens* (Trushina et al. 2013). Secretion of IAA or other compounds with auxin activity, which increases root surface area (Contreras-Cornejo et al. 2009), may also play a major role in colonization by making the plant cell wall more extensible. A common theme among these genes is the ability of their products to alter the physical properties of the host.

The plant-fungal molecular dialog can also impact colonization in less direct ways. Secreted and cell-surface proteins are likely to play an important role in colonization, through their role in reprogramming plant gene expression or metabolism. Hydrophobins, already mentioned, may alter surface properties and also act as signals to the plant. A given secreted protein might, in this way, affect the physical properties of the root surface or apoplast and, in addition, act as a MAMP to trigger or prime the plant immune system. Proteins secreted by a symbiont or pathogen that alter the functioning of plant cells are often referred to as effectors. Some are imported into plant cells. The paradigm of effector import into plant cells was first established for bacterial and oomycete pathogens and now seems to be of general importance in filamentous fungal pathogens; symbionts apparently use similar systems. An ectomycorrhizal symbiont, *Laccaria bicolor*, secretes a small protein designated MiSSP7. MiSSP7 is highly expressed during colonization of tree roots and is required for the symbiotic interaction (Plett et al. 2014, 2011). This effector protein is imported into plant cells, where it interacts with and stabilizes PtJAZ6, a negative regulator of jasmonic acid (JA)-induced gene regulation in the poplar tree host. In arbuscular mycorrhizae, likewise, a fungal effector interferes with plant gene expression, to establish the biotrophic interaction. A fungal effector targeted to the plant nucleus acts against the immune response. The mechanism is interaction with a specific plant transcription factor (Kloppholz et al. 2011). Small secreted proteins (SSPs) like the mycorrhizal effectors represent one class of candidates; hundreds of SSPs were predicted from an initial screen of the genomes of three sequenced *Trichoderma* species (Kubicek et al. 2011; Schmoll et al. 2016). *Trichoderma* generally does not cause plant disease; it is possible, however, that the effectors responsible for induced systemic resistance (ISR) or other types of systemic resistance are virulence factors that evolved secondarily to elicit resistance without triggering the full plant defense response. Sm1, discussed above, is a SSP. Loss of Sm1, however, does not lead to any obvious defect in colonization. Colonization and the induction of ISR are, at least in part, separate processes. Two considerations, though, lead us to the hypothesis that there is overlap between the factors required for colonization and the set of effector molecules triggering ISR. First, the initial ingress of *Trichoderma* exposes the plant to effectors that initiate defense responses. Second, effectors dampen the defense response to allow colonization to proceed. The precedent from mycorrhizae supports this view, though the identity of the molecules may be different. An examination of the three paralogs to SM1 in *T. virens* revealed that mutants lacking one, Sm2, were less able to colonize roots of maize seedlings (Crutcher et al. 2015). Thus, there is clear evidence for several classes of genes in different species of *Trichoderma* that encode proteins involved in colonization. An in-depth examination within one species would have the potential to discover other factors, as well as demonstrate their interrelationships and the timing at which they act.

1.5 Plant Genetic Determinants of Colonization

Plant genetic factors determining compatible or incompatible interactions with fungal pathogens are studied in detail, but much fewer studies address Trichoderma-plant interactions. Cell wall reinforcement is important to limit invasive growth. As mentioned above, an Arabidopsis mutant lacking isochorismate synthase 1, which cannot accumulate SA, showed that Trichoderma reaches the vascular cylinder of the mutant. Looking from the plant side, the main reason is that the mutant lacks the ability to respond to the fungus by cell wall reinforcement (Alonso-Ramírez et al. 2014). Loss of a single plant defense mechanism, the SA-induced pathway, thus resulted in a more compatible (where Trichoderma is the "pathogen") interaction, in which Trichoderma reached the vascular tissues. Working with *T. harzianum* and tomato, Martínez-Medina et al. (2017) showed that SA defenses limit Trichoderma colonization, while the JA pathway enhanced colonization, apparently by antagonizing the SA-dependent defenses.

1.6 Conclusion

As summarized in Sect. 1.4, a fair number of fungal genes involved in colonization have already been identified. The evidence is mainly from knockout or silencing mutants that have defects in colonization, and the targets of these approaches, for the most part, have been chosen based on candidate genes for predicted effectors, or enzymes and other proteins that could modulate plant cell wall physical properties. Nevertheless, finding cause-effect relationships is still difficult, because multiple genes and pathways are involved. A more systematic genetic analysis might finally identify central regulators of colonization in the fungal genome. For the plant side of the interaction, the full power of Arabidopsis genetics has yet to be employed.

Acknowledgements We thank Rinat Zaid for providing us with unpublished data and Dr. Nitsan Dahan at the Life Sciences & Engineering Center, Technion, for expert assistance with confocal microscopy. Studies of Trichoderma in the Horwitz Lab, including unpublished data shown here, were supported by the US-Israel Agricultural Research and Development Fund (BARD) and the US-Israel Binational Science Foundation (BSF), together with Charles M. Kenerley, Texas A&M University, USA.

References

Alonso-Ramírez A, Poveda J, Martín I, Hermosa R, Monte E, Nicolás C (2014) Salicylic acid prevents *Trichoderma harzianum* from entering the vascular system of roots. Mol Plant Pathol 15:823–831. https://doi.org/10.1111/mpp.12141

Bi F, Barad S, Ment D, Luria N, Dubey A, Casado V, Glam N, Mínguez JD, Espeso EA, Fluhr R, Prusky D (2016) Carbon regulation of environmental pH by secreted small molecules that modulate pathogenicity in phytopathogenic fungi. Mol Plant Pathol 17:1178–1195. https://doi.org/10.1111/mpp.12355

Białas A, Zess EK, De la Concepcion JC, Franceschetti M, Pennington HG, Yoshida K, Upson JL, Chanclud E, Wu C-H, Langner T, Maqbool A, Varden FA, Derevnina L, Belhaj K, Fujisaki K, Saitoh H, Terauchi R, Banfield MJ, Kamoun S (2018) Lessons in effector and NLR biology of plant-microbe systems. Mol Plant Microbe Interact 31:34–45. https://doi.org/10.1094/MPMI-08-17-0196-FI

Bonfante P, Genre A (2010) Mechanisms underlying beneficial plant–fungus interactions in mycorrhizal symbiosis. Nat Commun 1:1–11. https://doi.org/10.1038/ncomms1046

Brotman Y, Briff E, Viterbo A, Chet I (2008) Role of swollenin, an expansin-like protein from Trichoderma, in plant root colonization. Plant Physiol 147:779–789. https://doi.org/10.1104/pp.108.116293

Chacón MR, Rodríguez-Galán O, Benítez T, Sousa S, Rey M, Llobell A, Delgado-Jarana J (2007) Microscopic and transcriptome analyses of early colonization of tomato roots by Trichoderma harzianum. Int Microbiol 10:19–27

Contreras-Cornejo HA, Macias-Rodriguez L, Cortes-Penagos C, Lopez-Bucio J (2009) Trichoderma virens, a plant beneficial fungus, enhances biomass production and promotes lateral root growth through an auxin-dependent mechanism in arabidopsis. Plant Physiol 149:1579–1592. https://doi.org/10.1104/pp.108.130369

Crutcher FK, Moran-Diez ME, Ding S, Liu J, Horwitz BA, Mukherjee PK, Kenerley CM (2015) A paralog of the proteinaceous elicitor SM1 is involved in colonization of maize roots by Trichoderma virens. Fungal Biol 119:476–486. https://doi.org/10.1016/j.funbio.2015.01.004

Djonovic S, Vargas WA, Kolomiets MV, Horndeski M, Wiest A, Kenerley CM (2007) A proteinaceous elicitor Sm1 from the beneficial fungus trichoderma virens is required for induced systemic resistance in maize. Plant Physiol 145:875–889. https://doi.org/10.1104/pp.107.103689

van der Does HC, Duyvesteijn RGE, Goltstein PM, van Schie CN, Manders EMM, Cornelissen BJC, Rep M (2008) Expression of effector gene SIX1 of Fusarium oxysporum requires living plant cells. Fungal Genet Biol 45:1257–1264. https://doi.org/10.1016/j.fgb.2008.06.002

Dotson BR, Soltan D, Schmidt J, Areskoug M, Rabe K, Swart C, Widell S, Rasmusson AG (2018) The antibiotic peptaibol alamethicin from Trichoderma permeabilises Arabidopsis root apical meristem and epidermis but is antagonised by cellulase-induced resistance to alamethicin. BMC Plant Biol 18:165. https://doi.org/10.1186/s12870-018-1370-x

Druzhinina IS, Seidl-Seiboth V, Herrera-Estrella A, Horwitz BA, Kenerley CM, Monte E, Mukherjee PK, Zeilinger S, Grigoriev IV, Kubicek CP (2011) Trichoderma: the genomics of opportunistic success. Nat Rev Microbiol 9:749–759. https://doi.org/10.1038/nrmicro2637

Elad Y (2003) Biocontrol of foliar pathogens: mechanisms and application. Commun Agric Appl Biol Sci 68:17–24

Engelberth J, Koch T, Schüler G, Bachmann N, Rechtenbach J, Boland W (2001) Ion channel-forming alamethicin is a potent elicitor of volatile biosynthesis and tendril coiling Cross talk between jasmonate and salicylate signaling in lima bean. Plant Physiol 125:369–377

Franceschetti M, Maqbool A, Jiménez-Dalmaroni MJ, Pennington HG, Kamoun S, Banfield MJ (2017) Effectors of filamentous plant pathogens: commonalities amid diversity. Microbiol Mol Biol Rev 81:e00066-16. https://doi.org/10.1128/MMBR.00066-16

González-Pérez E, Ortega-Amaro MA, Salazar-Badillo FB, Bautista E, Douterlungne D, Jiménez-Bremont JF (2018) The Arabidopsis-Trichoderma interaction reveals that the fungal growth medium is an important factor in plant growth induction. Sci Rep 8:16427. https://doi.org/10.1038/s41598-018-34500-w

Guzmán-Guzmán P, Alemán-Duarte MI, Delaye L, Herrera-Estrella A, Olmedo-Monfil V (2017) Identification of effector-like proteins in Trichoderma spp and role of a hydrophobin in the plant-fungus interaction and mycoparasitism. BMC Genet 18:16. https://doi.org/10.1186/s12863-017-0481-y

Harman GE, Howell CR, Viterbo A, Chet I, Lorito M (2004) Trichoderma species — opportunistic, avirulent plant symbionts. Nat Rev Microbiol 2:43–56. https://doi.org/10.1038/nrmicro797

Karlsson M, Atanasova L, Jensen DF, Zeilinger S (2017a) Necrotrophic mycoparasites and their genomes. Microbiol Spectrosc 5:1005–1026. https://doi.org/10.1128/microbiolspec.FUNK-0016-2016

Karlsson M, Atanasova L, Jensen DF, Zeilinger S (2017b) Necrotrophic mycoparasites and their genomes. In: The fungal kingdom. American Society of Microbiology, Washington, DC, pp 1005–1026. https://doi.org/10.1128/microbiolspec.FUNK-0016-2016

Kloppholz S, Kuhn H, Requena N (2011) A secreted fungal effector of Glomus intraradices promotes symbiotic biotrophy. Curr Biol 21:1204–1209. https://doi.org/10.1016/j.cub.2011.06.044

Kubicek CP, Herrera-Estrella A, Seidl-Seiboth V, Martinez DA, Druzhinina IS, Thon M, Zeilinger S, Casas-Flores S, Horwitz BA, Mukherjee PK, Mukherjee M, Kredics L, Alcaraz LD, Aerts A, Antal Z, Atanasova L, Cervantes-Badillo MG, Challacombe J, Chertkov O, McCluskey K, Coulpier F, Deshpande N, von Döhren H, Ebbol DJ, Esquivel-Naranjo EU, Fekete E, Flipphi M, Glaser F, Gómez-Rodríguez EY, Gruber S, Han C, Henrissat B, Hermosa R, Hernández-Oñate M, Karaffa L, Kosti I, Le Crom S, Lindquist E, Lucas S, Lübeck M, Lübeck PS, Margeot A, Metz B, Misra M, Nevalainen H, Omann M, Packer N, Perrone G, Uresti-Rivera EE, Salamov A, Schmoll M, Seiboth B, Shapiro H, Sukno S, Tamayo-Ramos JA, Tisch D, Wiest A, Wilkinson HH, Zhang M, Coutinho PM, Kenerley CM, Monte E,

Bake SE, Grigoriev IV (2011) Comparative genome sequence analysis underscores mycoparasitism as the ancestral life style of Trichoderma. Genome Biol 12:R40. https://doi.org/10.1186/gb-2011-12-4-r40

Lahrmann U, Ding Y, Banhara A, Rath M, Hajirezaei MR, Döhlemann S, von Wirén N, Parniske M, Zuccaro A (2013) Host-related metabolic cues affect colonization strategies of a root endophyte. Proc Natl Acad Sci 110:13965–13970. https://doi.org/10.1073/pnas.1301653110

Lamdan N-L, Shalaby S, Ziv T, Kenerley CM, Horwitz BA (2015) Secretome of trichoderma interacting with maize roots: role in induced systemic resistance. Mol Cell Proteomics 14:1054–1063. https://doi.org/10.1074/mcp.M114.046607

Lorito M, Woo SL, Harman GE, Monte E (2010) Translational research on *Trichoderma*: from 'omics to the field. Annu Rev Phytopathol 48:395–417. https://doi.org/10.1146/annurev-phyto-073009-114314

Lu Z, Tombolini R, Woo S, Zeilinger S, Lorito M, Jansson JK (2004) In vivo study of trichoderma-pathogen-plant interactions, using constitutive and inducible green fluorescent protein reporter systems. Appl Environ Microbiol 70:3073–3081

Luo Y, Zhang D-D, Dong X-W, Zhao P-B, Chen L-L, Song X-Y, Wang X-J, Chen XL, Shi M, Zhang Y-Z (2010) Antimicrobial peptaibols induce defense responses and systemic resistance in tobacco against tobacco mosaic virus. FEMS Microbiol Lett 313:120–126. https://doi.org/10.1111/j.1574-6968.2010.02135.x

Martinez C, Blanc F, Le Clair E, Besnard O, Nicole M, Baccou JC (2001) Salicylic acid and ethylene pathways are differentially activated in melon cotyledons by active or heat-denatured cellulase from *Trichoderma longibrachiatum*. Plant Physiol 127:334–344

Martínez-Medina A, Appels FW, van Wees SCM (2017) Impact of salicylic acid- and jasmonic acid-regulated defences on root colonization by *Trichoderma harzianum* T-78. Plant Signal Behav 12:e1345404. https://doi.org/10.1080/15592324.2017.1345404

Masachis S, Segorbe D, Turrà D, Leon-Ruiz M, Fürst U, El Ghalid M, Leonard G, López-Berges MS, Richards TA, Felix G, Di Pietro A (2016) A fungal pathogen secretes plant alkalinizing peptides to increase infection. Nat Microbiol 1:16043. https://doi.org/10.1038/nmicrobiol.2016.43

Mathys J, De Cremer K, Timmermans P, Van Kerckhove S, Lievens B, Vanhaecke M, Cammue BA, De Coninck B (2012) Genome-wide characterization of ISR induced in *Arabidopsis thaliana* by *Trichoderma hamatum* T382 against *Botrytis cinerea* infection. Front Plant Sci 3:108. https://doi.org/10.3389/fpls.2012.00108

Minker KR, Biedrzycki ML, Kolagunda A, Rhein S, Perina FJ, Jacobs SS, Moore M, Jamann TM, Yang Q, Nelson R, Balint-Kurti P, Kambhamettu C, Wisser RJ, Caplan JL (2018) Semiautomated confocal imaging of fungal pathogenesis on plants: microscopic analysis of macroscopic specimens. Microsc Res Tech 81:141–152. https://doi.org/10.1002/jemt.22709

Morán-Diez E, Hermosa R, Ambrosino P, Cardoza RE, Gutiérrez S, Lorito M, Monte E (2009) The ThPG1 endopolygalacturonase is required for the *Trichoderma harzianum* – plant beneficial interaction. Mol Plant Microbe Interact 22:1021–1031. https://doi.org/10.1094/MPMI-22-8-1021

Morán-Diez ME, Trushina N, Lamdan N, Rosenfelder L, Mukherjee PK, Kenerley CM, Horwitz BA (2015) Host-specific transcriptomic pattern of Trichoderma virens during interaction with maize or tomato roots. BMC Genomics 16:8. https://doi.org/10.1186/s12864-014-1208-3

Nieto-Jacobo MF, Steyaert JM, Salazar-Badillo FB, Nguyen DV, Rostás M, Braithwaite M, De Souza JT, Jimenez-Bremont JF, Ohkura M, Stewart A, Mendoza-Mendoza A (2017) Environmental growth conditions of trichoderma spp. affects indole acetic acid derivatives, volatile organic compounds, and plant growth promotion. Front Plant Sci 8:102. https://doi.org/10.3389/fpls.2017.00102

Nogueira-Lopez G, Greenwood DR, Middleditch M, Winefield C, Eaton C, Steyaert JM, Mendoza-Mendoza A (2018) The apoplastic secretome of *Trichoderma virens* during interaction with maize roots shows an inhibition of plant defence and scavenging oxidative stress secreted proteins. Front Plant Sci 9:409. https://doi.org/10.3389/fpls.2018.00409

Oldroyd GED, Harrison MJ, Paszkowski U (2009) Reprogramming plant cells for endosymbiosis. Science 324:753–754. https://doi.org/10.1126/science.1171644

Parke JL (1991) Root colonization by indigenus and introduced microorganisms. In: The rhizosphere and plant growth. Springer, Dordrecht, pp 33–42. https://doi.org/10.1007/978-94-011-3336-4_4

Pelagio-Flores R, Esparza-Reynoso S, Garnica-Vergara A, López-Bucio J, Herrera-Estrella A (2017) Trichoderma-induced acidification is an early trigger for changes in arabidopsis root growth and determines fungal phytostimulation. Front Plant Sci 8:822. https://doi.org/10.3389/fpls.2017.00822

Peñalva MA, Tilburn J, Bignell E, Arst HN (2008) Ambient pH gene regulation in fungi: making connections. Trends Microbiol 16:291–300. https://doi.org/10.1016/j.tim.2008.03.006

Plett J, Martin F (2011) Blurred boundaries: lifestyle lessons from ectomycorrhizal fungal genomes. Trends Genet 27:14–22. https://doi.org/10.1016/j.tig.2010.10.005

Plett JM, Kemppainen M, Kale SD, Kohler A, Legué V, Brun A, Tyler BM, Pardo AG, Martin F (2011) A secreted effector protein of laccaria bicolor is required for symbiosis development. Curr Biol 21:1197–1203. https://doi.org/10.1016/j.cub.2011.05.033

Plett JM, Daguerre Y, Wittulsky S, Vayssieres A, Deveau A, Melton SJ, Kohler A, Morrell-Falvey JL, Brun A, Veneault-Fourrey C, Martin F (2014) Effector MiSSP7 of the mutualistic fungus Laccaria bicolor stabilizes the Populus JAZ6 protein and represses jasmonic acid (JA) responsive genes. Proc Natl Acad Sci 111:8299–8304. https://doi.org/10.1073/pnas.1322671111

Rippa S, Eid M, Formaggio F, Toniolo C, Béven L (2010) Hypersensitive-like response to the pore-former peptaibol alamethicin in *Arabidopsis thaliana*. Chembiochem 11:2042–2049. https://doi.org/10.1002/cbic.201000262

Rodriguez RJ, White JF Jr, Arnold AE, Redman RS (2009) Fungal endophytes: diversity and functional roles. New Phytol 182:314–330. https://doi.org/10.1111/j.1469-8137.2009.02773.x

Rotblat B, Enshell-Seijffers D, Gershoni JM, Schuster S, Avni A (2002) Identification of an essential component of the elicitation active site of the EIX protein elicitor. Plant J 32:1049–1055

Rubio MB, Dominguez S, Monte E, Hermosa R (2012) Comparative study of Trichoderma gene expression in interactions with tomato plants using high-density oligonucleotide microarrays. Microbiology 158:119–128. https://doi.org/10.1099/mic.0.052118-0

Ryder LS, Harris BD, Soanes DM, Kershaw MJ, Talbot NJ, Thornton CR (2012) Saprotrophic competitiveness and biocontrol fitness of a genetically modified strain of the plant-growth-promoting fungus *Trichoderma hamatum* GD12. Microbiology 158:84–97. https://doi.org/10.1099/mic.0.051854-0

Salas-Marina MA, Isordia-Jasso MI, Islas-Osun MA, Delgado-Sánchez P, Jiménez-Bremont JF, Rodríguez-Kessler M, Rosales-Saavedra MT, Herrera-Estrella A, Casas-Flores S (2015) The Epl1 and Sm1 proteins from Trichoderma atroviride and Trichoderma virens differentially modulate systemic disease resistance against different life style pathogens in *Solanum lycopersicum*. Front Plant Sci 6:77. https://doi.org/10.3389/fpls.2015.00077

Saloheimo M, Paloheimo M, Hakola S, Pere J, Swanson B, Nyyssönen E, Bhatia A, Ward M, Penttilä M (2002) Swollenin, a Trichoderma reesei protein with sequence similarity to the plant expansins, exhibits disruption activity on cellulosic materials. Eur J Biochem 269:4202–4211

Saravanakumar K, Fan L, Fu K, Yu C, Wang M, Xia H, Sun J, Li Y, Chen J (2016) Cellulase from *Trichoderma harzianum* interacts with roots and triggers induced systemic resistance to foliar disease in maize. Sci Rep 6:35543. https://doi.org/10.1038/srep35543

Sarrocco S, Matarese F, Baroncelli R, Vannacci G, Seidl-Seiboth V, Kubicek CP, Vergara M (2017) The constitutive endopolygalacturonase TvPG2 regulates the induction of plant systemic resistance by *Trichoderma virens*. Phytopathology 107:537–544. https://doi.org/10.1094/PHYTO-03-16-0139-R

Schmoll M, Dattenböck C, Carreras-Villaseñor N, Mendoza-Mendoza A, Tisch D, Alemán MI, Baker SE, Brown C, Cervantes-Badillo MG, Cetz-Chel J, Cristobal-Mondragon GR, Delay L, Esquivel-Naranjo EU, Frischmann A, Gallardo-Negrete JJ, García-Esquivel M, Gomez-Rodriguez EY, Greenwood DR, Hernández-Oñate M, Kruszewska JS, Lawry R, Mora-Montes HM, Muñoz-Centeno T, Nieto-Jacobo MF, Nogueira Lopez G, Olmedo-Monfil V, Osorio-Concepcion M, Piłsyk S, Pomraning K, Rodriguez-Iglesias A, Rosales-Saavedra MT, Sánchez-Arreguín JA, Seidl-Seiboth V, Stewart A, Uresti-Rivera EE, Wang C-L, Wang T-F, Zeilinger S, Casas-Flores S, Herrera-Estrella A (2016) The genomes of three uneven siblings: footprints of the lifestyles of three trichoderma species. Microbiol Mol Biol Rev 80:205–327. https://doi.org/10.1128/MMBR.00040-15

Shoresh M, Harman GE, Mastouri F (2010) Induced systemic resistance and plant responses to fungal biocontrol agents. Annu Rev Phytopathol 48:21–43. https://doi.org/10.1146/annurev-phyto-073009-114450

Thürich J, Meichsner D, Furch ACU, Pfalz J, Krüger T, Kniemeyer O, Brakhage A, Oelmüller R (2018) Arabidopsis thaliana responds to colonisation of Piriformospora indica by secretion of symbiosis-specific proteins. PLoS One 13:e0209658. https://doi.org/10.1371/journal.pone.0209658

Tisserant E, Malbreil M, Kuo A, Kohler A, Symeonidi A, Balestrin R, Charron P, Duensing N, Frei dit Frey N, Gianinazzi-Pearson V, Gilbert LB, Handa Y, Herr JR, Hijri M, Koul R, Kawaguchi M, Krajinski F, Lammers PJ, Masclaux FG, Murat C, Morin E, Ndikumana S, Pagni M, Petitpierre D, Requena N, Rosikiewicz P, Riley R, Saito K, San Clemente H, Shapiro H, van Tuinen D, Becard G, Bonfante P, Paszkowski U, Shachar-Hill YY, Tuskan GA, Young JPW, Sanders IR, Henrissat B, Rensing SA, Grigoriev IV, Corradi N, Roux C, Martin F (2013) Genome of an arbuscular mycorrhizal fungus provides insight into the oldest plant symbiosis. Proc Natl Acad Sci 110:20117–20122. https://doi.org/10.1073/pnas.1313452110

Trushina N, Levin M, Mukherjee PK, Horwitz BA (2013) PacC and pH–dependent transcriptome of the mycotrophic fungus *Trichoderma virens*. BMC Genomics 14:138. https://doi.org/10.1186/1471-2164-14-138

Viterbo A, Chet I (2006) TasHyd1, a new hydrophobin gene from the biocontrol agent Trichoderma asperellum, is involved in plant root colonization. Mol Plant Pathol 7:249–258. https://doi.org/10.1111/j.1364-3703.2006.00335.x

Viterbo A, Wiest A, Brotman Y, Chet I, Kenerley C (2007) The 18mer peptaibols from *Trichoderma virens* elicit plant defence responses. Mol Plant Pathol 8:737–746. https://doi.org/10.1111/j.1364-3703.2007.00430.x

Weaver MA, Kenerley CM (2008) Competitiveness of a genetically engineered strain of *Trichoderma virens*. Mycopathologia 166:51–59. https://doi.org/10.1007/s11046-008-9118-z

Yedidia II, Benhamou N, Chet II (1999) Induction of defense responses in cucumber plants (Cucumis sativus L.) By the biocontrol agent *Trichoderma harzianum*. Appl Environ Microbiol 65:1061–1070

Yu C, Dou K, Wang S, Wu Q, Ni M, Zhang T, Lu Z, Tang J, Chen J (2019) Elicitor hydrophobin Hyd1 interacts with Ubiquilin1-like to induce maize systemic resistance. J Integr Plant Biol:12796. https://doi.org/10.1111/jipb.12796

Chapter 2
The Vocabulary of Trichoderma-Plant Interactions

M. M. Monti, P. A. Pedata, L. Gualtieri, and M. Ruocco

Abstract In their natural environment, plants and microorganisms are persistently associated and communicating. This interaction is a finely tuned interplay triggered by chemical messages continuously exchanged by the actors. Recent studies confirm that the plant microbiome at root level is highly involved in many nutrient and biological processes. Commonly, soil and rhizosphere microbiome include species of *Trichoderma* genus, strongly influencing plant growth and differentiation and resistance to factors of abiotic and biotic stresses. Since we are imagining a molecular talk between the two interlocutors, here we produce the *Trichoderma* vocabulary, where we go through all its known "*words*". MAMPs will be described according to the classes listed below: microbial phytohormones (MP), proteins, secondary metabolites (SM) including microbial volatile organic compounds (MVOCs), other SM, and peptaibols. In the present vocabulary, we have included only those *Trichoderma*-secreted MAMPs for which a clear biological activity on plant has been demonstrated. In the present chapter, we aim at describing the factors known to participate in the multifaceted chatting between *Trichoderma* and its roommates, mainly the plant.

Keywords *Trichoderma* · Microbial phytohormones · Secondary metabolites · Microbial volatile organic compounds (MVOCs) · Peptaibols

2.1 Introduction

Species of the genus *Trichoderma* are among the most abundant biotrophic and saprotrophic fungi, commonly found as mycoparasites on other fungi, or as saprophytes on decaying wood and bark. Most *Trichoderma* species are microbial biological control agents (MBCAs) behaving as "generalists" since they are able to

M. M. Monti · P. A. Pedata · L. Gualtieri · M. Ruocco (✉)
Institute for Sustainable Plant Protection (IPSP-CNR), Portici, NA, Italy
e-mail: michelina.ruocco@ipsp.cnr.it

A. K. Sharma, P. Sharma (eds.), *Trichoderma*, Rhizosphere Biology,
https://doi.org/10.1007/978-981-15-3321-1_2

control a vast range of taxonomically unrelated pathogens by using a variety of mechanisms of action. *Trichoderma*-enriched biofertilizers may be used to enrich bioavailability to roots of phosphorous, crease soil-borne pathogens, and enhance soil health, increasing crop yield and lowering the need of fertilizers and pesticides. Moreover, recent studies highlighted that *Trichoderma* application on *Brassica rapa* increased biomass production, and potassium and secondary metabolite concentrations, with higher effects in not fertilized compared to fertilized soils, and increased calcium, magnesium, and iron levels in plants cultivated in low-fertility soils (Caporale et al. 2019). For its unique/distinctive lifestyle features, ability to colonize plant roots, and beneficial effect on plant physiology, *Trichoderma* can be considered as an opportunistic, avirulent plant symbiont (Harman et al. 2004).

The positive impact exerted by *Trichoderma* on the plant involves metabolic and phenotypic modifications at plant level (De Palma et al. 2019). This is the result of an intimate *Trichoderma*-plant interaction and crosstalk which modulate the behaviour of both partners. It is well known that microbes make the plant known their presence by producing molecules designed as microbe-associated molecular patterns (MAMPs). Even if few information are available on the molecular mechanisms eliciting signal transduction in the initial stages of the communication of plants with biocontrol fungi, during the past 10 years a growing literature has been shedding light on this aspect. In the present chapter we aim at describing the factors known to participate in the multifaceted chatting between *Trichoderma* and its roommates, mainly the plant. In the present vocabulary we have included only those *Trichoderma*-secreted MAMPs for which a clear biological activity on plant has been demonstrated. MAMPs will be described according to the classes listed below: microbial phytohormones (MP), proteins, secondary metabolites (SM) including microbial volatile organic compounds (MVOCs), other SM, and peptaibols.

2.2 Microbial Phytohormones

Plant hormones are signal molecules found across the plant kingdom, algae included, that are active in extremely low concentrations. Differently from animals, whose hormone production is confined to specialized glands, in plants each cell is able to produce phytohormones. These molecules are defined as "chemical regulators" produced by plants to adjust growth and development but also to respond to stresses. Also, microorganisms can produce and release phytohormones that we define as microbial phytohormones (MP).

IAA (indole-3-acetic acid): a diffusible phytohormone also secreted by several *Trichoderma* spp. even if its production is strongly influenced by the strain and the growing media. IAA biosynthesis pathway in fungi is dependent on tryptophan; therefore, adding this amino acid to the culture media enhances higher production of IAA (Nieto-Jacobo et al. 2017). Plant growth promotion by *T. virens* Gv29.8 and

T. atroviride IMI206040 has been explained with their ability to produce MP IAA (Contreras-Cornejo et al. 2009).

2.3 Proteins

Cellulase (EC 3.2.1.4): a 57-kDa endo-1,4-β-D-glucanase that catalyses the decomposition of cellulose and related polysaccharides. It is secreted by *T. longibrachiatum*, and due to the induction of multiple key defence processes and hypersensitive reaction-like symptoms, it has been recognized as a powerful elicitor of resistance in melon. Induction of defence mechanisms in melon plants occurs when either native or heat denatured protein is applied, even if the two forms differentially elicit ethylene and jasmonic acid (JA) pathways (Martinez et al. 2001).

Cellulase-Tv enzyme fraction obtained from *T. viride* containing pectinase (0.53 U/mg) and hemicellulase (1.1 U mg). The protein mix, when applied to plant cells, produces electrolyte leakage probably due to alterations in plasma membrane permeability and rupture, loss of cell viability, and strong alteration in phenolic metabolism accompanied by phenolic oxidation and consequent browning. As a consequence, application of this fraction of enzymes to grapevine cell cultures elicits production of phytoalexins, benzoic acid, and resveratrol, thus increasing resistance to *Botrytis cinerea* (Calderon et al. 1993).

EIX (Cellulysin): a 22-kDa endoxylanase catalyzing xylanolysis. This protein is commonly produced by *Trichoderma* and *Gliocladium* species growing on xylan (Dean et al. 1989). In the plant, translocation of EIX occurs via the vascular system. In tobacco, it elicits a hypersensitive response (HR) with many associated symptoms: elicitation of ethylene biosynthesis, electrolyte leakage, pathogenesis-related (PR) protein expression, and limited necrosis (Bailey et al. 1990). In tomato (*Solanum lycopersicum*), it is sensed by the plants by a endoxylanase receptor protein (EIX) (Bailey et al. 1993; Ron et al. 2000). Ethylene biosynthesis is correlated to the accumulation of 1-aminocyclopropane-1-carboxylic acid (*acc*) synthase and *acc* oxidase transcripts (Avni et al. 1994). It induces the biosynthesis of VOCs in *Nicotiana plumbaginifolia*, *Phaseolus lunatus*, and *Zea mays* with a pattern largely resembling the one induced by herbivore damage or JA treatment, suggesting the triggering of the octadecanoid signalling pathway (Piel et al. 1997).

Hydrophobins small proteins of less than 100 amino acids, presenting eight cysteine residues in conserved positions, found on the outer surfaces of cell walls of hyphae and conidia of all filamentous asco- and basidiomycetes (Seidl-Seiboth et al. 2011). Hydrophobins drive interactions between the fungus and its environment. Conventionally, they are separated into two classes (class I and II) on the basis of their solubility in solvents, hydropathy profiles, and spacing between the conserved cysteines.

HFB7 a class II hydrophobin with a taxonomically restricted presence in only two clades of *Trichoderma* (*Harzianum* and *Virens*). In *T. virens*, it has been found strongly upregulated in the initial stages of interaction of fungus with tomato roots, whereas it declined during a prolonged interaction. Expression of *hfb7* gene increases when *T. virens* interacts with other fungi and in response to a diversity of abiotic stress conditions, as well as during growth of aerial hyphae in standing liquid culture (Przylucka et al. 2017).

HYTLO1 a 7.2-kDa class II hydrophobin-secreted *T. longibrachiatum* MK1, a strain used for biocontrol. This hydrophobin has antifungal activity and plays a role as a MAMP and plant growth promotion (PGP), mainly stimulating root formation. It can increase resistance to pathogens and activate defence-related responses mediated by reactive oxygen species (ROS), superoxide dismutase, oxylipin, phytoalexins, and PR proteins (Ruocco et al. 2015). HYTLO1 plays a key role in the initial stages of the plant-fungus interaction, causing a rapid cytosolic Ca^{2+} increase in *Lotus japonicus* suspension cell cultures resulting in the activation of the defence-related genes *mpk3*, *wrk33*, and *cp450*. By immunocytochemical analyses, it has been demonstrated that this fungal hydrophobin forms a protein film covering the plant cell wall (Moscatiello et al. 2018).

PKS/NRPS a polyketide synthase of *T. virens* that is known to induce systemic resistance (ISR) response in maize. Mutation in PKS/NRPS hybrid gene prevented *T. virens* from inducing the defence response gene *pal* (phenylalanine ammonia lyase) but not the defence response gene *aos* (allene oxide synthase) (Mukherjee et al. 2012a).

QID74 a *T. harzianum* CECT 2413 74-kDa cysteine-rich cell wall protein showing a surprising homology to a salivary protein of *Chironomus*, a dipteran insect. QIDd74 plays a central role in adherence to hydrophobic surfaces and cellular protection (Rosado et al. 2007) and plays a role in root architecture by inducing lateral root formation in cucumber and tomato. QID74 expression could contribute to secretion of higher levels of growth-promoting compounds that enhance root hair proliferation. As a consequence of modification in root architecture, the total absorbing surface increases, promoting uptake and translocation of NPK and micronutrients in the sprouts, overall enhancing the plant biomass (Samolski et al. 2012).

Serine Proteinase an 18-kDa *T. virens*-secreted glycosylated protein, significantly able to elicit terpenoid and peroxidase activity in cotton radicles when applied either in its native or denatured form (Hanson and Howell 2004).

SM1 a 12.60 kDa *T. virens*-secreted proteinaceous elicitor designated as small protein 1 member of the cerato-platanin family (Djonovic et al. 2006, 2007), not glycosylated, which may interact with oligosaccharides or other small molecules. Its monomeric form is active and able to elicit defence responses in the plant (Crutcher and Kenerley 2019). SM1 orthologs have been purified and characterized from *T. harzianum* (Freitas et al. 2014), *T. atroviride* (eliciting plant response EPL1)

(Seidl et al. 2006), and *T. asperellum* (EPL1-Tas) (Yu et al. 2018). Purified SM1 and its orthologs trigger production of ROS in the seedlings of several plant species and induce, both locally and systemically, the expression of defence-related genes. Its action has been associated with significant induction of many genes as those involved in JA, and green leaf volatile biosynthesis, and those coding for a peroxidase and an α-dioxygenase. Deletion and overexpression of SM1 result in a significant reduction and enhancement of disease protection, respectively. In contrast, neither deletion nor overexpression of SM1 influences normal growth or development of *T. virens*, germination of conidia, production of gliotoxin, hyphal coiling, hydrophobicity, or competence to colonize maize roots (Djonovic et al. 2007).

SM2 a 12.6-kDa *T. virens* cerato-platanin protein family elicitor, paralog of SM1. Gene expression of *Sm2* and of its ortholog *Epl2* is related to the formation and maturation of conidia in *T. virens* and *T. atroviride*, respectively (Frischmann et al. 2013), and has been found important for promoting plant protection in the maize-*Cochliobolus heterostrophus* pathosystem (Gaderer et al. 2015).

SSCP a group of small secreted cysteine-rich proteins from *T. virens*. Following gene knockout experiments, the role of four SSCPs (77560, 71692, 92810, 111486) in ISR in plants has been investigated. Deleted gene lines showed improved ISR activity compared to a control strain, demonstrating them as negative effectors. These results suggest that these SSCPs may be crucial for the fine-tuning of ISR operated by *Trichoderma*. Indeed, maize plants colonized by the mutants are more resistant to *C. heterostrophus* than plants interacting with control strain (Lamdan et al. 2015).

Tas-ACDS 1-Aminocyclopropane-1-carboxylate deaminase from *T. asperellum* T203 involved in plant root growth promotion of canola (*Brassica napus*) (Viterbo et al. 2010).

TASHYD1 a *T. asperellum* class I hydrophobin contributing to the interaction of *Trichoderma* with the plant. Deleted mutants for *TasHyd1* gene showed a reduction of both adhesion of germinating spores on root surfaces and intercellular colonization of the root. It has been postulated that TASHYD1 might also offer protection to the growing hyphae of *T. asperellum* from plant defence compounds locally synthesized during the early stages of plant colonization, permitting this fungus to invade the intercellular spaces of the plant root (Viterbo and Chet 2006).

TasSwo Swollenin, a *T. asperellum* 55-kDa expansin-like protein. This is a group of non-enzymatic proteins, playing a crucial role in deagglomeration of cellulose and other cell wall polysaccharides by loosening macrofibrils, leading to plant biomass degradation. Swollenin has a bimodular structure, composed of an N-terminal carbohydrate-binding module family-1 domain (CBD) connected by a linker region to an expansin-like domain, with significant similarity to an EXPB-like protein precursor from plants. Study of *T. asperellum*-cucumber (*Cucumis sativus*) root interaction showed that the swollenin CBD peptide is the active part of the protein because in cucumber it stimulates local defence responses in roots and leaves, providing local protection toward infection by *B. cinerea* and *Pseudomonas syringae* pv *lachrymans*, with a role of MAMP in the *T. asperellum*-plant interaction (Brotman et al. 2008).

Thioredoxin TV81963 and TV216458 are two thioredoxins of *T. virens* involved in the modification of arrangement of -S-S- bonds in proteins. *T. virens* secretes these enzymes into the apoplast following colonization of maize roots, where they may function as putative effector-like proteins, decreasing ROS activity and reducing the plant defence reactions (Nogueira-Lopez et al. 2018).

ThPG1 a 38.3-kDa endopolygalacturonase from *T. harzianum* T34. Its primary structure presents a putative signal peptide of 21 amino acids with the conserved glycosyl-hydrolases family-28 domain correlated to its catalytic activity of cell wall degradation. ThPG1 is needed for an efficient root colonization by *T. harzianum* and to activate ISR-like defence in *Arabidopsis* (Moran-Diez et al. 2009).

ThPH1 and ThPH2 two hydrolases (exo-1,4-β-D-glucanase and endo-1,4-β-D-glucanase) from *T. harzianum* triggering the transitory production of ROS. ThPH1 and ThPH2 are necessary to activate ISR in maize elevating free cytosolic calcium level in maize leaf and upregulating genes associated to the jasmonate/ethylene signalling pathway. Deletion mutants of *T. harzianum* for these two hydrolases could not trigger the immune defence-related genes in maize, thus protecting it against leaf disease. These hydrolases thus could be defined as *Trichoderma* MAMPs (Saravanakumar et al. 2016).

TV11061 a disulphideisomerase from *T. virens* involved in the rearrangement of -S-S- bonds in proteins. *T. virens* secretes this enzyme into the apoplast following colonization of maize roots, where it may function as putative effector-like protein, decreasing ROS activity and reducing the plant defence reactions (Nogueira-Lopez et al. 2018).

TvHYDII1 a class II hydrophobin family member from *T. virens*, which enhances its antagonistic activity against *Rhizoctonia solani* and is involved in plant root colonization. In fact, *T. virens* strains deleted for *tvhydii1* gene have a diminished ability to colonize tomato roots, while those overexpressing it show enhanced root colonization (Guzmán-Guzmán et al. 2017).

TvINV a *T. virens* invertase of glycosyl hydrolase family-32. A homologous sequence has been identified in *T. atroviride* genome. Expression of the gene coding for TvINV was detected only in medium containing sucrose or in the presence of plants. TvINV expression is fundamental for the control of root colonization, as *T. virens* mutant strains with reduced production of TvINV and unable to metabolize sucrose present higher ability in colonizing roots (Vargas et al. 2009).

2.4 Secondary Metabolites

Secondary metabolites (SMs) are small bioactive molecules apparently not directly essential for fungal growth, but proved to play a role in signalling, development, and interaction of microorganisms with the environment. One hypothesis is that micro-organisms have evolved SM production for communication with, or defence against, other microbes or multicellular organisms, as plants, in order to build their own habitat (Brakhage and Schroeckh 2011). Fungi genome sequencing is revealing a huge number of potential gene clusters for the production of putative SMs. *Trichoderma* spp. have a huge repertoire of genes putatively implicated in the biosynthesis of SMs, such as non-ribosomal peptides, polyketides, terpenoids, and pyrones, but at the moment few of them have been characterized as MAMPs.

2.4.1 Microbial Volatile Organic Compounds

Volatile organic compounds (VOCs) are composite blends of low-molecular-weight compounds, responsible for inter-and intra-individual communication involving plants, antagonists, and mutualistic symbionts both below and above ground. MVOCs are commonly produced by a vast range of microbes, from bacteria to fungi, through all the stages of their life. An increasing amount of evidence suggests that MVOCs are intercepted by the plant and can contribute both positively or negatively to plant fitness (Kanchiswamy et al. 2015), and therefore, MVOCs can be considered part of the words that *Trichoderma* spp. use to communicate with plants.

Butanal (C_4H_8O) an aldehyde with MW of 72.11 emitted by *Trichoderma* spp. inhibiting *Arabidopsis* seed germination (Lee et al. 2019).

1-Decene ($C_{10}H_{20}$) an alkene with a MW of 140.27 emitted by *Trichoderma* spp. inhibiting *Arabidopsis* seed germination but promoting plant development (Lee et al. 2019).

2-Heptanone ($C_7H_{14}O$) a ketone with MW of 114.19 emitted by *Trichoderma* spp. inhibiting Arabidopsis seed germination but promoting plant development (Lee et al. 2019).

2-Heptylfuran ($C_{11}H_{18}O$) a heteroaromatic with MW of 166.26 emitted by *Trichoderma* spp. involved in the promotion of Arabidopsis plant development (Lee et al. 2019).

(D)-Limonene ($C_{10}H_{16}$) a monoterpene with a MW 136.23 emitted by *Trichoderma* spp. involved in the promotion of *Arabidopsis* plant development (Lee et al. 2019).

(*S*)-Limonene ($C_{10}H_{16}$) a monoterpene with a MW 136.23 emitted by *Trichoderma* spp. inhibiting *Arabidopsis* seed germination but promoting plant development (Lee et al. 2019).

2-Methyl butanal and 3-methyl butanal ($C_5H_{10}O$) aldehydes with MW of 86.13 emitted by *Trichoderma* spp. inhibiting *Arabidopsis* seed germination (Lee et al. 2019).

2-Methyl-1-butanol and 3-methyl-1-butanol ($C_5H_{12}O$) are amyl alcohols with MW of 88.15 emitted by *Trichoderma* spp. inhibiting *Arabidopsis* seed germination but promoting plant development (Lee et al. 2019).

2-Methyl-1-propanol ($C_4H_{10}O$) an alcohol with MW of 74.12 emitted by *Trichoderma* spp. involved in the promotion of *Arabidopsis* plant development (Lee et al. 2019).

Nonanal ($C_9H_{18}O$) an aldehyde with MW of 142.24 emitted by *Trichoderma* spp. inhibiting *Arabidopsis* seed germination and plant development (Lee et al. 2019).

Octanoic acid ($C_8H_{16}O_2$) an ester with MW of 144.21 emitted by *Trichoderma* spp. involved in the promotion of *Arabidopsis* plant development (Lee et al. 2019).

2-Octenal ($C_8H_{14}O$) an aldehyde with a MW of 126.20 emitted by *Trichoderma* spp. inhibiting *Arabidopsis* seed germination and plant development (Lee et al. 2019).

3-Octanone ($C_8H_{16}O$) a ketone with MW of 128.21 emitted by *Trichoderma* spp. inhibiting *Arabidopsis* seed germination (Lee et al. 2019).

1-Octanol ($C_8H_{18}O$) an alcohol with MW of 130.23 emitted by *Trichoderma* spp. inhibiting *Arabidopsis* seed germination (Lee et al. 2019).

1-Octen-3-ol ($C_8H_{16}O$) an alcohol with MW of 128.21 emitted by *Trichoderma* spp. inhibiting *Arabidopsis* seed germination and plant development (Lee et al. 2019).

1-Octen-3-one ($C_8H_{14}O$) a ketone with MW of 126.20 emitted by *Trichoderma* spp. inhibiting *Arabidopsis* seed germination and plant development (Lee et al. 2019).

1-Pentanol ($C_5H_{12}O$) an alcohol with MW of 88.15 emitted by *Trichoderma* spp. inhibiting *Arabidopsis* seed germination and plant development (Lee et al. 2019).

2-Pentanone ($C_5H_{10}O$) a ketone with MW of 86.13 emitted by *Trichoderma* spp. inhibiting *Arabidopsis* seed germination (Lee et al. 2019).

6PP (6-n-pentyl-6H-pyran-2-one): ($C_{10}H_{14}O_2$) a monoterpene with MW of 166.22. It is a major secondary metabolite produced by several *Trichoderma* spp., mainly by *T. atroviride* and *T. asperellum*. Its effects on plants are multiple: 6PP isolated from *T. atroviride* P1 (Vinale et al. 2008) is directly implicated in the triggering of plant defence genes and plant growth promotion (Marra et al. 2019). A decrease of disease symptoms caused on tomato and canola by *B. cinerea* and *Leptosphaeria maculans*, respectively, was obtained on seedlings treated with 6PP (Vinale et al. 2008). As well, 6PP from *T. atroviride* controls *Arabidopsis thaliana* root morphogenesis via auxin signalling, modulating in a specific and dose-dependent manner in primary roots the expression of PIN auxin-transport proteins (Garnica-Vergara et al. 2016). On the contrary, it has been reported that germination of *Arabidopsis* seeds is not influenced by 6PP application (Lee et al. 2019).

Sesquiterpenes a mix of different molecules (β-caryophyllene, (−)-β-elemene, germacrene D, τ-cadinene, δ-cadinene, α-amorphene, and τ-selinene and monoterpenes β-myrcene, trans-β-ocimene, and cis-β-ocimene) produced by several *Trichoderma* species. Sesquiterpene activity on plants has been correlated with root branching promotion and plant biomass increase, probably as a result of an increased nutrient uptake augmenting root absorptive capacity (Contreras-Cornejo et al. 2014), even if β-ocimene has been correlated with inhibition of *Arabidopsis* plant development (Lee et al. 2019).

TD (trichodiene): ($C_{15}H_{24}$) with MW of 204.35. TD synthesis is catalysed by a sesquiterpene synthase produced by several *Trichoderma* species. TD strongly induces the expression of salicylic acid (SA)-responsive genes and reduces the growth of lateral tomato root (Malmierca et al. 2015).

2.4.2 Other SM

Anthraquinone one of the major secondary metabolites produced by *T. harzianum* T22 and A6 strains (Vinale et al. 2008). It can promote tomato and canola growth (Vinale et al. 2008).

Ferricrocin a siderophore produced by *T. virens*. Ferricrocins normally play a crucial role in iron homeostasis of filamentous fungi and are involved in several Fe-related metabolic processes. Intracellular ferricrocin produced by *T. virens* is implicated in growth, conidiation, and ISR in maize. *T. virens* mutants, lacking ferricrocin, colonize more aggressively the roots than the wild type. For ISR plant assays, seeds were planted in *Trichoderma*-inoculated soil and, subsequently, leaves were inoculated with the necrotrophic leaf pathogen *C. heterostrophus*. Differently from the wild strain, the mutants failed to confer resistance against the invading pathogen, suggesting a loss of ISR (Mukherjee et al. 2018).

HA (Harzianic acid): a tetramic acid derivative isolated from several strains of *T. harzianum* recently described as a novel siderophore, for its iron binding ability (Vinale et al. 2013). It can promote the growth of cucumber plants and reduce disease symptoms caused by *Pythium* spp., *P. irregulare*, and *R. solani* both in the laboratory and glasshouse conditions (Vinale et al. 2009).

HA (Harzianum A): a non-phytotoxic trichothecene from *T. arundinaceum* IBT 40837. It plays an important function in priming tomato plants against *B. cinerea*, inducing a faster and higher levels of expression of the genes involved in the defence response to the pathogen attack (Malmierca et al. 2012).

Harzianolide a major SM of the group of butenolides, produced by several strains of *T. harzianum*. It significantly promotes tomato seedling growth and tomato plant biomass production both in a hydroponic culture and soil (Cai et al. 2013) and growth of wheat and canola seedlings, in addition to an auxin-like activity on etiolated pea stems (Vinale et al. 2008).

Harzianopyridone one of the major SM produced by *T. harzianum* T39. Application of this SM can induce ISR on tomato and canola vs. *B. cinerea* or *L. maculans*, respectively. Harzianopyridone significantly inhibits and/or changes the pathogenicity and virulence-related protein PMR1 of *B. cinerea* (Saravanakumar et al. 2018; Vinale et al. 2008).

Iso-HA (Isoharzianic acid): a stereoisomer of harzianic acid also isolated from *T. harzianum*. Applications of iso-HA in vitro can inhibit the radial growth mycelium of *Sclerotina sclerotiorum* and *R. solani*. It is also able to induce disease resistance and improve the germination of tomato seeds. Moreover, the presence of tomato plant tissue in the growing liquid medium may affect the production of HA and iso-HA. Indeed, plant tissue stimulates the production of HA by negatively modulating, in parallel, the biosynthesis of its analog (Vinale et al. 2014).

T22 Azaphilone one of the major secondary metabolites produced by *T. harzianum* T22. Application of this secondary metabolite can induce ISR on canola vs. *L. maculans* (Vinale et al. 2008).

T39 Butenolide one of the major secondary metabolites produced by *T. harzianum* T22 and T39. It can induce tomato and canola growth promotion (Vinale et al. 2008).

2.4.2.1 Peptaibols

Peptaibols are short peptides of non-ribosomal origin containing α-aminoisobutyric acid (Aib) and other non-proteinogenic amino acids and a C-terminal 1,2-amino alcohol (Mukherjee et al. 2012b). They form voltage-gated channels with selectivity toward monovalent cations, which could explain their antibiotic function (Duclohier and Wroblewski 2001). Their ecological relevance is due not only to their antimicrobial properties but also to their ability to interact with plants activating ISR. In *Trichoderma*, genomes have been found in two peptaibol synthetases organized in 18 and 14 modules (Hermosa et al. 2014) and some *Trichoderma* peptaibols have been deeply studied and can be referred as MAMPs.

ALA (Alamethicin): a 20-mer peptide from *T. viride* (Brewer et al. 1987). In lima bean (*P. lunatus*), ALA potently elicits the biosynthesis of VOCs mainly via the octadecanoid-signalling pathway, with a 20-fold increase of JA. ALA can also upregulate SA biosynthesis and, in several species like *Bryonia*, *Lathyrus*, and *Pisum*, induces tendril coiling; however, in this case the response seems to be independent from octadecanoid biosynthesis (Engelberth et al. 2001).

18-Residue Peptaibols in *T. virens* Gv29-8, the 18-residue peptaibols act as elicitors of non-cultivar-specific defence responses, playing a critical role in the chemical communication between *Trichoderma* and cucumber plants (Viterbo et al. 2007).

TK VI (Trichokonin VI): the main trichokonin produced by *T. longibrachiatum* SMF2. In *Arabidopsis* primary root TK VI acts as an inhibitor of both cell division and cell elongation, disturbing the maintenance of the root stem cell niche, in a dose-dependent manner, within a concentration range of 0.5–10 μM. In the root tip, it can also modulate local auxin biosynthesis and polar auxin transport, increasing the auxin level and disrupting the auxin response gradients. TK VI is also an elicitor of systemic resistance in Chinese cabbage and tobacco and an inductor of programmed cell death in tumour cells (Shi et al. 2016).

Trichokonins 20-mer antimicrobial peptaibols isolated from *T. pseudokoningii* SMF2 able to induce defence responses and ISR in tobacco (*Nicotiana tabacum* var. Samsun NN) against infection of tobacco mosaic virus (TMV). Trichokonin application can also significantly increase activities of PR enzymes PAL and POD and upregulate the expression of several plant defence genes (Luo et al. 2010).

2.5 Conclusion

The world of microbial-plant interactions is amazing but still in the early beginning of its discovery. Among microorganisms, *Trichoderma* spp. are some of the most extensively studied, mainly due to their potential commercial applications worldwide. *Trichoderma* spp. are considered as model organisms, so results obtained studying this genus can also be extended to other fungi. Despite the huge number of secreted proteins by *Trichoderma* spp., only a few have been functionally characterized. In the last decade researchers have invested considerable efforts to unravel the molecular mechanisms that modulate and regulate the interaction between *Trichoderma* and the plant, mainly in order to identify molecules with potential applications in agriculture.

In spite of the growing number of papers on "*Trichoderma* biochemistry", very few are on well-characterized *Trichoderma* MAMPs. Indeed, the dialogue between *Trichoderma* and plants at the root level is still far from being deciphered, and we still do not have data on how plants can perceive the signals emitted by *Trichoderma*, as only the receptor for EIX having been identified.

Proteinaceous MAMPs are the most studied and analysed, probably because proteins are among the first molecules to be investigated under different perspectives. However, thanks to the use of new and sophisticated facilities for VOC detection, also the number of identified volatiles is growing very fast, even if their characterization is still poor and controversial, often leading to argumentative results.

In this chapter the reader can find the vocabulary of the known "*words*" that *Trichoderma* spp. use to chat with plants. This first vocabulary aims at being a stimulus for the definition of an increasingly more complete one, to be upgraded over time. Looking through the references reported in this chapter, it can be noted that even if research in this field is in its very beginning, many published results could give new insights if interpreted considering the interaction *Trichoderma*-plants as a dynamic system based on continuous signalling from both sides.

References

Avni A, Bailey BA, Mattoo AK, Anderson JD (1994) Induction of ethylene biosynthesis in Nicotiana tabacum by a *Trichoderma viride* xylanase is correlated to the accumulation of 1-aminocyclopropane-1-carboxylic acid (ACC) synthase and oxidase transcripts. Plant Physiol 106:1049–1055

Bailey BA, Dean JFD, Anderson JD (1990) An Ethylene biosynthesis inducing Endoxylanase elicits electrolyte leakage and necrosis in nicotiana tabacum cv xanthi leaves. Plant Physiol 94:1849–1854

Bailey BA, Korcak RF, Anderson JD (1993) Sensitivity to an Ethylene Biosynthesis inducing Endoxylanase in Nicotiana tabacum cv Xanthi is controlled by a single dominant gene. Plant Physiol 101:1081–1088

Brakhage AA, Schroeckh V (2011) Fungal secondary metabolites - strategies to activate silent gene clusters. Fungal Genet Biol 48:15–22

Brewer D, Mason FG, Taylor A (1987) The production of alamethicins by *Trichoderma* spp. Can J Microbiol 33:619–625

Brotman Y, Briff E, Viterbo A, Chet I (2008) Role of swollenin, an expansin-like protein from *Trichoderma*, in plant root colonization. Plant Physiol 147:779–789

Cai F, Yu G, Wang P, Wei Z, Fu L, Shen Q, Chen W (2013) Harzianolide, a novel plant growth regulator and systemic resistance elicitor from *Trichoderma harzianum*. Plant Physiol Biochem 73:106–113

Calderon AA, Zapata JM, Munoz R, Pedreno MA, Barcelo AR (1993) Resveratrol production as a part of the hypersensitive like response of grapevine cells to an elicitor from *Trichoderma viride*. New Phytol 124:455–463

Caporale AG, Vitaglione P, Troise AD, Pigna M, Ruocco M (2019) Influence of three different soil types on the interaction of two strains of *Trichoderma harzianum* with Brassica rapa subsp sylvestris cv esculenta, under soil mineral fertilization. Geoderma 350:11–18

Contreras-Cornejo AH, Macias-Rodriguez L, Cortes-Penagos C, Lopez-Bucio J (2009) *Trichoderma virens*, a plant beneficial fungus, enhances biomass production and promotes lateral root growth through an auxin-dependent mechanism in arabidopsis. Plant Physiol 149:1579–1592

Contreras-Cornejo AH, Macias-Rodriguez L, Herrera-Estrella A, Lopez-Bucio J (2014) The 4-phosphopantetheinyl transferase of *Trichoderma virens* plays a role in plant protection against Botrytis cinerea through volatile organic compound emission. Plant and Soil 379:261–274

Crutcher FK, Kenerley CM (2019) Analysis of a putative glycosylation site in the *Trichoderma virens* elicitor SM1 reveals no role in protein dimerization. Biochem Biophys Res Commun 509:817–821

De Palma M, Salzano M, Villano C, Aversano R, Lorito M, Ruocco M, Docimo T, Piccinelli AL, D'Agostino N, Tucci M (2019) Transcriptome reprogramming, epigenetic modifications and alternative splicing orchestrate the tomato root response to the beneficial fungus *Trichoderma harzianum*. Hortic Res 6:5

Dean JFD, Gamble HR, Anderson JD (1989) The Ethylene biosynthesis-inducing xylanase- its induction in *Trichoderma viride* and certain plant pathogens. Phytopathology 79:1071–1078

Djonovic S, Pozo MJ, Dangott LJ, Howell CR, Kenerley CM (2006) Sm1, a proteinaceous elicitor secreted by the biocontrol fungus *Trichoderma virens* induces plant defense responses and systemic resistance. Mol Plant Microbe Interact 19:838–853

Djonovic S, Vargas WA, Kolomiets MV, Horndeski M, Wiest A, Kenerley CM (2007) A proteinaceous elicitor Sm1 from the beneficial fungus *Trichoderma virens* is required for induced systemic resistance in maize. Plant Physiol 145:875–889

Duclohier H, Wroblewski H (2001) Voltage-dependent pore formation and antimicrobial activity by alamethicin and analogues. J Membr Biol 184:1–12

Engelberth J, Koch T, Schuler G, Bachmann N, Rechtenbach J, Boland W (2001) Ion channel-forming alamethicin is a potent elicitor of volatile biosynthesis and tendril coiling Cross talk between jasmonate and salicylate signaling in lima bean. Plant Physiol 125:369–377

Freitas RS, Steindorff AS, Ramada MH, Linhares de Siqueira SJ, Noronha EF, Ulhoa CJ (2014) Cloning and characterization of a protein elicitor Sm1 gene from *Trichoderma harzianum*. Biotechnol Lett 36:783–788

Frischmann A, Neudl S, Gaderer R, Bonazza K, Zach S, Gruber S, Spadiut O, Friedbacher G, Grothe H, Seidl-Seiboth V (2013) Self-assembly at air/water interfaces and carbohydrate binding properties of the small secreted protein EPL1 from the fungus *Trichoderma atroviride*. J Biol Chem 288:4278–4287

Gaderer R, Lamdan NL, Frischmann A, Sulyok M, Krska R, Horwitz BA, Seidl-Seiboth V (2015) Sm2, a paralog of the *Trichoderma* cerato-platanin elicitor Sm1, is also highly important for plant protection conferred by the fungal-root interaction of *Trichoderma* with maize. BMC Microbiol 15:2

Garnica-Vergara A, Barrera-Ortiz S, Munoz-Parra E, Raya-Gonzalez J, Mendez-Bravo A, Macias-Rodriguez L, Francisco Ruiz-Herrera L, Lopez-Bucio J (2016) The volatile 6-pentyl-2H-pyran-2-one from *Trichoderma atroviride* regulates Arabidopsis thaliana root morphogenesis via auxin signaling and ETHYLENE INSENSITIVE 2 functioning. New Phytol 209:1496–1512

Guzmán-Guzmán P, Alemán-Duarte MI, Delaye L, Herrera-Estrella A, Olmedo-Monfil V (2017) Identification of effector-like proteins in *Trichoderma* spp and role of a hydrophobin in the plant-fungus interaction and mycoparasitism. BMC Genet 18:16

Hanson LE, Howell CR (2004) Elicitors of plant defense responses from biocontrol strains of *Trichoderma virens*. Phytopathology 94:171–176

Harman GE, Howell CR, Viterbo A, Chet I, Lorito M (2004) *Trichoderma* species - opportunistic, avirulent plant symbionts. Nat Rev Microbiol 2:43–56

Hermosa R, Elena Cardoza R, Belen Rubio M, Gutierrez S, Monte E (2014) Secondary metabolism and antimicrobial metabolites of *Trichoderma*. In: Gupta VK, Schmoll M, Herrera Estrella A, Upadhyay RS, Druzhinina I, Tuohy MG (eds) Biotechnology and biology of trichoderma. Newnes, Oxford. https://doi.org/10.1016/B978-0-444-59576-8.00010-2

Kanchiswamy CN, Mainoy M, Maffei ME (2015) Chemical diversity of microbial volatiles and their potential for plant growth and productivity. Front Plant Sci 6:151

Lamdan N-L, Shalaby S, Ziv T, Kenerley CM, Horwitz BA (2015) Secretome of *Trichoderma* interacting with maize roots: role in induced systemic resistance. Mol Cell Proteomics 14:1054–1063

Lee S, Behringer G, Hung R, Bennett J (2019) Effects of fungal volatile organic compounds on Arabidopsis thaliana growth and gene expression. Fungal Ecol 37:1–9

Luo Y, Zhang D-D, Dong X-W, Zhao P-B, Chen L-L, Song X-Y, Wang X-J, Chen X-L, Shi M, Zhang Y-Z (2010) Antimicrobial peptaibols induce defense responses and systemic resistance in tobacco against tobacco mosaic virus. FEMS Microbiol Lett 313:120–126

Malmierca MG, Cardoza RE, Alexander NJ, McCormick SP, Hermosa R, Monte E, Gutierrez S (2012) Involvement of *Trichoderma Trichothecenes* in the biocontrol activity and induction of plant defense-related genes. Appl Environ Microbiol 78:4856–4868

Malmierca MG, McCormick SP, Cardoza RE, Alexander NJ, Monte E, Gutierrez S (2015) Production of trichodiene by *Trichoderma harzianum* alters the perception of this biocontrol strain by plants and antagonized fungi. Environ Microbiol 17:2628–2646

Marra R, Lombardi N, d'Errico G, Troisi J, Scala G, Vinale F, Woo S, Lonanomi G, Lorito M (2019) Application of *Trichoderma* strains and metabolites enhances soybean productivity and nutrient content. J Agric Food Chem 67:1814

Martinez C, Blanc F, Le Claire E, Besnard O, Nicole M, Baccou JC (2001) Salicylic acid and ethylene pathways are differentially activated in melon cotyledons by active or heat-denatured cellulase from *Trichoderma longibrachiatum*. Plant Physiol 127:334–344

Moran-Diez E, Hermosa R, Ambrosino P, Cardoza RE, Gutierrez S, Lorito M, Monte E (2009) The ThPG1 endopolygalacturonase is required for the *Trichoderma harzianum*-plant beneficial interaction. Mol Plant Microbe Interact 22:1021–1031

Moscatiello R, Sello S, Ruocco M, Barbulova A, Cortese E, Nigris S, Baldan B, Chiurazzi M, Mariani P, Lorito M, Navazio L (2018) The hydrophobin HYTLO1 secreted by the biocontrol fungus *Trichoderma longibrachiatum* triggers a NAADP-mediated calcium signalling pathway in lotus japonicus. Int J Mol Sci 19:E2596

Mukherjee PK, Buensanteai N, Moran-Diez ME, Druzhinina IS, Kenerley CM (2012a) Functional analysis of non-ribosomal peptide synthetases (NRPSs) in *Trichoderma virens* reveals a polyketide synthase (PKS)/NRPS hybrid enzyme involved in the induced systemic resistance response in maize. Microbiology 158:155–165

Mukherjee PK, Horwitz BA, Kenerley CM (2012b) Secondary metabolism in *Trichoderma* - a genomic perspective. Microbiology 158:35–45

Mukherjee PK, Hurley JF, Taylor JT, Puckhaber L, Lehner S, Druzhinina I, Schumacher R, Kenerley CM (2018) Ferricrocin, the intracellular siderophore of *Trichoderma virens*, is involved in growth, conidiation, gliotoxin biosynthesis and induction of systemic resistance in maize. Biochem Biophys Res Commun 505:606–611

Nieto-Jacobo MF, Steyaert JM, Salazar-Badillo FB, Nguyen DV, Rostas M, Braithwaite M, De Souza JT, Jimenez-Bremont JF, Ohkura M, Stewart A, Mendoza-Mendoza A (2017) Environmental growth conditions of *Trichoderma* spp. affects indole acetic acid derivatives, volatile organic compounds, and plant growth promotion. Front Plant Sci 8:102

Nogueira-Lopez G, Greenwood DR, Middleditch M, Winefield C, Eaton C, Steyaert JM, Mendoza-Mendoza A (2018) The apoplastic secretome of *Trichoderma virens* during interaction with maize roots shows an inhibition of plant defence and scavenging oxidative stress secreted proteins. Front Plant Sci 9:509

Piel J, Atzorn R, Gabler R, Kuhnemann F, Poland W (1997) Cellulysin from the plant parasitic fungus *Trichoderma viride* elicits volatile biosynthesis in higher plants via the octadecanoid signalling cascade. FEBS Lett 416:143–148

Przylucka A, Akcapinar GB, Chenthamara K, Cai F, Grujic M, Karpenko J, Livoi M, Shen Q, Kubicek CP, Druzhinina IS (2017) HFB7-A novel orphan hydrophobin of the *Harzianum* and *Virens* clades of *Trichoderma*, is involved in response to biotic and abiotic stresses. Fungal Genet Biol 102:63–76

Ron M, Kantety R, Martin GB, Avidan N, Eshed Y, Zamir D, Avni A (2000) High-resolution linkage analysis and physical characterization of the EIX-responding locus in tomato. Theor Appl Genet 100:184–189

Rosado IV, Rey M, Codon AC, Govantes J, Moreno-Mateos MA, Benitez T (2007) QID74 Cell wall protein of *Trichoderma harzianum* is involved in cell protection and adherence to hydrophobic surfaces. Fungal Genet Biol 44:950–964

Ruocco M, Lanzuise S, Lombardi N, Woo SL, Vinale F, Marra R, Varlese R, Manganiello G, Pascale A, Scala V, Turra D, Scala F, Lorito M (2015) Multiple roles and effects of a novel *Trichoderma hydrophobin*. Mol Plant Microbe Interact 28:167–179

Samolski I, Rincon AM, Mary Pinzon L, Viterbo A, Monte E (2012) The qid74 gene from *Trichoderma harzianum* has a role in root architecture and plant biofertilization. Microbiology 158:129–138

Saravanakumar K, Fan L, Fu K, Yu C, Wang M, Xia H, Sun J, Li Y, Chen J (2016) Cellulase from *Trichoderma harzianum* interacts with roots and triggers induced systemic resistance to foliar disease in maize. Sci Rep 6:35543

Saravanakumar K, Lu Z, Xia H, Wang M, Sun J, Wang S, Wang Q-q, Li Y, Chen J (2018) Triggering the biocontrol of Botrytis cinerea by *Trichoderma harzianum* through inhibition of pathogenicity and virulence related proteins. Front Agric Sci Eng 5:271–279

Seidl V, Marchetti M, Schandl R, Allmaier G, Kubicek CP (2006) Epl1, the major secreted protein of Hypocrea atroviridis on glucose, is a member of a strongly conserved protein family comprising plant defense response elicitors. FEBS J 273:4346–4359

Seidl-Seiboth V, Gruber S, Sezerman U, Schwecke T, Albayrak A, Neuhof T, von Doehren H, Baker SE, Kubicek CP (2011) Novel hydrophobins from *Trichoderma* define a new hydrophobin subclass: protein properties, evolution, regulation and processing. J Mol Evol 72:339–351

Shi W-L, Chen X-L, Wang L-X, Gong Z-T, Li S, Li C-L, Xie BB, Zhang W, Shi M, Li C, Zhang Y-Z, Son X-Y (2016) Cellular and molecular insight into the inhibition of primary root growth of Arabidopsis induced by peptaibols, a class of linear peptide antibiotics mainly produced by *Trichoderma* spp. J Exp Bot 67:2191–2205

Vargas WA, Mandawe JC, Kenerley CM (2009) Plant-derived sucrose is a key element in the symbiotic association between *Trichoderma virens* and maize plants. Plant Physiol 151:792–808

Vinale F, Sivasithamparam K, Ghisalberti EL, Marra R, Barbetti MJ, Li H, Woo SL, Lorito M (2008) A novel role for *Trichoderma* secondary metabolites in the interactions with plants. Physiol Mol Plant Pathol 72:80–86

Vinale F, Flematti G, Sivasithamparam K, Lorito M, Marra R, Skelton BW, Ghisalberti EL (2009) Harzianic acid, an antifungal and plant growth promoting metabolite from *Trichoderma harzianum*. J Nat Prod 72:2032

Vinale F, Nigro M, Sivasithamparam K, Flematti G, Ghisalberti EL, Ruocco M, Varlese R, Marra R, Lanzuise S, Eid A, Woo SL, Lorito M (2013) Harzianic acid: a novel siderophore from *Trichoderma harzianum*. FEMS Microbiol Lett 347:123–129

Vinale F, Manganiello G, Nigro M, Mazzei P, Piccolo A, Pascale A, Ruocco M, Marra R, Lombardi N, Lanzuise S, Varlese R, Cavallo P, Lorito M, Woo SL (2014) A novel fungal metabolite with beneficial properties for agricultural applications. Molecules 19:9760–9772

Viterbo A, Chet I (2006) TasHyd1, a new hydrophobin gene from the biocontrol agent *Trichoderma asperellum*, is involved in plant root colonization. Mol Plant Pathol 7:249–258

Viterbo A, Wiest A, Brotman Y, Chet I, Kenerley C (2007) The 18mer peptaibols from *Trichoderma virens* elicit plant defence responses. Mol Plant Pathol 8:737–746

Viterbo A, Landau U, Kim S, Chernin L, Chet I (2010) Characterization of ACC deaminase from the biocontrol and plant growth-promoting agent *Trichoderma asperellum* T203. FEMS Microbiol Lett 305:42–48

Yu W, Mijiti G, Huang Y, Fan H, Wang Y, Liu Z (2018) Functional analysis of eliciting plant response protein Epl1-Tas from *Trichoderma asperellum* ACCC30536. Sci Rep 8:7974

Chapter 3
Could *Trichoderma* Be a Plant Pathogen? Successful Root Colonization

Jorge Poveda, Daniel Eugui, and Patricia Abril-Urias

Abstract Root colonization by *Trichoderma* requires a complex molecular dialogue between fungus and plant. This colonization is limited to the outermost layers of the root and does not penetrate the plant vascular bundle. It is possible for a symbiotic relationship to be established in which *Trichoderma* improves plant growth and development through increasing systemic resistance against future possible attacks from pests and/or pathogens, increasing tolerance to abiotic stresses (i.e. salinity, drought, low temperatures), improving the capacity to absorb nutrients and actively stimulating plant growth. Various studies have shown that the ability to activate plant defences via salicylic acid (SA) is essential for the proper development of root colonization. Otherwise, *Trichoderma* would cross the thin line between symbiotic microorganism and opportunistic pathogen, reaching the root vascular bundle and causing the death of the plant tissues.

Keywords *Trichoderma* · Root colonization · Salicylic acid · Benefits · Plant pathogen

---

All authors contributed equally to this work.

J. Poveda (✉)
Spanish-Portuguese Institute for Agricultural Research (CIALE), University of Salamanca, Salamanca, Spain

Biological Mission of Galicia (CSIC), Pontevedra, Spain
e-mail: jorgepoveda@usal.es

D. Eugui
Spanish-Portuguese Institute for Agricultural Research (CIALE), University of Salamanca, Salamanca, Spain

Blue Agro Bioscience, Noain, Navarra, Spain

P. Abril-Urias
Spanish-Portuguese Institute for Agricultural Research (CIALE), University of Salamanca, Salamanca, Spain

Institute of Environmental Sciences of Castilla-La Mancha (ICAM), University of Castilla-La Mancha, Toledo, Spain

A. K. Sharma, P. Sharma (eds.), *Trichoderma*, Rhizosphere Biology,
https://doi.org/10.1007/978-981-15-3321-1_3

3.1 Introduction

The use of species of the genus *Trichoderma* as agents of biological control in agriculture has been extensively studied by numerous authors. There are several direct mechanisms of antagonism through which *Trichoderma* is able to reduce attacks on plants by plant pathogenic agents. These include mycoparasitism (as well as parasitism of nematodes), antibiosis and competition for space and resources. These strategies produce a series of direct benefits for plants owing to the absence of attacks and to metabolic investments in plant defences.

In addition, the symbiotic relationship between *Trichoderma* and plant has a series of indirect benefits that result from the process of root colonization undertaken by the fungus. In this chapter, we will consider the process of root colonization by *Trichoderma*, the indirect benefits obtained by plants through their symbiotic relationship with *Trichoderma* and the thin line that separates mutualism from parasitism in this interaction.

3.2 Root Colonization by *Trichoderma*

Root colonization by rhizosphere beneficial microorganisms is key in facilitating mechanisms that determine the efficacy in increasing plant growth, biocontrol potential and the ability to induce tolerance against abiotic stresses (Harman 2011; Hermosa et al. 2012; Guzmán-Guzmán et al. 2018). Comparative analysis of the genomes of *Trichoderma reesei*, *T. atroviride* and *T. virens* has shown that mycoparasitism is the ancestral way of life of the genus *Trichoderma* and that colonization of the rhizosphere subsequently arose, something that would have been facilitated by the presence in the soil of pathogens and exudates from the roots of plants (Kubicek et al. 2011) as a result of a shift towards a more generalist way of life (Kubicek et al. 2019). *Trichoderma* strains are therefore found in many root ecosystems.

In 1999, Yedidia et al. conducted an electron microscopy of ultrathin sections from *Trichoderma*-treated cucumber roots. They observed that penetration was mainly restricted to the root's epidermis and outer cortex. They also observed a strengthening of the epidermal and cortical cell walls through localized accumulation of callose and cellulose. Furthermore, in 2000, Yedidia et al. observed the formation of typical fungal structures previously associated with mycoparasitic interactions of *T. harzianum* strains that colonize cucumber roots, because these are characterized by hyphal coiling and formation of appressoria.

Later, in 2007, Chacón et al. studied the root-colonization process of tomato plants by *T. harzianum* (Fig. 3.1). They observed the profuse adhesion of hyphae to the plant roots as well as colonization of the root epidermis and cortex. The transformation of *T. harzianum* with green fluorescent protein (GFP) allowed analysis by confocal microscopy that revealed intercellular hyphal growth (cellular

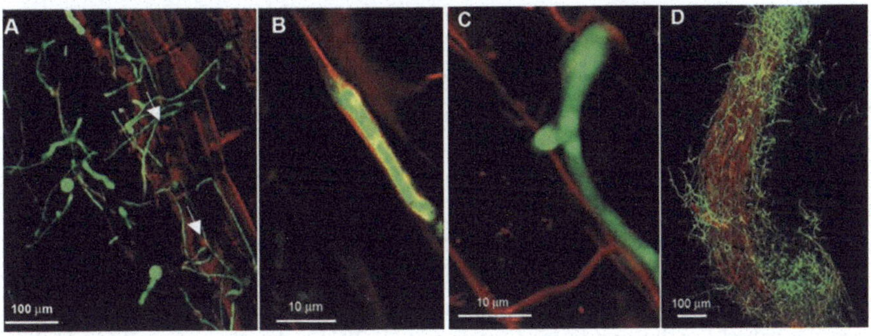

Fig. 3.1 Confocal microscope view of the colonization of tomato roots by *T. harzianum* GFP22. The green fluorescence caused by the expression of GFP indicates the presence of the fungus, while the red fluorescence caused by propidium iodide shows the plant cells. (**a**) The hyphae have entered the roots and colonized the intercellular space of the epidermis 10 h after the inoculation of the fungus in 3-week-old plants. (**b**) The hyphae have grown between the cells of the epidermis and are surrounded by the plant's cell walls. (**c**) In some cases, the hyphae exhibited a swollen tip, which was identified as a non-septate vesicle. (**d**) 24 h after inoculation, large-scale colonization of the root surface by the fungus is observed. (Extracted from Chacón et al. 2007)

apoplast) and the formation of plant-induced papilla-like hyphal tips. Moreover, they observed that the contact between *T. harzianum* and the roots persisted over a long period of time, and they reported the presence of yeast-like cells in the fungi. This form of root colonization was subsequently reported in cucumber roots in a similar study (Samolski et al. 2012) and later still in olive tree roots (Carrero-Carrón et al. 2018).

In the case of *Trichoderma*, the root-colonization process is encouraged by and is a direct evolutionary consequence of the exudates of the plant's roots and the protection of the niche to be colonized (Woo and Lorito 2007). In this regard, it has been observed that sucrose from plants is an important intracellular nutritive resource for *Trichoderma*: some fungal molecular mechanisms that are involved in the acquisition of root exudates have been described as a "di/tripeptide transporter and a permease/intracellular invertase system" (Vizcaíno et al. 2006; Vargas et al. 2009). In addition, a hydroponic experiment conducted by Zhang et al. (2014) demonstrated the positive effects of organic acids on the colonization of cucumber roots by *T. harzianum*, and the authors reported that certain organic acids could be used as nutritional sources for *Trichoderma*.

Colonization involves an ability to recognize and adhere to roots, penetrate the plant and withstand toxic metabolites produced by the plant in response to invasion (Hermosa et al. 2012). The plant's recognition of its invaders and vice versa implies numerous molecular responses. Fungus-root adhesion requires the action of hydrophobins, small molecules made of hydrophobic proteins from the outermost cell wall layer that coat the fungal cell surface, as well as of expansion-like proteins related to cell wall development. The class I hydrophobin produced by *T. asperellum* has been described as being involved in the process of root colonization by fostering the *Trichoderma*-root's adherence and protecting the hyphae of the plant's defensive compounds (Viterbo and Chet 2006).

To establish colonies, *Trichoderma* requires different molecular tools—for example, swollenin, a protein that contains an N-terminal carbohydrate-binding module family 1 domain with cellulose-binding function and a C-terminal expansion-like domain and that plays a key role in a fungus's root-colonization capacity (Brotman et al. 2008). This is also the case for other proteins with cellulase activity (Saravanakumar et al. 2016) and even for elicitors of induced systemic resistance (ISR) as the secondary metabolite SM1 from *T. virens* (Djonović et al. 2007). The action of "an effector-like protein in colonization of roots by *Trichoderma*" has also been described for the first time (Crutcher et al. 2015). Another example is ThPG1 endopolygalacturonase in *T. harzianum*, which is required for active root colonization and plant defence induction in tomato plants (Morán-Diez et al. 2009).

A pioneering study showed that the colonization of the root by *Trichoderma* was accompanied by the elicitation of systemic responses, in this case against attack from the necrotrophic fungus *Botrytis cinerea* (de Meyer et al. 1998). Later studies linked these responses to the induction of defence genes through the jasmonic acid (JA) and ethylene (ET) pathways with the inoculation of cucumber roots with *T. asperellum* against the leaf pathogen *Pseudomonas syringae* pv. *lachrymans* (Shoresh et al. 2005; Korolev et al. 2008). Therefore, it is clear that root colonization by *Trichoderma* entails the activation of plant defences, both locally and systemically (Shoresh et al. 2010; Hermosa et al. 2013) and even for relatively long periods of time (Tucci et al. 2011). Moreover, in 2019, Yuan et al. observed an activation of salicylic acid (SA)-related genes, as well as of JA- and ET-related genes, as a result of colonization of the roots of cucumber plants by *T. longibrachiatum*, which improved these plant's defences against attacks by *B. cinerea* on their leaves. SA seems to be the key hormone in the activation of defence responses at the root level after fungal-beneficial colonization (de Palma et al. 2019). All this suggests that both delays and overlaps in the expression of genes related to different plant defence routes may occur (Salas-Marina et al. 2011), with these adopting a wave-like response dynamic to *Trichoderma* stimuli (Rubio et al. 2014).

Therefore, it seems that the plant may perceive *Trichoderma* as hostile and that the activation of the plant's defences may limit penetration by the fungus to the root's outermost cell layers. Transcriptomic studies have shown that during the first few hours of interaction between the fungus and the plant, a broad genetic reprogramming occurs. This is preceded by a transient decrease in the plant's immune response during the first 24 h, which is probably what allows colonization of the root to take place (Morán-Diez et al. 2012). Once *Trichoderma* manages to avoid the plant's defences, the root's defensive position normalizes and it refocuses on growth, which thereafter is further promoted by the beneficial fungus (Brotman et al. 2013).

However, the ability of *Trichoderma* to colonize roots has been described in terms of mycoparasitism of other symbiotic plant microorganisms. Arbuscular mycorrhizal fungi (AMF) are ubiquitous soil microorganisms that colonize the roots of nearly 80% of terrestrial plants. They provide water and nutrients (i.e. phosphorus) in exchange for carbon photosynthetic products. They promote plant growth and productivity, and they have been shown to induce plant defence responses against pathogens (de Jaeger et al. 2010). In this regard, *Trichoderma* has been described as an effective parasite of AMF species. It increases conidia germination in the presence of these fungi (Filion et al. 1999), or it actively parasitizes them, as is the case of *T. atroviride* in its interaction with *Gigaspora gigantea* and that of *G. margarita* with *Medicago truncatula* (Lace et al. 2015). It does so without killing other plant-beneficial microorganisms, but it does obtain important nutrients from these fungi. de Jaeger et al. (2010) describe how *T. harzianum* is able to mycoparasitize the extraradical mycelium hyphae and spores of *Glomus intraradices* (now *Rhizophagus intraradices*) and *G. irregulare* (now *R. irregularis*). But they also describe for the first time how *T. harzianum* is able to mycoparasitize the intraradical mycelium hyphae and vesicles in *Solanum tuberosum* roots. This finding not only indicates the ability of *Trichoderma* to use AMF as an entry bridge into roots but also opens up the possibility that other microorganisms, possibly including plant pathogens, can also parasitize AMF and colonize the root's insides more easily. Similarly, Velmourougane et al. (2017) have observed how, through root biofilms formed by bacteria *Azotobacter chroococcum*, *T. viride* is able to colonize chickpea roots to a greater or lesser extent.

3.3 Benefits Obtained by the Plant from Successful Colonization

After *Trichoderma* has concluded the difficult process of establishing a root colony, the plant begins to receive the indirect benefits of symbiosis.

3.3.1 Induction of Systemic Resistance

The induction of ISR in plants is a consequence of the interactions between the different triggers released by microorganisms and plant's receptors (Nawrocka and Małolepsza 2013). It begins with the distinction, by plant-cell surface pattern recognition receptors (PRR), of specific components of microorganisms' cell surfaces, which are known as microorganism-associated molecular patterns (MAMPs), if they are beneficial microorganisms, or pathogen-associated molecular patterns (PAMPs), if they are molecules associated with harmful microorganisms (Hermosa et al. 2013; Contreras-Cornejo et al. 2014). The systemic stimulation of defence mechanisms prepares the plant's immune response, allowing it to act more quickly against a subsequent attack from any kind of pathogen. As a result, the possibility that the disease will spread is reduced (Hermosa et al. 2012; Bisen et al. 2016; Mendoza-Mendoza et al. 2018).

Similarly, PRRs also respond to molecules such as cell wall oligomers or cuticle fragments released by microbial hydrolytic activity in plant tissues, which are known as damage-associated molecular patterns (DAMPs). Stimulation of these PRR leads to a cascade of signal transduction and to activation of MAMP- or PAMP-triggered immunity (MTI or PTI) (Hermosa et al. 2013; Mendoza-Mendoza et al. 2018). This process, in turn, involves the fortification of the cell wall through the deposit of structural substances such as callose and lignin, the production of secondary antimicrobial metabolites like phytoalexins, the deposit of chitinases and glucanases—pathogenesis-related (PR) proteins that degrade pathogens' cell walls—and the accumulation of phytohormones such as SA, JA and ET, whose signal is transmitted systemically through neighbouring cells (Contreras-Cornejo et al. 2014; Mendoza-Mendoza et al. 2018).

Biological control agents (BCAs) can act directly through mycoparasitism, antibiosis and competition with the pathogen, among other mechanisms, as well as indirectly through the induction of ISR in plants, which is the most viable strategy when direct action is not effective against pathogens. In addition, this is a durable and effective response against a broad spectrum of pathogens that can last for several months, so a subsequent reinduction of plant defences is necessary (Nawrocka and Małolepsza 2013; Bisen et al. 2016).

Calderón et al. (1993) were the first to demonstrate the induction of systemic resistance mediated by *Trichoderma* in grapevine. *Trichoderma* not only prepares the plant's defences against bacteria and fungi but also its defences against other plant pathogens (Elsharkawy et al. 2013; Coppola et al. 2017). To date, there are few studies that determine the control that *Trichoderma* can exert on different insects. Those which have been conducted are limited in their focus to aphids, thrips and caterpillars. Coppola et al. (2019) demonstrate how *T. atroviride* induces defensive responses in the plant that decrease the development of two pests that have different eating habits and harm tomato plants: the moth (noctuid) *Spodoptera littoralis* and the aphid *Macrosiphum euphorbiae*. In addition, the presence of this fungus produces metabolic alterations in the tomato plant that result in the production of volatile organic compounds (VOCs) capable of improving the attraction of the parasitoid *Aphidius ervi*, thus reinforcing this plant's defences. Studies have also described how *Trichoderma* is able to induce ISR against viruses such as the cucumber mosaic virus, in the case of tomato plants (Vitti et al. 2016), and even against nematodes (Sharon et al. 2001; Martínez-Medina et al. 2017a; de Medeiros et al. 2017).

It must be borne in mind that the induction of resistance may vary due to the influence of abiotic environmental factors, as well as to the presence of pathogens and symbionts in the rhizosphere, since it is known that there are strains whose positive impact on plants has been observed only under unfavourable environmental conditions (Nawrocka and Małolepsza 2013). For example, *T. harzianum* T22 supposes a greater benefit for the plant under conditions of abiotic stress and in the presence of the pathogen *Pythium ultimum* (Mastouri et al. 2010). Similarly, research conducted by Sriram et al. (2009) and Morán-Diez et al. (2009) shows changes associated with resistance induced by *Trichoderma* in plants inoculated with pathogens such as *Rhizoctonia solani* or *Phytophthora capsici*.

It has been observed that *Trichoderma* is able to induce resistance in a wide variety of species; this leads to colonization of the roots and to a transcriptomic, proteomic and metabolomic modification in the plant (Mukherjee et al. 2012). Plants react instantaneously to the invasion of *Trichoderma* through physical changes that include the deposition of callose or the thickening of the cell wall and through biochemical changes such as the local and systemic accumulation of reactive oxygen species (ROS). The induction of signalling pathways related to SA and JA/ET phytohormones then occurs, resulting in the induction of resistance responses against attack by pathogens throughout the plant (Nawrocka and Małolepsza 2013).

The traditional description of *Trichoderma*-plant interaction—for example, that of Brotman et al. (2012)—holds that plant defence response occurs via ISR, which depends on JA/ET signalling pathways. However, the defence induced by *Trichoderma* may be ISR, SAR (systemic acquired resistance) or both of these (Mendoza-Mendoza et al. 2018). At high inoculum doses, *Trichoderma* can trigger a SAR response mediated by SA that is similar to that found in the case of necrotrophic pathogens (Mukherjee et al. 2012; Bisen et al. 2016). On the other hand, recent studies maintain that colonization of the *Arabidopsis* root by *T. asperellum* produces an ISR resistance through a cascade of SA signalling, and both SA and JA/ET signalling pathways are combined in an ISR resistance that is activated by cell-free culture filtrates of *Trichoderma*, probably because more than one elicitor triggers plant defences (Yoshioka et al. 2011; Hermosa et al. 2013). Likewise, this ISR induced through JA and SA pathways was observed in a study conducted by Jogaiah et al. (2018), in which the role of *T. virens* in tomato plant's defence against *Fusarium oxysporum* f. sp. *lycopersici* was analysed.

Therefore, *Trichoderma*-related induction of the SA, ET and JA pathways in the same plant may indicate the presence of alternative induced resistance mechanisms and a more complicated signalling network that connects the SAR and ISR defence response pathways (Brotman et al. 2012; Nawrocka and Małolepsza 2013). This variation depends on the species of plant, the strain of *Trichoderma* and the pathogen against which the defence response is directed, since one analysis of the mechanisms involved in the ISR induced by *T. harzianum* in tomato plants against *B. cinerea* shows how the systemic resistance induced by *Trichoderma* requires not only JA signalling routes, but also SA and abscisic acid (ABA) ones (although JA plays the central role). This finding demonstrates that this type of resistance is a flexible process that includes different signalling pathways that depend on the pathogen's mode of action (Martínez-Medina et al. 2013). Likewise, a recent study by Yuan et al. (2019) describes how the inoculation of cucumber roots performed by *T. longibrachiatum* H9 significantly inhibits infection by *B. cinerea* by activating the signalling pathways associated with JA/ET and SA phytohormones.

On the other hand, *Trichoderma* can produce effectors—which are also volatile—that activate the ETI (effector-triggered immunity) response, predisposing the plant to respond more quickly and more intensely to subsequent attack by a pathogen (Tucci et al. 2011; Mendoza-Mendoza et al. 2018). For example, the production of harzianolide and 6-n-pentyl-6H-pyran-2-one (6PP) by *Trichoderma* spp. has been correlated with tomato plant's resistance against *B. cinerea* (Contreras-Cornejo et al. 2014).

During plant colonization, *Trichoderma* undergoes a transcriptomic reprogramming that depends on the species of both the *Trichoderma* and the plant. Secretion and perception of peptides are two mechanisms that fungi have developed to perceive and communicate with their surroundings (Mendoza-Mendoza et al. 2018). Sucrose derived from plants is an important resource that *Trichoderma* uses to facilitate colonization of the root and coordination of defence mechanisms (Hermosa et al. 2012). It has been observed that *Trichoderma* is capable of secreting soluble proteins (via the classical type II secretion system) that include cell wall-degrading enzymes (CWDEs), effector enzymes, host-surface adhesion and recognition enzymes, adhesion-only proteins (adhesins and hydrophobins) and proteins involved in secondary metabolism, among others. All these proteins assist in the root's entry into the root tissues and prevent the plant's activation of defence responses. And at the same time, they prepare the plant for potential attacks from pathogens. The presence of CWDEs in *Trichoderma* secretome alerts the plant immune system of the presence of this fungus. In addition, the activity of these proteins assists in degrading plant cell walls and directly facilitates colonization by increasing the plasticity of the cell wall, though perhaps this activity only takes place when penetration will otherwise not be successful, given that interaction between the plant and the fungus is fundamentally beneficial. On the other hand, studies have also shown how CWDEs, specifically the β-glucanases and chitinases of *T. asperellum*, are able to induce defence responses against *Corynespora cassiicola* and *Curvularia aeria*, which cause leaf spot disease in lettuce. These proteins therefore also play an important role in improving plant resistance (Baiyee et al. 2019).

Even so, it is important to bear in mind that root colonization and induction of ISR are not synonymous: the hydrolysis of the pectin, in particular the middle lamina, should facilitate admission and apoplastic colonization. However, ISR may be activated before, after or during penetration by proteins or soluble secreted metabolites into root tissues (Mendoza-Mendoza et al. 2018).

In general, plants detect the presence of fungi by recognizing MAMPs, such as ergosterol and chitin oligomers that are released from the cell walls of these microorganisms. This recognition occurs through the LysM CERK1 and CEBiP receptor proteins, which have been identified in *Arabidopsis* and rice, respectively (Contreras-Cornejo et al. 2014). The structural proteins and the secondary metabolites of *Trichoderma* act as MAMPs, while its enzymes generate DAMPs when they act against other fungi and against the cell walls of plants (Hermosa et al. 2013). The different strains of *Trichoderma* produce a great variety of MAMPs, which to date are among the most extensively described within the beneficial fungi. The first *Trichoderma* MAMP that was identified was an ET-inducing xylanase (Xyn2/Eix), a potent elicitor of the defensive response in tobacco and tomato plants that promotes ET biosynthesis and then leaf ISR by inducing the expression of the gene that encodes ACC synthase (Hermosa et al. 2012, 2013; Contreras-Cornejo et al. 2014).

Some *Trichoderma* elicitors that can induce a wide variety of changes in plants are similar to MAMPs. The elicitors released by the fungi are classified into different types, the most important of which are proteins or peptides with low molecular weights that range from 6 to 42 kDa. Examples include enzymes such as glucanases, cellulases, chitinases, serine proteases or xylanases, indole compounds, fatty acids and their derivatives such as glycosphingolipids, and oligosaccharides or polysaccharides, encompassing chitin and chitin-like compounds (Nawrocka and Małolepsza 2013).

The elicitor produced by *Trichoderma* that has been most effectively characterized is Sm1/Epl1, a protein from the cerato-platanin (CP) family that is rich in cysteine and is abundantly secreted. The monomeric form of Sm1 is in a glycosylated state, which is essential for all elicitors. It has been suggested that the monomeric form in the non-glycosylated state is susceptible to dimerization driven by oxidation in plants, making Sm1 inactive as an ISR inducer (Mukherjee et al. 2012; Hermosa et al. 2012; Nawrocka and Małolepsza 2013). It acts as a JA elicitor, so it induces ISR in cotton and corn (Hermosa et al. 2013; Lamdan et al. 2015).

Other *Trichoderma* elicitors that have been fully described include: the polygalacturonase ThPG1, a cell wall-degrading enzyme that degrades pectin and produces oligogalacturonides that then act as DAMPs, which ultimately causes SAR activation, and peptaibols, which are peptides that are produced as secondary metabolites with antibiotic activity against fungi and bacteria. An example here is trichokonins, which are produced by *T. pseudokoningii*, which induces systemic resistance against the tobacco mosaic virus through the upregulation of genes related to SA, JA and ET (Hermosa et al. 2013; Contreras-Cornejo et al. 2014), and swollenin that have carbohydrate-binding domains (Swo1) and LysM-type chitin-binding domains and are therefore capable of dampening plant defences while the fungus colonizes the root; in the presence of fungal plant pathogens, the activity of these chitinases may cause the release of chitooligosaccharides, contributing to induction of defence (Lamdan et al. 2015); 4-phosphopantetheinyl transferases (PPTases) from *T. virens* related with the induction of SA and camalexin-dependent plant defence responses in *A. thaliana* (Velázquez-Robledo et al. 2011); and 6PP, a common *Trichoderma* secondary metabolite that is responsible for the coconut aroma and yellow pigmentation associated with some strains of this genus and that inhibits the growth of pathogens such as *Fusarium oxysporum* (Hermosa et al. 2013; Contreras-Cornejo et al. 2014) and *Spodoptera frugiperda* (Contreras-Cornejo et al. 2018).

3.3.2 *Increased Tolerance to Abiotic Stress*

Numerous studies have shown that fungi of the genus *Trichoderma* increase the ability of plants to tolerate various abiotic stresses, such as drought, salinity or extreme temperatures, even in different soil types and at different pH levels (Harman 2000; Adams et al. 2007; Bae et al. 2009). In 2008, Donoso et al. inoculated two wheat genotypes submitted to water deficit with a strain of *T. harzianum*. They observed that the fungus produced a positive effect on the aerial and root growth of plants under stress.

This effect on plants is the result of various factors that arise in the interaction between *Trichoderma* and the plant, and the stimulation of root growth and development described above plays a fundamental role (Harman 2000; Harman et al. 2004a). Root colonization of the plant by these fungi increases root growth, which means an increase in productivity. In turn, as the plant grows to be stronger, it is better able to tolerate various stress situations (Harman et al. 2004b). Harman (2000) showed that inoculation of corn plants with the T22 strain of *T. harzianum* produced an increase in foliar and root biomass and a reduction in the plant's sensitivity to drought. He suggests that these outcomes may be a result of increased root development.

Under stress conditions, so-called ROS occur in plants, which can cause cell damage. Although plants have the capacity to detoxify and block these compounds, under conditions of severe or prolonged stress, they can accumulate (Mittler 2002). *Trichoderma* reduces oxidative stress in plants and helps them to tolerate it (Shoresh et al. 2010). Shoresh and Harman observed that tomato plants inoculated with *T. harzianum* expressed genes that encode antioxidant enzymes (Shoresh and Harman 2008). In 2010, Mastouri et al. proved that under conditions of osmotic saline stress or extreme temperatures, tomato seeds treated with *T. harzianum* T22 germinated earlier and more evenly than did non-inoculated seeds. They found that inoculated seeds exhibited a lower accumulation of ROS relative to non-inoculated ones, and they observed that an exogenous antioxidant resulted in an effect similar to that produced by fungus inoculation. They proposed that these findings were caused by induction of plant physiological protection against oxidative damage (Mastouri et al. 2010). These studies show the ability of *Trichoderma* to increase the detoxifying capacity of ROS in plants, as well as its correlation with the induction of tolerance against abiotic stresses (Shoresh et al. 2010), which was later confirmed (Mastouri et al. 2012). More recently, in 2018, Yasmeen and Siddiqui observed that *T. harzianum* Th-6 improved the tolerance of corn and rice plants to saline conditions in hydroponic cultivation, improving stomatic conductivity and increasing water content and antioxidant activity.

Another mechanism by which plants acquire a greater tolerance to abiotic stresses due to interaction with *Trichoderma* is the reprogramming of their gene expression (Shoresh et al. 2010). Once a relationship between the plant and the microorganism is established, a response is produced whereby the plant's expression of genes related to antioxidant response (such as peroxidase activity), those related to tolerance of extreme temperatures (such as chaperones) and those related to the osmoprotective effect (such as osmotines) is increased (Alfano et al. 2007; Segarra et al. 2007; Shoresh and Harman 2008; Shoresh et al. 2010). In 2013, Brotman et al. observed an improvement in germination in *Arabidopsis* and cucumber plants inoculated with *Trichoderma* under conditions of saline stress, and they describe an increase in osmoregulation and antioxidant gene expression. In addition, they observed transcription factors involved in abiotic stress response (MYB51 or

MYB15, for example). ACC deaminase activity was also enhanced, allowing a reduction in the levels of ethylene, which had increased under stress. Another effect on plant gene expression caused by interaction with *Trichoderma* was observed by Bae et al. in 2009. After inoculating cocoa plants with a strain of *T. hamatum*, they observed that the expression of seven expressed sequence tags (ESTs) involved in drought response was delayed for at least 3 days relative to that in non-inoculated plants. This retarded the physiological changes produced by the drought and increased water content and plant growth.

Other mechanisms by which *Trichoderma* indirectly increases plant's tolerance to abiotic stress have also been studied. Shoresh and Harman observed in 2008 that starch content in the leaves of plants inoculated with *Trichoderma* was higher than that of non-inoculated plants. This effect may provide a greater carbon reserve for the plant when prolonged drought conditions and the stomatal closure that results from them cause a carbon deficit. The potassium (K) content of the inoculated plants is also higher; this can help plants to tolerate saline stress, since K interferes with the assimilation of the nutrient (Yedidia et al. 2001; Shabala and Cuin 2008). More recently, researchers have studied possible future uses of fungi of the genus *Trichoderma* as a source of genes for the construction of transgenic plants that could solve various problems. For example, the *cht42* and *cht33* genes of a strain of *T. harzianum* were successfully expressed in tobacco plants, increasing their tolerance to salinity and heavy metals (de las Mercedes Dana et al. 2006). The *hsp70* gene was also expressed in *Arabidopsis*, increasing its tolerance to various abiotic stresses (Montero-Barrientos et al. 2010).

3.3.3 Increase in Nutrient Intake

The microorganisms that inhabit the root zone of plants have a great influence on the dynamic balance in which soil nutrients are found (López-Bucio et al. 2015). The ability of various fungi of the genus *Trichoderma* to promote plant growth has been well known for some time. In 1992, Kleifeld and Chet observed the positive effect of a strain of *T. harzianum* in a greenhouse trial with different species of cultivated plants.

Many strains of fungi of the genus *Trichoderma* increase the root biomass of the plants with which they interact, improving plant's ability to capture nutrients from the soil solution. *T. virens* and *T. atroviride* increase the growth of *Arabidopsis* seedlings, both in the aerial part and in the radicular part, by stimulating the formation of lateral roots (Contreras-Cornejo et al. 2009). In the case of cucumber plants, this modification of the root architecture increases biomass by increasing the absorption of NPK nutrients and micronutrients (Samolski et al. 2012).

In addition to its influence on root architecture, *Trichoderma* secretes exudates that increase the solubility of various nutrients, making it easier for plants to absorb them. Siderophores are one of these substances, and they have a great influence on iron nutrition (López-Bucio et al. 2015). This element is very important to plant physiology: it is part of numerous enzymatic systems, and it is found in cytochromes, chloroplasts, mitochondria and the Calvin cycle, among others (Terry and Zayed 1995; Mengel and Kirkby 2001).

In the presence of oxygen, free ferrous iron (Fe^{2+}) is rapidly oxidized to become insoluble ferric iron (Fe^{3+}). To support this situation in which there is a lower availability of this micronutrient, some bacteria and fungi have the ability to synthesize molecules called siderophores, which have great affinity for iron. These are released from the cells and into the surrounding environment, where they become attached to iron molecules. They are then reabsorbed by specific transporters (Searle et al. 2015). This mechanism can also benefit crops, since they are able to acquire this chelated iron by plasma membrane reductases (Welch et al. 1993).

In addition, as Martínez-Medina et al. have previously described, the interaction of the fungus with the plant activates the plant's systemic induced resistance by the transcription factor MYB72, which also has a function in the assimilation of iron by favouring such assimilation during fungus-plant interaction (Martínez-Medina et al. 2017a, b). In another study conducted in 2014, a strain of siderophore-producing *T. asperellum* raised the available iron content in sterilized soil and increased cucumber plant's ability to absorb the element (Zhao et al. 2014).

The release of organic acids and Fe^{3+}-reducing enzymes seems to contribute to reducing this element to the soluble form Fe^{2+}, which, together with the production of siderophores, enhances plant's iron absorption capacity (López-Bucio et al. 2015).

The solubilization of phosphate is one of the most important benefits that can be obtained from organisms used as biofertilizers. This is because the gradual depletion of non-renewable sources of this element is progressively increasing the price of phosphate fertilizer (López-Bucio et al. 2015). Numerous studies have demonstrated the ability of fungi of the genus *Trichoderma* to solubilize phosphate, even in soils contaminated with heavy metals (Rawat and Tewari 2011). In another study, a strain of *T. harzianum* increased the content of phosphorus and other nutrients in tomato seedlings, as well as the height and diameter of the stem and the fresh and dry weights (Azarmi et al. 2012). Along with the action of different organic acids in the solubilization of these elements, another possible explanation for this increase in phosphorous is an increase in the expression of phosphorus transport genes in the plant as a consequence of interaction with *Trichoderma* (López-Bucio et al. 2015).

It has been demonstrated that various fungi of the *Trichoderma* genus improve plant's capacity to capture nitrogen, the most important nutrient in agricultural production (Harman 2011). The inoculation of *Trichoderma* in several field trials allowed the nitrogen supply to be reduced by 30–50% without causing a fall in crop production (Shoresh et al. 2010; Harman 2011). The mechanisms by which the fungus acts to produce this effect appear to be an increase in the plant's root volume, which increases success in finding nutrients during soil exploration (Harman 2011), and an increase in the expression of nitrogen transport genes in the plant, which makes it more efficient in assimilating the nutrient (Singh et al. 2018).

The data obtained in various studies show an increase in the nutritional content of horticultural and cultivated species when they are inoculated with fungi of the genus

Trichoderma. In 1999, it was observed that a strain of *T. harzianum* increased the availability of various nutrients—for example, phosphorus, iron, manganese and zinc—in an in vitro nutrient solution (Altomare et al. 1999). Then, in 2001, Yedidia et al. demonstrated the ability of a strain of *T. harzianum* to increase the biomass of cucumber plants inoculated under hydroponic conditions and to raise phosphorus content by 90% and iron content by 30%, as well as levels of other micronutrients such as copper, zinc, manganese and sodium. A study conducted in 2015 by Li et al. showed the ability of a strain of *T. harzianum* to solubilize phosphorus in the form of phytate, iron, copper and zinc by releasing organic acids detected by HPLC such as lactic, citric or tartaric acid, or by releasing siderophores or reducing iron.

3.3.4 Biostimulant Capacity

It is known that the ability of numerous *Trichoderma* species to promote plant growth is due, to a large extent, to their stimulating the root system of plants. For example, in a study completed in 2001, inoculation of cucumber plants with the T-203 strain of *T. harzianum* in a hydroponic system produced a 95% increase in plant's root area and a 75% increase in their root length relative to non-inoculated plants. This in turn resulted in a 25% increase in total plant biomass in dry weight. According to the authors, this was possibly due to the fungus' production of plant hormones (Yedidia et al. 2001). Harman et al. observed, both in research trials and in commercial practice, that the inoculation with the T22 strain of *T. harzianum* produced an increase in the development of roots in various cultivated plants, with corn having received the most attention. In addition, they observed that the effect lasted throughout the life cycle of annual plants and that this effect could be achieved with a reduced inoculum via seed treatment (Harman 2000; Harman et al. 2004a, b).

As research in this field has progressed, it has been established that this effect on root architecture is related to auxins. In a 2009 study involving *T. virens*, *T. atroviride* and the plant *A. thaliana*, it was observed that plants inoculated with said fungi had a phenotype related to these hormones. In addition, these fungi were producers of compounds from this family of hormones—for example, indole-3-acetic acid (IAA) (Contreras-Cornejo et al. 2009). Subsequently, in 2012, it was shown that the *qid74* gene of the CECT 2413 strain of *T. harzianum* induces modifications in the root architecture of plants during interaction, producing an increase in absorption volume. This promotes absorption and transport of nutrients in the plant, which in turn produces an increase in the overall biomass of the inoculated plants (Samolski et al. 2012). More recently, in 2017, Pelagio-Flores et al. proposed, as an explanation of this phenomenon, that *Trichoderma* acidifies the rhizosphere, causing root sensitivity to pH to reprogramme root architecture through the redistribution of auxins in columella root cap cells.

In addition to the ability of *Trichoderma* to promote plant growth, the genus' capacity to produce phytohormones has been demonstrated in numerous studies (Contreras-Cornejo et al. 2013). Analogous auxin hormones related to root growth and development, and particularly IAA, have received the most attention from researchers (Hoyos-Carvajal et al. 2009). In 2009, Contreras-Cornejo et al. showed that promotion of the growth of *A. thaliana* by a strain of *T. virens* was the result of the metabolic pathway response to auxins, which was activated by the production of IAA and other auxin analogues by the fungus. The authors hypothesized that the fungus produces these hormones to promote root growth in the plant and thus facilitate colonization of it.

Martínez-Medina et al. (2014) studied the levels of this hormone in plants inoculated with strains of *T. ghanese*, *T. hamatum* and *T. harzianum*. Increased auxin levels were found in plants inoculated with these growth-promoting strains. Although the authors did not confirm whether these hormones had been produced by the plant or by the fungus, it could at least be established that they were the product of interaction between the two.

Studies have also confirmed that, besides IAA, some strains of *Trichoderma* produce ACC deaminase. This hormone is an inhibitor of ethylene, which is the phytohormone growth inhibitor par excellence. In 2010, two different studies investigated this hormone in the context of interaction between fungi of the genus *Trichoderma* and plants. Viterbo et al. demonstrated that the T203 strain of *T. asperellum* is capable of producing the said hormone, and Martínez-Medina et al. verified that plants inoculated with *T. harzianum* showed higher levels of IAA and ACC deaminase in their aerial parts compared to uninoculated plants, with both hormones acting in a synergistic way to promote growth of the plants (Martínez-Medina et al. 2010; Viterbo et al. 2010).

3.4 *Trichoderma* as a Plant Pathogen: The Importance of SA

SA is a plant hormone involved in plant resistance against different organisms at both systemic and local levels, as well as in plant's responses to abiotic stresses during growth and development (Rivas-San Vicente and Plasencia 2011). The mechanisms by which SA is able to activate defences at the root level and prevent infection by pathogenic microorganisms are well known, but the way in which it acts in the context of beneficial microorganisms has recently been reconsidered. In a similar way to what occurs with phytopathogens, beneficial microorganisms also need to evade, at least partially, plant's defensive responses to establish a mutualistic association with them (Zamioudis and Pieterse 2012). SA plays a fundamental role in such responses to biotrophic pathogens, which normally live in intercellular spaces (Glazebrook 2005). In the same vein, a transitory accumulation is required during the first phases of mutualistic microorganism's colonization of the root (Medina et al. 2003).

The rhizosphere is full of microorganisms, many of which are neither advantageous nor harmful to plants. However, some of these microorganisms are pathogens, while others are beneficial. Beneficial bacteria represent a very small proportion of soil's whole bacterial set, and the plant must be very precise when it chooses whether it is worth beginning the process of forming symbiotic complexes. It has been determined that SA is the hormone responsible for this microbial selection, since mutants deficient in the said hormone do not effectively benefit from symbiotic relationships with the beneficial bacteria present (Lebeis et al. 2015). In addition, it has been proven that rhizobacteria require a partial suppression of plant responses to SA to nodulate (Peleg-Grossman et al. 2009; Stacey et al. 2006), although it may be that SA plays an important role only in nodulations of an indeterminate type (van Spronsen et al. 2003).

The application of SA in plants inoculated with mycorrhizal fungi promotes a synergistic effect on tolerance to salinity, improving root activity and structure (Ghazanfar et al. 2015). However, it has been shown that root colonization by these fungi promotes local and systemic accumulation of phenolic defence compounds against pathogens, a response mediated by SA (Zhu et al. 2015). By contrast, it has been found that for the association between mycorrhizae and plants to occur, the partial suppression of SA-dependent responses is necessary (Pozo and Azcón-Aguilar 2007; López-Ráez et al. 2010). Similar findings were observed in a study focused on the beneficial interaction between the endophytic fungus *Serendipita indica* (or *Piriformospora indica*) and *A. thaliana* (Jacobs et al. 2011).

The beneficial microorganism can colonize the root to create its own microenvironment in the rhizosphere and rhizoplane. It is clear that all species of endophytic microorganisms compete intensely for nutrients and space in the rhizosphere via allelopathy and autotoxicity. However, some microorganisms are able to penetrate plants and colonize their roots. Therefore, it is possible that fungus-plant symbiotic interactions of this type start with the fungus acting as an opportunist plant pathogen in an initial phase that involves the partial suppression of plant defence responses in the roots, the fungus's aim being to avoid the adversities found in the rhizosphere (Fesel and Zuccaro 2016; Banik 2019).

As far as *Trichoderma* is concerned, in 2014, Alonso-Ramírez et al. described the importance of SA in the ability of *Trichoderma* to colonize roots. They demonstrated that the absence of a defensive response mediated by this plant hormone caused a massive colonization of plant tissues by the fungi and the death of the plant. Several studies indicate that for *Trichoderma*-plant interaction to take place, it is necessary for a plant's defences to be partially suppressed. This suppression produces a decrease in the expression levels of defence genes regulated by SA, both locally and systemically, after the penetration of *Trichoderma* into the root's tissues (Morán-Diez et al. 2012; Brotman et al. 2013). The use of a mutant line of *A. thaliana* that is unable to synthesize SA (*sid2*) showed that root colonization by *T. harzianum* can reach the plant's vascular bundles. In addition, the presence of the fungus at leaf level could be verified endophytically, as a result of massive plant colonization undertaken by it. The same study revealed the possibility that the absence of a defensive response mediated by SA, such as the accumulation of callose in the intercellular spaces, may be what causes *Trichoderma* to be able to colonize vascular bundles. Finally, by means of a study involving hydroponic culture, these authors determined that not only did colonization of the plant by the

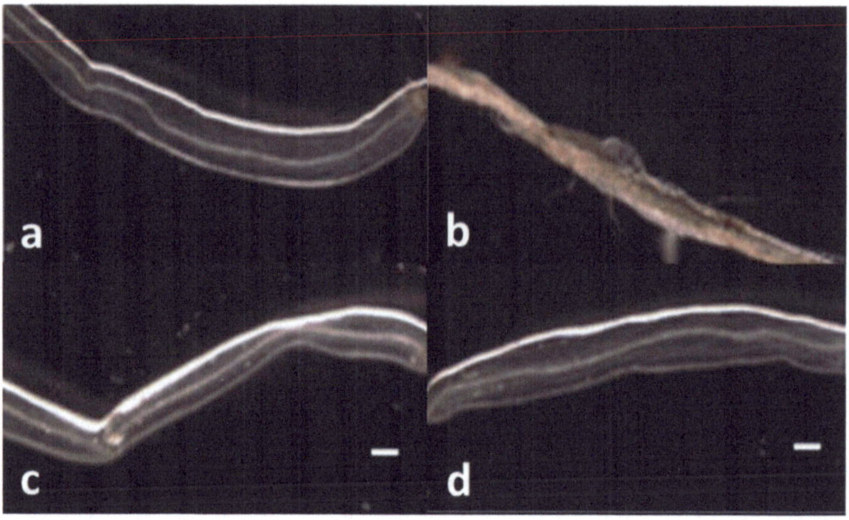

Fig. 3.2 Effect of *T. harzianum* (T34) on *Arabidopsis* Col-0 (**a**, **b**) and *salicylic acid induction deficient 2* (*sid2*) mutant (**c**, **d**) roots grown in hydroponic culture, treated with the fungus (**b**, **d**) or untreated (**a**, **c**). Representative sections of roots of both *Arabidopsis* lines that were treated with T34 or were not treated were photographed with a Leica stereomicroscope. Scale bars, 0.15 mm. (Extracted from Alonso-Ramírez et al. 2014)

fungus occur on a massive scale in the absence of SA, but also that as a consequence of that colonization, the plant collapsed and died, exhibiting symptoms such as root rot (Fig. 3.2) (Alonso-Ramírez et al. 2014). Despite this, other authors have not observed these phenotypes in *sid2* after colonization by *T. hamatum* on land (Mathys et al. 2012). This discrepancy is possibly due to the use of a hydroponic culture, since on land the symptoms of plant death are not always observed (Alonso-Ramírez et al. 2014).

Many authors have corroborated these results and examined them in more depth in their studies. After they carried out similar trials with tomato plants, the same results were corroborated by Fiorini et al. (2016) and Martínez-Medina et al. (2017a), as well as in transcriptomic studies of tomato roots (de Palma et al. 2019). In addition, other studies involving other fungi—for example, the endophytic fungus *Clonostachys rosea* in tomato plants and *A. thaliana* (Alvarez-Nordström 2014) and *T. virens* in corn (Morán-Diez et al. 2015)—have produced the same findings. It has even been possible to describe some of the compounds possibly released by *Trichoderma* that are capable of activating defence responses mediated by SA and preventing root colonization—for example, terpenes such as trichodermol (Cardoza et al. 2015) and squalene (Malmierca et al. 2015).

As a result of the studies on root colonization by *Trichoderma* described in this chapter, interesting future lines of research focused on studying this symbiotic process from an evolutionary point of view have opened up. There are currently several studies on the synthesis and accumulation of interesting and useful secondary metabolites by various species of marine-alga-endophytic *Trichoderma* that potentially have a pharmacological use (Su et al. 2018). Examples here include trichocitrin from *T. citrinoviride* in the brown alga *Dictyopteris prolifera* (Liang et al. 2016) or in the green alga *Cladophora* spp. (Marra et al. 2019), or the bisabolane, cyclonerane and harziane derivatives from *T. asperellum* in the red alga *Sargassum* sp. (Song et al. 2018). These even develop epiphytically and produce trichocarotins A-H and trichocadinin A from *T. virens* in the case of *Gracilaria vermiculophylla* (Shi et al. 2018). But this process of tissue colonization and the plant defence response involved in algal colonization are still being studied, as they are in the case of other plants at intermediate evolutionary steps up to higher vascular plants.

3.5 Conclusions

In summary, effective root colonization by *Trichoderma* provides the plant with a series of benefits that are derived from the symbiotic relationship, for example, an induction of systemic resistance against possible attacks from pests and/or pathogens, an increase in tolerance against abiotic stresses, a promotion of plant growth by physiological mechanisms and an improvement in the acquisition of nutrients through the roots.

Summarizing what has been described in this chapter, Fig. 3.3 shows the role that SA plays in effective root colonization by *Trichoderma*.

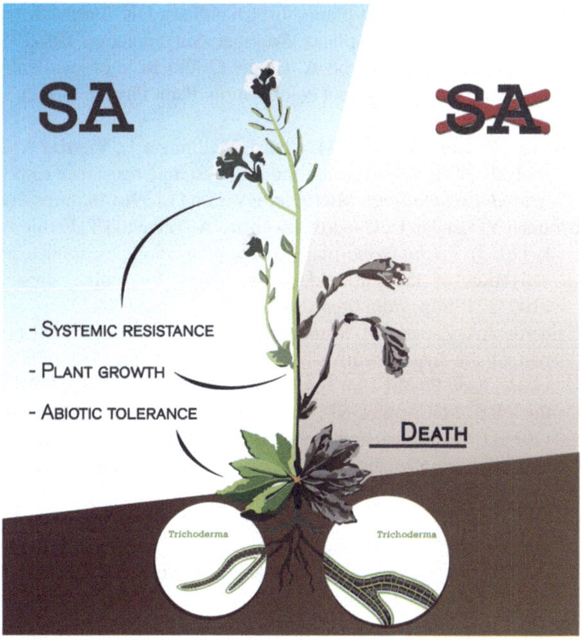

Fig. 3.3 Schematization of root colonization by *Trichoderma* in *A. thaliana* plants. Initially, the plant defends itself against being colonized by *Trichoderma*. The plant prevents *Trichoderma* from penetrating the vascular bundle and causing the expected indirect benefits. In the absence of an SA-mediated response, *Trichoderma* massively colonizes the roots, reaching the vascular bundle, producing plant death and possibly becoming an opportunistic pathogen. Infographic made by Darío Rodríguez-Prieto

References

Adams P, De-Leij FAAM, Lynch JM (2007) *Trichoderma harzianum* Rifai 1295-22 mediates growth promotion of crack willow (*Salix fragilis*) saplings in both clean and metal-contaminated soil. Microb Ecol 54(2):306–313. https://doi.org/10.1007/s00248-006-9203-0

Alfano G, Ivey MLL, Cakir C, Bos JIB, Miller SA, Madden LV, Hoitink HAJ (2007) Systemic modulation of gene expression in tomato by *Trichoderma hamatum* 382. Phytopathology 97 (4):429–437. https://doi.org/10.1094/phyto-97-4-0429

Alonso-Ramírez A, Poveda J, Martín I, Hermosa R, Monte E, Nicolás C (2014) Salicylic acid prevents *Trichoderma harzianum* from entering the vascular system of roots. Mol Plant Pathol 15(8):823–831. https://doi.org/10.1111/mpp.12141

Altomare C, Norvell WA, Björkman T, Harman GE (1999) Solubilization of phosphates and micronutrients by the plant-growth-promoting and biocontrol fungus. Appl Environ Microbiol 65(7):2926–2933

Alvarez-Nordström S (2014) Endophytic growth of *Clonostachys rosea* in tomato and *Arabidopsis*. Dissertation, The Swedish University of Agricultural Sciences, Uppsala, Suecia (Sweden)

Azarmi R, Hajieghrari B, Giglou A (2012) Effect of *Trichoderma* isolates on tomato seedling growth response and nutrient uptake. Afr J Biotechnol 10(31):5850–5855. https://doi.org/10.5897/ajb10.1600

Bae H, Sicher RC, Kim MS, Kim SH, Strem MD, Melnick RL, Bailey BA (2009) The beneficial endophyte *Trichoderma hamatum* isolate DIS 219b promotes growth and delays the onset of the drought response in *Theobroma cacao*. J Exp Bot 60(11):3279–3295. https://doi.org/10.1093/jxb/erp165

Baiyee B, Ito SI, Sunpapao A (2019) *Trichoderma asperellum* T1 mediated antifungal activity and induced defense response against leaf spot fungi in lettuce (*Lactuca sativa* L.). Physiol Mol Plant Pathol 106:96–101. https://doi.org/10.1016/j.pmpp.2018.12.009

Banik A (2019) Plant endophytes: true symbiont or opportunistic pathogens? EC Microbiol 15:533–535

Bisen K, Keswani C, Patel JS, Sarma BK, Singh HB (2016) *Trichoderma* spp.: efficient inducers of systemic resistance in plants. In: Chaudhary DK, Verma A (eds) Microbial-mediated induced systemic resistance in plants. Springer, Singapore, pp 185–195

Brotman Y, Briff E, Viterbo A, Chet I (2008) Role of swollenin, an expansin-like protein from *Trichoderma*, in plant root colonization. Plant Physiol 147(2):779–789. https://doi.org/10.1104/pp.108.116293

Brotman Y, Lisec J, Méret M, Chet I, Willmitzer L, Viterbo A (2012) Transcript and metabolite analysis of the *Trichoderma*-induced systemic resistance response to *Pseudomonas syringae* in *Arabidopsis thaliana*. Microbiology 158(1):139–146. https://doi.org/10.1099/mic.0.052621-0

Brotman Y, Landau U, Cuadros-Inostroza A, Takayuki T, Fernie AR, Chet I, Viterbo A, Willmitzer L (2013) *Trichoderma*-plant root colonization: escaping early plant defense responses and activation of the antioxidant machinery for saline stress tolerance. PLoS Pathog 9(3): e1003221. https://doi.org/10.1371/journal.ppat.1003221

Calderón AA, Zapata JM, Muñoz R, Pedreño MA, Barceló AR (1993) Resveratrol production as a part of the hypersensitive-like response of grapevine cells to an elicitor from *Trichoderma viride*. New Phytol 124(3):455–463. https://doi.org/10.1111/j.1469-8137.1993.tb03836.x

Cardoza RE, McCormick SP, Malmierca MG, Olivera ER, Alexander NJ, Monte E, Gutiérrez S (2015) Effects of trichothecene production on the plant defense response and fungal physiology: overexpression of the *Trichoderma arundinaceum tri4* gene in *T. harzianum*. Appl Environ Microbiol 81(18):6355–6366. https://doi.org/10.1128/AEM.01626-15

Carrero-Carrón I, Rubio MB, Niño-Sánchez J, Navas-Cortés JA, Jiménez-Díaz RM, Monte E, Hermosa R (2018) Interactions between *Trichoderma harzianum* and defoliating *Verticillium dahliae* in resistant and susceptible wild olive clones. Plant Pathol 67(8):1758–1767. https://doi.org/10.1111/ppa.12879

Chacón MR, Rodríguez Galán O, Benítez Fernández CT, Sousa S, Rey M, Llobell González A, Delgado Jarana J (2007) Microscopic and transcriptome analyses of early colonization of tomato roots by *Trichoderma harzianum*. Int Microbiol 10(1):19–27. https://doi.org/10.2436/20.1501.01.4

Contreras-Cornejo HA, Macias-Rodriguez L, Cortes-Penagos C, Lopez-Bucio J (2009) *Trichoderma virens*, a plant beneficial fungus, enhances biomass production and promotes lateral root growth through an auxin-dependent mechanism in *Arabidopsis*. Plant Physiol 149 (3):1579–1592. https://doi.org/10.1104/pp.108.130369

Contreras-Cornejo HA, Ortiz-Castro R, López-Bucio J, Mukherjee PK (2013) Promotion of plant growth and the induction of systemic defence by *Trichoderma*: physiology, genetics and gene expression. In: Mukherjee PK, Horwitz BA, Singh US, Mala M, Schmoll M (eds) Trichoderma: biology and applications. CABI, Wallingford, pp 173–194

Contreras-Cornejo HA, Macías-Rodríguez L, López-Bucio J (2014) Enhanced plant immunity using *Trichoderma*. In: Gupta VK (ed) Biotechnology and biology of *Trichoderma*. Elsevier, Oxford, pp 495–504

Contreras-Cornejo HA, Macías-Rodríguez L, del-Val E, Larsen J (2018) The root endophytic fungus *Trichoderma atroviride* induces foliar herbivory resistance in maize plants. Appl Soil Ecol 124:45–53. https://doi.org/10.1016/j.apsoil.2017.10.004

Coppola M, Cascone P, Chiusano ML, Colantuono C, Lorito M, Pennacchio F, Rao R, Woo SL, Guerrieri E, Digilio MC (2017) *Trichoderma harzianum* enhances tomato indirect defense against aphids. Insect Sci 24(6):1025–1033. https://doi.org/10.1111/1744-7917.12475

Coppola M, Cascone P, Di Lelio I, Woo SL, Lorito M, Rao R, Pennacchio F, Guerrieri E, Digilio MC (2019) *Trichoderma atroviride* P1 colonization of tomato plants enhances both direct and indirect defence barriers against insects. Front Physiol 10:813. https://doi.org/10.3389/fphys.2019.00813

Crutcher FK, Moran-Diez ME, Ding S, Liu J, Horwitz BA, Mukherjee PK, Kenerley CM (2015) A paralog of the proteinaceous elicitor SM1 is involved in colonization of maize roots by *Trichoderma virens*. Fungal Biol 119(6):476–486. https://doi.org/10.1016/j.funbio.2015.01.004

Djonović S, Vargas WA, Kolomiets MV, Horndeski M, Wiest A, Kenerley CM (2007) A proteinaceous elicitor Sm1 from the beneficial fungus *Trichoderma virens* is required for induced systemic resistance in maize. Plant Physiol 145(3):875–889. https://doi.org/10.1104/pp.107.103689

Donoso EP, Bustamante RO, Caru M, Niemeyer HM (2008) Water deficit as a driver of the mutualistic relationship between the fungus *Trichoderma harzianum* and two wheat genotypes. Appl Environ Microbiol 74(5):1412–1417. https://doi.org/10.1128/AEM.02013-07

Elsharkawy MM, Shimizu M, Takahashi H, Ozaki K, Hyakumachi M (2013) Induction of systemic resistance against Cucumber mosaic virus in *Arabidopsis thaliana* by *Trichoderma asperellum* SKT-1. Plant Pathol J 29(2):193. https://doi.org/10.5423/PPJ.SI.07.2012.01

Fesel PH, Zuccaro A (2016) Dissecting endophytic lifestyle along the parasitism/mutualism continuum in *Arabidopsis*. Curr Opin Microbiol 32:103–112. https://doi.org/10.1016/j.mib.2016.05.008

Filion M, St-Arnaud M, Fortin JA (1999) Direct interaction between the arbuscular mycorrhizal fungus *Glomus intraradices* and different rhizosphere microorganisms. New Phytol 141 (3):525–533

Fiorini L, Guglielminetti L, Mariotti L, Curadi M, Picciarelli P, Scartazza A, Sarrocco S, Vannacci G (2016) *Trichoderma harzianum* T6776 modulates a complex metabolic network to stimulate tomato *cv.* Micro-Tom growth. Plant and Soil 400(1-2):351–366. https://doi.org/10.1007/s11104-015-2736-6

Ghazanfar B, Cheng Z, Ahmad I, Khan AR, Hanqiang L, Haiyan D, Fang C (2015) Synergistic and individual effect of *Glomus etunicatum* root colonization and acetyl salicylic acid on root activity and architecture of tomato plants under moderate NaCl stress. Pak J Bot 47 (6):2047–2054

Glazebrook J (2005) Contrasting mechanisms of defense against biotrophic and necrotrophic pathogens. Annu Rev Phytopathol 43:205–227. https://doi.org/10.1146/annurev.phyto.43.040204.135923

Guzmán-Guzmán P, Porras-Troncoso MD, Olmedo-Monfil V, Herrera-Estrella A (2018) *Trichoderma* species: versatile plant symbionts. Phytopathology 109(1):6–16. https://doi.org/10.1094/PHYTO-07-18-0218-RVW

Harman GE (2000) Myths and dogmas of biocontrol changes in perceptions derived from research on *Trichoderma harzianum* T-22. Plant Dis 84(4):377–393. https://doi.org/10.1094/PDIS.2000.84.4.377

Harman GE (2011) *Trichoderma*-not just for biocontrol anymore. Phytoparasitica 39(2):103–108. https://doi.org/10.1007/s12600-011-0151-y

Harman GE, Lorito M, Chet I, Howell CR, Viterbo A (2004a) *Trichoderma* species- opportunistic, avirulent plant symbionts. Nat Rev Microbiol 2(1):43–56. https://doi.org/10.1038/nrmicro797

Harman GE, Petzoldt R, Comis A, Chen J (2004b) Interactions between *Trichoderma harzianum* strain T22 and maize inbred line Mo17 and effects of these interactions on diseases caused by *Pythium ultimum* and *Colletotrichum graminicola*. Phytopathology 94(2):147–153. https://doi.org/10.1094/phyto.2004.94.2.147

Hermosa R, Viterbo A, Chet I, Monte EG (2012) Plant-beneficial effects of *Trichoderma* and of its genes. Microbiology 158(1):17–25. https://doi.org/10.1099/mic.0.052274-0

Hermosa R, Rubio MB, Cardoza RE, Nicolás C, Monte E, Gutiérrez S (2013) The contribution of *Trichoderma* to balancing the costs of plant growth and defense. Int Microbiol 16(2):69–80. https://doi.org/10.2436/20.1501.01.181

Hoyos-Carvajal L, Orduz S, Bissett J (2009) Growth stimulation in bean (*Phaseolus vulgaris* L.) by *Trichoderma*. Biol Control 51(3):409–416. https://doi.org/10.1016/j.biocontrol.2009.07.018

Jacobs S, Zechmann B, Molitor A, Trujillo M, Petutschnig E, Lipka V, Kogel KH, Schäfer P (2011) Broad-spectrum suppression of innate immunity is required for colonization of *Arabidopsis* roots by the fungus *Piriformospora indica*. Plant Physiol 156(2):726–740. https://doi.org/10.1104/pp.111.176446

de Jaeger N, Declerck S, de La Providencia IE (2010) Mycoparasitism of arbuscular mycorrhizal fungi: a pathway for the entry of saprotrophic fungi into roots. FEMS Microbiol Ecol 73(2):312–322. https://doi.org/10.1111/j.1574-6941.2010.00903.x

Jogaiah S, Abdelrahman M, Tran LSP, Ito SI (2018) Different mechanisms of *Trichoderma virens*-mediated resistance in tomato against *Fusarium* wilt involve the jasmonic and salicylic acid pathways. Mol Plant Pathol 19(4):870–882. https://doi.org/10.1111/mpp.12571

Kleifeld O, Chet I (1992) *Trichoderma harzianum*-interaction with plants and effect on growth response. Plant and Soil 144:267–272. https://doi.org/10.1007/BF00012884

Korolev N, David DR, Elad Y (2008) The role of phytohormones in basal resistance and *Trichoderma*-induced systemic resistance to *Botrytis cinerea* in *Arabidopsis thaliana*. Biol Control 53(4):667–683. https://doi.org/10.1007/s10526-007-9103-3

Kubicek CP, Herrera-Estrella A, Seidl-Seiboth V, Martinez DA, Druzhinina IS, Thon M, Zeilinger S, Casas-Flores S, Horwitz BA, Mukherjee PK, Mukherjee M, László Kredics L, Luis D, Alcaraz LD, Aerts A, Zsuzsanna Antal Z, Atanasova L, Cervantes-Badillo MG, Challacombe J, Chertkov O, McCluskey K, Coulpier F, Deshpande N, von Döhren H, Ebbole DJ, Esquivel-Naranjo EU, Fekete E, Flipphi M, Glaser F, Gómez-Rodríguez EY, Gruber S, Han C, Henrissa B, Hermosa R, Miguel Hernández-Oñate M, Levente Karaffa L, Idit Kosti I, Le Crom S, Lindquist E, Lucas S, Lübeck M, Lübeck PS, Margeot A, Metz B, Misra M, Nevalainen H, Omann MN, Perrone G, Uresti-Rivera EE, Salamov A, Schmoll M, Seiboth B, Shapiro H, Sukno S, Tamayo-Ramos JA, Tisch D, Wiest A, Wilkinson HH, Michael Zhang M, Coutinho PM, Kenerley CM, Monte E, Baker SE, Grigoriev IV (2011) Comparative genome sequence analysis underscores mycoparasitism as the ancestral life style of *Trichoderma*. Genome Biol 12:R40. https://doi.org/10.1186/gb-2011-12-4-r40

Kubicek CP, Steindorff AS, Chenthamara K, Manganiello G, Henrissat B, Zhang J, Cai F, Kopchinskiy AG, Kubicek EM, Kuo A, Baroncelli R, Sarrocco S, Noronha EF, Vannacci G, Shen Q, Grigoriev IV, Druzhinina IS (2019) Evolution and comparative genomics of the most common *Trichoderma* species. BMC Genomics 20(1):485. https://doi.org/10.1186/s12864-019-5680-7

Lace B, Genre A, Woo S, Faccio A, Lorito M, Bonfante P (2015) Gate crashing arbuscular mycorrhizas: in vivo imaging shows the extensive colonization of both symbionts by *Trichoderma atroviride*. Environ Microbiol Rep 7(1):64–77. https://doi.org/10.1111/1758-2229.12221

Lamdan NL, Shalaby S, Ziv T, Kenerley CM, Horwitz BA (2015) Secretome of *Trichoderma* interacting with maize roots: role in induced systemic resistance. Mol Cell Proteomics 14 (4):1054–1063. https://doi.org/10.1074/mcp.M114.046607

de las Mercedes Dana M, Pintor-Toro JA, Cubero B (2006) Transgenic tobacco plants overexpressing chitinases of fungal origin show enhanced resistance to biotic and abiotic stress agents. Plant Physiol 142(2):722–730. https://doi.org/10.1104/pp.106.086140

Lebeis SL, Paredes SH, Lundberg DS, Breakfield N, Gehring J, McDonald M, Malfatti S, Glavina del Rio T, Jones CD, Tringe SG, Dangl JL (2015) Salicylic acid modulates colonization of the root microbiome by specific bacterial taxa. Science 349(6250):860–864. https://doi.org/10.1126/science.aaa8764

Li RX, Cai F, Pang G, Shen QR, Li R, Chen W (2015) Solubilisation of phosphate and micronutrients by *Trichoderma harzianum* and its relationship with the promotion of tomato plant growth. PLoS One 10(6):1–16. https://doi.org/10.1371/journal.pone.0130081

Liang XR, Miao FP, Song YP, Guo ZY, Ji NY (2016) Trichocitrin, a new fusicoccane diterpene from the marine brown alga-endophytic fungus *Trichoderma citrinoviride* cf-27. Nat Prod Res 30(14):1605–1610. https://doi.org/10.1080/14786419.2015.1126264

López-Bucio J, Pelagio-Flores R, Herrera-Estrella A (2015) *Trichoderma* as biostimulant: exploiting the multilevel properties of a plant beneficial fungus. Sci Hortic 196:109–123. https://doi.org/10.1016/j.scienta.2015.08.043

López-Ráez JA, Verhage A, Fernández I, García JM, Azcón-Aguilar C, Flors V, Pozo MJ (2010) Hormonal and transcriptional profiles highlight common and differential host responses to arbuscular mycorrhizal fungi and the regulation of the oxylipin pathway. J Exp Bot 61 (10):2589–2601. https://doi.org/10.1093/jxb/erq089

Malmierca MG, McCormick SP, Cardoza RE, Monte E, Alexander NJ, Gutiérrez S (2015) Trichodiene production in a *Trichoderma harzianum erg1*-silenced strain provides evidence of the importance of the sterol biosynthetic pathway in inducing plant defense-related gene expression. Mol Plant Microbe Interact 28(11):1181–1197. https://doi.org/10.1094/MPMI-06-15-0127-R

Marra R, Nicoletti R, Pagano E, DellaGreca M, Salvatore MM, Borrelli F, Lombardi N, Vinale F, Woo SL, Andolfi A (2019) Inhibitory effect of trichodermanone C, a sorbicillinoid produced by *Trichoderma citrinoviride* associated to the green alga *Cladophora* sp., on nitrite production in LPS-stimulated macrophages. Nat Prod Res 33:3389–3397. https://doi.org/10.1080/14786419.2018.1479702

Martínez-Medina A, Roldán A, Albacete A, Pascual JA (2010) The interaction with arbuscular mycorrhizal fungi or *Trichoderma harzianum* alters the shoot hormonal profile in melon plants. Phytochemistry 72(2–3):223–229. https://doi.org/10.1016/j.phytochem.2010.11.008

Martínez-Medina A, Fernández I, Sánchez-Guzmán MJ, Jung SC, Pascual JA, Pozo MJ (2013) Deciphering the hormonal signalling network behind the systemic resistance induced by *Trichoderma harzianum* in tomato. Front Plant Sci 4:206. https://doi.org/10.3389/fpls.2013.00206

Martínez-Medina A, Del Mar Alguacil M, Pascual JA, van Wees SCM (2014) Phytohormone profiles induced by *Trichoderma* isolates correspond with their biocontrol and plant growth-promoting activity on melon plants. J Chem Ecol 40(7):804–815. https://doi.org/10.1007/s10886-014-0478-1

Martínez-Medina A, Appels FV, van Wees SC (2017a) Impact of salicylic acid-and jasmonic acid-regulated defences on root colonization by *Trichoderma harzianum* T-78. Plant Signal Behav 12(8):e1345404. https://doi.org/10.1080/15592324.2017.1345404

Martínez-Medina A, van Wees SCM, Pieterse CMJ (2017b) Airborne signals from *Trichoderma* fungi stimulate iron uptake responses in roots resulting in priming of jasmonic acid dependent defences in shoots of *Arabidopsis thaliana* and *Solanum lycopersicum*. Plant Cell Environ 40 (11):2691–2705. https://doi.org/10.1111/pce.13016

Mastouri F, Björkman T, Harman GE (2010) Seed treatment with *Trichoderma harzianum* alleviates biotic, abiotic, and physiological stresses in germinating seeds and seedlings. Phytopathology 100(11):1213–1221. https://doi.org/10.1094/PHYTO-03-10-0091

Mastouri F, Björkman T, Harman GE (2012) *Trichoderma harzianum* enhances antioxidant defense of tomato seedlings and resistance to water deficit. Mol Plant Microbe Interact 25 (9):1264–1271. https://doi.org/10.1094/mpmi-09-11-0240

Mathys J, De Cremer K, Timmermans P, van Kerkhove S, Lievens B, Vanhaecke M, Cammue BP, De Coninck B (2012) Genome-wide characterization of ISR induced in *Arabidopsis thaliana* by *Trichoderma hamatum* T382 against *Botrytis cinerea* infection. Front Plant Sci 3:108. https://doi.org/10.3389/fpls.2012.00108

de Medeiros HA, de Araújo Filho JV, De Freitas LG, Castillo P, Rubio MB, Hermosa R, Monte E (2017) Tomato progeny inherit resistance to the nematode *Meloidogyne javanica* linked to plant growth induced by the biocontrol fungus *Trichoderma atroviride*. Sci Rep 7:40216. https://doi.org/10.1038/srep40216

Medina MJH, Gagnon H, Piché Y, Ocampo JA, Garrido JMG, Vierheilig H (2003) Root colonization by arbuscular mycorrhizal fungi is affected by the salicylic acid content of the plant. Plant Sci 164(6):993–998. https://doi.org/10.1016/S0168-9452(03)00083-9

Mendoza-Mendoza A, Zaid R, Lawry R, Hermosa R, Monte E, Horwitz BA, Mukherjee PK (2018) Molecular dialogues between *Trichoderma* and roots: role of the fungal secretome. Fungal Biol Rev 32(2):62–85. https://doi.org/10.1016/j.fbr.2017.12.001

Mengel K, Kirkby EA (2001) Principles of plant nutrition. Springer, Dordrecht, p 635

de Meyer G, Bigiriman J, Elad Y, Höfte M (1998) Induced systemic resistance in *Trichoderma harzianum* T39 biocontrol of *Botrytis cinerea*. Eur J Plant Pathol 104(3):279–286. https://doi.org/10.1023/A:1008628806616

Mittler R (2002) Oxidative stress, antioxidants and stress tolerance. Trends Plant Sci 7(9):405–410. https://doi.org/10.1016/S1360-1385(02)02312-9

Montero-Barrientos M, Hermosa R, Cardoza RE, Gutiérrez S, Nicolás C, Monte E (2010) Transgenic expression of the *Trichoderma harzianum hsp70* gene increases *Arabidopsis* resistance to heat and other abiotic stresses. J Plant Physiol 167(8):659–665. https://doi.org/10.1016/j.jplph.2009.11.012

Morán-Diez E, Hermosa R, Ambrosino P, Cardoza RE, Gutiérrez S, Lorito M, Monte E (2009) The ThPG1 endopolygalacturonase is required for the *Trichoderma harzianum* – plant beneficial interaction. Mol Plant Microbe Interact 22(8):1021–1031. https://doi.org/10.1094/MPMI-22-8-1021

Morán-Diez E, Rubio B, Domínguez S, Hermosa R, Monte E, Nicolás C (2012) Transcriptomic response of *Arabidopsis thaliana* after 24 h incubation with the biocontrol fungus *Trichoderma harzianum*. J Plant Physiol 169(6):614–620. https://doi.org/10.1016/j.jplph.2011.12.016

Morán-Diez ME, Trushina N, Lamdan NL, Rosenfelder L, Mukherjee PK, Kenerley CM, Horwitz BA (2015) Host-specific transcriptomic pattern of *Trichoderma virens* during interaction with maize or tomato roots. BMC Genomics 16(1):8. https://doi.org/10.1186/s12864

Mukherjee M, Mukherjee PK, Horwitz BA, Zachow C, Berg G, Zeilinger S (2012) *Trichoderma* – plant–pathogen interactions: advances in genetics of biological control. Indian J Microbiol 52 (4):522–529. https://doi.org/10.1007/s12088-012-0308-5

Nawrocka J, Małolepsza U (2013) Diversity in plant systemic resistance induced by *Trichoderma*. Biol Control 67(2):149–156. https://doi.org/10.1016/j.biocontrol.2013.07.005

de Palma M, Salzano M, Villano C, Aversano R, Lorito M, Ruocco M, Docimo T, Piccinelli AL, D'Agostino N, Tucci M (2019) Transcriptome reprogramming, epigenetic modifications and alternative splicing orchestrate the tomato root response to the beneficial fungus *Trichoderma harzianum*. Hortic Res 6(1):5. https://doi.org/10.1038/s41438-018-0079-1

Pelagio-Flores R, Esparza-Reynoso S, Garnica-Vergara A, López-Bucio J, Herrera-Estrella A (2017) *Trichoderma*-induced acidification is an early trigger for changes in *Arabidopsis* root growth and determines fungal phytostimulation. Front Plant Sci 8:822. https://doi.org/10.3389/fpls.2017.00822

Peleg-Grossman S, Golani Y, Kaye Y, Melamed-Book N, Levine A (2009) NPR1 protein regulates pathogenic and symbiotic interactions between *Rhizobium* and legumes and non-legumes. PLoS One 4(12):e8399. https://doi.org/10.1371/journal.pone.0008399

Pozo MJ, Azcón-Aguilar C (2007) Unraveling mycorrhiza-induced resistance. Curr Opin Plant Biol 10(4):393–398. https://doi.org/10.1016/j.pbi.2007.05.004

Rawat R, Tewari L (2011) Effect of abiotic stress on phosphate solubilization by biocontrol fungus *Trichoderma* sp. Curr Microbiol 62(5):1521–1526. https://doi.org/10.1007/s00284-011-9888-2

Rivas-San Vicente M, Plasencia J (2011) Salicylic acid beyond defence: its role in plant growth and development. J Exp Bot 62(10):3321–3338. https://doi.org/10.1093/jxb/err031

Rubio MB, Quijada NM, Pérez E, Domínguez S, Monte E, Hermosa R (2014) Identifying beneficial qualities of *Trichoderma parareesei* for plants. Appl Environ Microbiol 80(6):1864–1873. https://doi.org/10.1128/AEM.03375-13

Salas-Marina MA, Silva-Flores MA, Uresti-Rivera EE, Castro-Longoria E, Herrera-Estrella A, Casas-Flores S (2011) Colonization of *Arabidopsis* roots by *Trichoderma atroviride* promotes growth and enhances systemic disease resistance through jasmonic acid/ethylene and salicylic acid pathways. Eur J Plant Pathol 131(1):15–26. https://doi.org/10.1007/s10658-011-9782-6

Samolski I, Rincón AM, Pinzón LM, Viterbo A, Monte E (2012) The *qid74* gene from *Trichoderma harzianum* has a role in root architecture and plant biofertilization. Microbiology 158 (1):129–138. https://doi.org/10.1099/mic.0.053140-0

Saravanakumar K, Fan L, Fu K, Yu C, Wang M, Xia H, Sun J, Yaqian L, Chen J (2016) Cellulase from *Trichoderma harzianum* interacts with roots and triggers induced systemic resistance to foliar disease in maize. Sci Rep 6:35543. https://doi.org/10.1038/srep35543

Searle LJ, Méric G, Porcelli I, Sheppard SK, Lucchini S (2015) Variation in siderophore biosynthetic gene distribution and production across environmental and faecal populations of *Escherichia coli*. PLoS One 10(3):1–14. https://doi.org/10.1371/journal.pone.0117906

Segarra G, Casanova E, Bellido D, Odena MA, Oliveira E, Trillas I (2007) Proteome, salicylic acid, and jasmonic acid changes in cucumber plants inoculated with *Trichoderma asperellum* strain T34. Proteomics 7(21):3943–3952. https://doi.org/10.1002/pmic.200700173

Shabala S, Cuin TA (2008) Potassium transport and plant salt tolerance. Physiol Plant 133 (4):651–669. https://doi.org/10.1111/j.1399-3054.2007.01008.x

Sharon E, Bar-Eyal M, Chet I, Herrera-Estrella A, Kleifeld O, Spiegel Y (2001) Biological control of the root-knot nematode *Meloidogyne javanica* by *Trichoderma harzianum*. Phytopathology 91(7):687–693. https://doi.org/10.1094/PHYTO.2001.91.7.687

Shi ZZ, Fang ST, Miao FP, Yin XL, Ji NY (2018) Trichocarotins A–H and trichocadinin A, nine sesquiterpenes from the marine-alga-epiphytic fungus *Trichoderma virens*. Bioorg Chem 81:319–325. https://doi.org/10.1016/j.bioorg.2018.08.027

Shoresh M, Harman GE (2008) The molecular basis of shoot responses of maize seedlings to *Trichoderma harzianum* T22 inoculation of the root: a proteomic approach. Plant Physiol 147 (4):2147–2163. https://doi.org/10.1104/pp.108.123810

Shoresh M, Yedidia I, Chet I (2005) Involvement of jasmonic acid/ethylene signaling pathway in the systemic resistance induced in cucumber by *Trichoderma asperellum* T203. Phytopathology 95(1):76–84. https://doi.org/10.1094/PHYTO-95-0076

Shoresh M, Harman GE, Mastouri F (2010) Induced systemic resistance and plant responses to fungal biocontrol agents. Annu Rev Phytopathol 48:21–43. https://doi.org/10.1146/annurev-phyto-073009-114450

Singh BN, Dwivedi P, Sarma BK, Singh GS, Singh HB (2018) *Trichoderma asperellum* T42 reprograms tobacco for enhanced nitrogen utilization efficiency and plant growth when fed with N nutrients. Front Plant Sci 9:163. https://doi.org/10.3389/fpls.2018.00163

Song YP, Liu XH, Shi ZZ, Miao FP, Fang ST, Ji NY (2018) Bisabolane, cyclonerane, and harziane derivatives from the marine-alga-endophytic fungus *Trichoderma asperellum* cf44-2. Phytochemistry 152:45–52. https://doi.org/10.1016/j.phytochem.2018.04.017

van Spronsen PC, Tak T, Rood AM, van Brussel AA, Kijne JW, Boot KJ (2003) Salicylic acid inhibits indeterminate-type nodulation but not determinate-type nodulation. Mol Plant Microbe Interact 16(1):83–91. https://doi.org/10.1094/MPMI.2003.16.1.83

Sriram S, Manasa SB, Savitha MJ (2009) Potential use of elicitors from *Trichoderma* in induced systemic resistance for the management of *Phytophthora capsici* in red pepper. J Biol Control 23(4):449–456

Stacey G, McAlvin CB, Kim SY, Olivares J, Soto MJ (2006) Effects of endogenous salicylic acid on nodulation in the model legumes *Lotus japonicus* and *Medicago truncatula*. Plant Physiol 141(4):1473–1481. https://doi.org/10.1104/pp.106.080986

Su D, Ding L, He S (2018) Marine-derived *Trichoderma* species as a promising source of bioactive secondary metabolites. Mini Rev Med Chem 18(20):1702–1713. https://doi.org/10.2174/1389557518666180727130826

Terry N, Zayed AM (1995) Physiology and biochemistry of leaves under iron deficiency. In Iron nutrition in soils and plants. In: Badìa J (ed) Iron nutrition in soils and plants. Kluwer Academic Publishers, Dordrecht, pp 283–294

Tucci M, Ruocco M, De Masi L, De Palma M, Lorito M (2011) The beneficial effect of *Trichoderma* spp. on tomato is modulated by the plant genotype. Mol Plant Pathol 12 (4):341–354. https://doi.org/10.1111/j.1364-3703.2010.00674.x

Vargas WA, Mandawe JC, Kenerley CM (2009) Plant-derived sucrose is a key element in the symbiotic association between *Trichoderma virens* and maize plants. Plant Physiol 151 (2):792–808. https://doi.org/10.1104/pp.109.141291

Velázquez-Robledo R, Contreras-Cornejo HA, Macías-Rodríguez L, Hernández-Morales A, Aguirre J, Casas-Flores S, López-Bucio J, Herrera-Estrella A (2011) Role of the 4-phosphopantetheinyl transferase of *Trichoderma virens* in secondary metabolism and induction of plant defense responses. Mol Plant Microbe Interact 24(12):1459–1471. https://doi.org/10.1094/MPMI-02-11-0045

Velmourougane K, Prasanna R, Singh S, Chawla G, Kumar A, Saxena AK (2017) Modulating rhizosphere colonisation, plant growth, soil nutrient availability and plant defense enzyme activity through *Trichoderma viride-Azotobacter chroococcum* biofilm inoculation in chickpea. Plant and Soil 421(1-2):157–174. https://doi.org/10.1007/s11104-017-3445-0

Viterbo ADA, Chet I (2006) *TasHyd1*, a new hydrophobin gene from the biocontrol agent *Trichoderma asperellum*, is involved in plant root colonization. Mol Plant Pathol 7 (4):249–258. https://doi.org/10.1111/j.1364-3703.2006.00335.x

Viterbo A, Landau U, Kim S, Chernin L, Chet I (2010) Characterization of ACC deaminase from the biocontrol and plant growth-promoting agent *Trichoderma asperellum* T203. FEMS Microbiol Lett 305(1):42–48. https://doi.org/10.1111/j.1574-6968.2010.01910.x

Vitti A, Pellegrini E, Nali C, Lovelli S, Sofo A, Valerio M, Scopa A, Nuzzaci M (2016) *Trichoderma harzianum* T-22 induces systemic resistance in tomato infected by Cucumber mosaic virus. Front Plant Sci 7:1520. https://doi.org/10.3389/fpls.2016.01520

Vizcaíno JA, Cardoza RE, Hauser M, Hermosa R, Rey M, Llobell A, Becker JM, Gutiérrez S, Monte E (2006) *ThPTR2*, a di/tri-peptide transporter gene from *Trichoderma harzianum*. Fungal Genet Biol 43(4):234–246. https://doi.org/10.1016/j.fgb.2005.12.003

Welch RM, Norvell WA, Schaefer SC, Shaff JE, Kochian LV (1993) Induction of iron (III) and copper(II) reduction in pea (*Pisum sativum* L.) roots by Fe and Cu status: does the root-cell plasmalemma Fe(III)-chelate reductase perform a general role in regulating cation uptake? Planta 190(4):555–561. https://doi.org/10.1007/BF00224795

Woo SL, Lorito M (2007) Exploiting the interactions between fungal antagonists, pathogens and the plant for biocontrol. In: Vurro M, Gressel J (eds) Novel biotechnologies for biocontrol agent enhancement and management. Springer, Dordrecht, pp 107–130

Yasmeen R, Siddiqui ZS (2018) Ameliorative effects of *Trichoderma harzianum* on monocot crops under hydroponic saline environment. Acta Physiol Plant 40(1):1–14. https://doi.org/10.1007/s11738-017-2579-2

Yedidia I, Benhamou N, Chet I (1999) Induction of defense responses in cucumber plants (*Cucumis sativus* L.) by the biocontrol agent *Trichoderma harzianum*. Appl Environ Microbiol 65 (3):1061–1070. 0099-2240/99/$04.0010

Yedidia I, Benhamou N, Kapulnik Y, Chet I (2000) Induction and accumulation of PR proteins activity during early stages of root colonization by the mycoparasite *Trichoderma harzianum* strain T-203. Plant Physiol Biochem 38(11):863–873. https://doi.org/10.1016/S0981-9428(00)01198-0

Yedidia I, Srivastva AK, Kapulnik Y, Chet I (2001) Effect of *Trichoderma harzianum* on micro-element concentrations and increased growth of cucumber plants. Plant and Soil 235 (2):235–242. https://doi.org/10.1023/A:1011990013955

Yoshioka Y, Ichikawa H, Naznin HA, Kogure A, Hyakumachi M (2011) Systemic resistance induced in *Arabidopsis thaliana* by *Trichoderma asperellum* SKT-1, a microbial pesticide of seedborne diseases of rice. Pest Manag Sci 68(1):60–66. https://doi.org/10.1002/ps.2220

Yuan M, Huang Y, Ge W, Jia Z, Song S, Zhang L, Huang Y (2019) Involvement of jasmonic acid, ethylene and salicylic acid signaling pathways behind the systemic resistance induced by *Trichoderma longibrachiatum* H9 in cucumber. BMC Genomics 20(1):144. https://doi.org/10.1186/s12864-019-5513-8

Zamioudis C, Pieterse CM (2012) Modulation of host immunity by beneficial microbes. Mol Plant Microbe Interact 25(2):139–150. https://doi.org/10.1094/MPMI-06-11-0179

Zhang F, Meng X, Yang X, Ran W, Shen Q (2014) Quantification and role of organic acids in cucumber root exudates in *Trichoderma harzianum* T-E5 colonization. Plant Physiol Biochem 83:250–257. https://doi.org/10.1016/j.plaphy.2014.08.011

Zhao L, Wang F, Zhang Y, Zhang J (2014) Involvement of *Trichoderma asperellum* strain T6 in regulating iron acquisition in plants. J Basic Microbiol 54(S1):115–124. https://doi.org/10.1002/jobm.201400148

Zhu H, Zhang R, Chen W, Gu Z, Xie X, Zhao H, Yao Q (2015) The possible involvement of salicylic acid and hydrogen peroxide in the systemic promotion of phenolic biosynthesis in clover roots colonized by arbuscular mycorrhizal fungus. J Plant Physiol 178:27–34. https://doi.org/10.1016/j.jplph.2015.01.016

Chapter 4
Harnessing the Perception of *Trichoderma* Signal Molecules in Rhizosphere to Improve Soil Health and Plant Health

Sevugapperumal Nakkeeran, Suppaiah Rajamanickam, Murugavel Vanthana, Perumal Renukadevi, and Malaiyandi Muthamilan

Abstract In the fungal kingdom, *Trichoderma* is the most exploited biocontrol agent. It is known for its ability to colonize plant root, suppress many plant pathogens, and act as inducer of systemic resistance in plant and is also involved in growth promotional activities. The knowledge on rhizosphere signalling between plant root–*Trichoderma* interaction is highly essential to trigger the immune response and to improve plant health. There is a bidirectional perception of signal molecules for the successful establishment of its association. The interaction of *Trichoderma* starts with contact to the root surface, followed by attachment, penetration, and colonization. The signal molecules from root exudates and secretomes of *Trichoderma* are essential for successful accomplishment of a symbiotic relationship. Among the complex composition of root exudates, sucrose plays a vital role in the attraction of *Trichoderma* to the plant root system. The hydrophobin and swollenins and cysteine-rich proteins secreted by *Trichoderma* plays a role in each step towards successful plant root interaction. This review enlightens our knowledge on root–*Trichoderma* interaction and its ability to overcome plant's defense mechanism to prove itself as a "true friend" to the plant system and for improvement of soil and plant health.

Keywords *Trichoderma* · Signal molecules · Rhizosphere specificity · Hydrophobin · Swollenins

4.1 Introduction

Microorganisms are the first living organisms established in the earth followed by the plant kingdom. Microbes have coevolved with plants. The continuous association and evolution decide the relationship of the microbes with plants either as

S. Nakkeeran (✉) · S. Rajamanickam · M. Vanthana · P. Renukadevi · M. Muthamilan
Department of Plant Pathology, Tamil Nadu Agricultural University, Coimbatore, India
e-mail: nakkeeranayya@tnau.ac.in

A. K. Sharma, P. Sharma (eds.), *Trichoderma*, Rhizosphere Biology,
https://doi.org/10.1007/978-981-15-3321-1_4

biotroph, hemibiotroph, necrotroph, or saprotrophs. These relationships are decided based on the molecular communications between both partners, which may be either beneficial or harmful. Hence, it plays a dual role, both positive (beneficial microorganisms) and negative (pathogenic microorganism). Communications are mediated through signal molecules from the microbes to the plants or vice versa. The photosynthates diverted to the rhizosphere decide the colonization of microbes including *Trichoderma*. The rich biological niche of the rhizosphere represents the health of the plant system. So it is of paramount importance to understand the plant–microbe interaction in the rhizosphere. The plant root exudate composition is highly complex and complete characterization of the root exudate is not possible. The root exudates are grouped into low-molecular-weight compounds and high-molecular-weight compounds. The low-molecular-weight compounds like sugars, amino acids, and secondary metabolites are highly diverse and have various functions. The high-molecular-weight compounds include polysaccharides and proteins which are less diverse but are released in high proportion. These compounds are transported to the environment through an active or passive process (Badri and Vivanco 2009). Transporters like the ABC transporter, MATE exporter (multidrug and toxic compound extrusion family), major facilitator superfamily (MFS), and aluminum-activated malate transporter family (ALMT) are involved in the active transportation of high-molecular-weight compounds. These root secretions recruit specific microorganisms in the rhizosphere. The exact mechanism behind the recruitment of microorganisms is still unclear (Reddy et al. 2012; Takanashi et al. 2014).

The plant root–*Trichoderma* interaction resembles the interaction of mycorrhiza with symbionts of the plant root. Though *Trichoderma* is an opportunistic symbiont, it can survive saprophytically and remain in close association with the plant system (Harman et al. 2004). The *Trichoderma* species secretes CAZymes (carbohydrate-active enzymes), cysteine-rich proteins, hydrophobins, and many other enzymes to suppress the plant immunity against their colonization. The ISR trigger provides a priming effect against plant pathogenic organisms. The SAR is triggered during initial interaction of the plant–*Trichoderma* surpassing SAR through increased JA/ET which are antagonists of SA. However, our knowledge on *Trichoderma*–root interactions is only primitive.

Hence, understanding of the cross talk between microbes and plants is essential to define the relationship. The role of chemical molecules involved in the molecular dialogues at different trophic levels is not well explored. In this review, we describe how molecular communication occurs between *Trichoderma* and the rhizosphere to improve plant health and soil health.

4.2 Root Exudates and Rhizosphere Colonization by *Trichoderma*

Interaction and communication between the rhizomicrobiome and plants are basically determined by the root colonization of the host. Colonization is mediated by chemical cues in the rhizosphere, which occur at the cost of carbon budgeting of the photosynthates produced by plants. Around 20% of the photosynthates are diverted to the rhizosphere as root exudates and are released via the roots to the rhizosphere. The complex nutrient profile in the root exudates serves either as an attractant or repellant for *Trichoderma*. The nutrients of low and high molecular weight mediate tritrophic interactions between *Trichoderma*, plant, and pathogens in the rhizosphere. Sugars, organic acids, amino acids, phenols, and other secondary metabolites are low-molecular-weight compounds, while proteins and polysaccharides are high-molecular-weight compounds. These exudates vary according to age and stage of the crop (Chaparro et al. 2014). The carbohydrates and amino acids serve as a chemoattractant for diverse microbes in soil. However, other compounds mediate highly specific interactions (Abdel-Lateif et al. 2012). Secondary metabolites and plant hormones attract beneficial soil microbes and aid in defending plants against pathogens. Root exudates like lectins are involved in symbiotic interactions and defense responses. Root caps and root tips release mucilage rich in polysaccharides into the rhizosphere, which contain protein rich in arabinogalactan. These proteins are used to attract beneficial microorganisms like *Trichoderma* and beneficial bacteria. Thus, *Trichoderma* grows on the rhizosphere and sometimes enter as endophytes into the crop plants.

4.3 Rhizosphere Specificity of *Trichoderma*

Colonization of the roots by *Trichoderma* on tomato upregulated the genes associated with the formation of structures for infection on root tissues (Samolski et al. 2009). Initiation of *Trichoderma* attachment on the root system suppresses the uptake of nutrients and carbohydrate metabolism due to the plant defense mechanism (Rubio et al. 2014). Analysis of molecular communication between tomato roots and *Trichoderma* interaction in hydroponic system indicated that the genes associated with the creation of building blocks, nutrient exchange, and carbohydrate synthesis were induced (Rubio et al. 2012). The secreted quinoprotein glucose dehydrogenase gene is expressed more preferentially on tomato roots after interaction with *Trichoderma*. However, a group of scientists emphasize that *Trichoderma* has a wide host range. If so, there may be any specific adaptation by *Trichoderma* to the hosts. But the answer behind the issue was: the response is regulated not by the *Trichoderma* but it is the host plant that decides the regulation. Interaction of *T. virens* strain with tomato and maize clearly indicated that the subset of 35 genes was upregulated differently in tomato and maize (Moran-Diez et al. 2015). Similarly, interaction of *Trichoderma* with maize roots induced the expression of genes associated with the synthesis of small secreted proteins, short-chain oxidoreductases, glycosyl hydrolases, and symbiosis-related invertases (Vargas et al. 2009). Hence, though *Trichoderma* has wide adaptability, the selective expression of different genes varies depending upon the host, thus explaining that the response is mediated by the host and not by *Trichoderma*. However, to have in-depth knowledge on the biochemical interaction of *Trichoderma* with the rhizosphere of different hosts, complete transcriptome profiling of rhizosphere interaction is imperative.

4.4 *Trichoderma* CWDE and MAMP Molecules on Root Colonization

The basic platform associated with plant–microbe interactions is the plant cell walls. Colonization of the plants is enabled by cell wall digestion followed by penetration. Like that of the soilborne pathogens, *Trichoderma* also produce cell wall-degrading enzymes (CWDE) like glycosyl hydrolases (GHs), which facilitate *Trichoderma* to access the root tissues and also serve as a source of secreted proteins which facilitate communication through the production of signal molecules. Carbohydrate-active enzymes (CAZymes) characterized in eight *Trichoderma* genomes belongs to 136 GH families. In particular, *T. reesei* QM6a contains 68 families and *T. reesei* RUTC30 contains 75 families. The genome of *T. harzianum* has 130 and *T. guizhouense* has 136 GH families. Variations in the CAZymes reflect on the ability of the species and isolates of *Trichoderma* to colonize the substrate and host plants (Li et al. 2017; Taniguchi et al. 2011). CWDE of *Trichoderma* secretome not only favors colonization of the roots but also triggers the immune response of the host plant. Inactive cellulase and xylanase proteins were identified as the first MAMP molecule from *Trichoderma* to trigger innate immune response on host plants (Avni et al. 1994; Martinez et al. 2001; Rotblat et al. 2002). The CWDE does not cause extensive damage to the cell wall, but also results in the production of DAMP signal molecules. Oligogalacturonides were released as a DAMP signal molecule, due to the enzymatic action of endopolygalacturonase ThPG1 of *Trichoderma* with the roots of *Arabidopsis* and tomato. It was the first identified DAMP molecule involved in ISR (Moran-Diez et al. 2009). Enzymatic action of chitinases from *Trichoderma* released DAMP molecule chitooligosaccharides from the chitin polymers associated with ISR (Woo et al. 2006). Though CWDE releases MAMP and DAMP signal molecules, it increases the plasticity of the root cells and thus facilitates efficient colonization. Root colonization in tomato by *T. harzianum* was enhanced by the production of endopolygalacturonase ThPG1. It was reflected by an eightfold increase of *T. harzianum* DNA in the root tissues through qPCR, which was not the case with the *T. harzianum* silenced for ThPG1 expression (Moran-Diez et al. 2009). Thus, the molecular communication between endopolygalacturonase of *Trichoderma* and the roots of the host plant favors colonization. The endophytic *T. virens* and *T. harzianum* produce cellulases, endopolygalacturonases, and xylanases (Moran-Diez et al. 2015; Chacon et al. 2007). Coculturing of these endophytes along with maize or tomato upregulated eight genes out of which seven genes were involved in polysaccharide metabolism favoring root colonization (Moran-Diez et al. 2015). Thus, initiatives made to overexpress the genes associated with polysaccharide metabolism would favor microbial communication network between *Trichoderma* spp. and host plant, which would enhance the level of colonization of roots and thus protect the plants from biotic and abiotic stress.

4.5 MAMP and DAMP Molecules of *Trichoderma*-Mediated Systemic Resistance

The molecular signals from *Trichoderma* include structural MAMPs; secreted MAMPs like "xylanases, cellulases, cerato-platanins, swollenins, avirulence proteins, lysin motif (LysM) domains, peptaibols, 6-pentyl pyrones, trichothecenes," phytohormones; and cell-wall-degrading polygalacturonases (Hermosa et al. 2013). Fungal elicitors are generally deposited in the root cell apoplast region. The effectors may be a protein or secondary metabolite. The MAMP molecule xylanase–ET-inducing xylanase Eix/Xyn2 is an elicitor produced by *T. viride*. It elicits ET biosynthesis and induces ISR in plant (Avni et al. 1994; Martinez et al. 2001; Rotblat et al. 2002; Ron and Avni 2004). Martinez et al. (2001) reported that *Trichoderma* induced cellulases, salicylic acid, and ethylene signalling pathways. Ceratoplatanin proteins are small secreted proteins rich in cysteines and trigger root colonization and ISR induced by JA in cotton and maize (Hermosa et al. 2012). Swollenin protein identified in *T. reesei* was similar to plant expansin and colonized root, but it does not trigger ISR (Brotman et al. 2008). The avirulence proteins avr4 and avr9 reported from *T. atroviride* and *T. harzianum* are SSCP homologues that stimulate SAR through activation of EDS1 (enhanced disease susceptibility 1) gene against *Rhizoctonia solani* and *B. cinerea* (Marra et al. 2006; Harman et al. 2004). These avirulence genes also protect the fungal cell wall from plant chitinase enzyme (Stergiopoulos and De Wit 2009). Thus, colonization of roots of the host plant by *Trichoderma* spp. triggers innate immunity against plant pathogens (Moran-Diez et al. 2009; De Lorenzo et al. 2011). Woo et al. (2006) reported the DAMP activity of chitooligosaccharides by chitinase activity of plant or *Trichoderma* on rhizospheric fungi.

4.6 Trichoderma Proteins Involved in Root Colonization

The protein domain with eight cysteines is present in the fungal extracellular membrane (CFEM) of *Trichoderma*. The family proteins of CFEM contain TMD proteins, which have diverse roles in sensing of plant surface, asexual development, iron assimilation, and colonization as that of the fungal pathogens (Nasser et al. 2016; Kou et al. 2017; Sabnam and Roy Barman 2017). PTH11, the GPCR with CFEM domain in *Trichoderma*, is essential for protein activity (Kou et al. 2017). Besides, the *Trichoderma* genome had 50 soluble secreted proteins with CFEM domain. Hence, a thorough understanding of the secreted proteins with CFEM domain would help in tinkering the genes to enhance rhizosphere colonization by *Trichoderma*.

4.7 Molecular Signals of *Trichoderma* and ISR

Rhizosphere colonization of *Trichoderma* activates the ISR and SAR through the salicylic acid (SA) and jasmonate/ethylene (JA/ET) pathways. *Trichoderma* colonization induces rapid ion fluxes and oxidative burst along with callose deposition and synthesis of polyphenols (Shoresh et al. 2010). It is followed by the coordination of signals with NPR genes, causing varying degrees of resistance to the invading pathogens (Shoresh et al. 2010). The host plant contact and increased population of *Trichoderma* induce SA-mediated SAR (Segarra et al. 2007; Salas-Marina et al. 2011; Yoshioka et al. 2012). MAPK from *T. virens* interacting with MAPK from cucumber resulted in induced downstream defense in cucumber (Viterbo et al. 2005; Shoresh et al. 2006). Compounds like xylanase, peptaibols, alpha amino isobutyric acid, trichovirin II, and alamethicin from *Trichoderma* spp. induce the plants' defense mechanism (Luo et al. 2010; Druzhinina et al. 2011). The gene Sm1/Epl1 from *Trichoderma* that belongs to the cerato-platanin family secretes a large amount of cysteine-rich hydrophobin-like proteins and induced ISR in maize plants (Djonovic et al. 2006, 2007; Seidl et al. 2006; Mukherjee et al. 2012). The glycosylated monomers of Sm1 induced ISR. However, non-glycosylated monomers of Sm1 are prone to oxidative-driven dimerization and could not act as an elicitor to induce ISR. Sm1 coding for the oligomer with binding properties of *N*-acetyl glucosamine is highly conserved and could be responsible for ISR (de Oliveira et al. 2011). Besides, defense responses in maize were also triggered by a PKS/NRPS hybrid enzyme (Mukherjee et al. 2012).

Root colonization by *Trichoderma* damages the cell wall and produces DAMP molecules, which are responsible for the ISR and conferred resistance to foliar and soilborne pathogens. The seeds of tomato plants treated with *T. harzianum* decreased the intensity of leaf blight caused by *Botrytis cinerea* (Moran-Diez et al. 2009). *Trichoderma* hydrolyze the pectin in the middle lamella, penetrating and colonizing the apoplastic region. During the process of penetration and colonization of root tissues, metabolites and soluble secreted proteins are produced and trigger ISR. Interaction of *T. virens* I10 with roots of tomato or pectin induced the expression of TvPg1 coding for endopolygalacturonase, while Tvpg2 was constitutively expressed and was responsible for the induction of ISR (Baroncelli et al. 2016; Sarrocco et al. 2017). Besides, the plant defense elicitors, viz., endo-1,4-β-D-glucanase and exo-1,4-β-D-glucanase, induce ISR activated through MAMP and DAMP. These enzymes were responsible for inducing a resistance mechanism in maize (Saravanakumar et al. 2016). *T. virens* colonization in maize roots induced 27 different glycosyl hydrolases. Due to substrate specificity, 17 were over expressed in the maize root tissues and ten were secreted by *T. virens* alone and not by the host (Lamdan et al. 2015). *T. harzianum* colonization on maize roots was promoted by enzymes of cellulose, conserved cellulose binding domains present in the C-terminus of *Thph1* and N-terminus of *Thph2*. Treatment with mutant strains of *T. harzianum* (*Thph1* and *Thph2*) expressed the reduced level of cellobiose production on maize roots. However, application of cellulase-like proteins secreted by the genes *Thph1* and *Thph2* induced overexpression of cellobiose. LC-MS analysis indicated that JA content in maize leaves and roots was increased in plants treated with cellulase protein from *T. harzianum* than the wild type. Further, increased mRNA levels of Opr7, Pr4, Aoc1, and Erf1 were observed with the expression of defense-related genes that increased with days of *Trichoderma* treatments (Saravanakumar et al. 2016). Thus, it indicated that the production of cellobiose is dependent on genes *Thph1* and *Thph2* that are rational for the triggered ISR and defense genes.

In addition, it also increased shoot weight and height and root weights in plants treated with the wild strain compared with the plants treated with the mutant strain of *Trichoderma*. Hence, cellulase-like proteins secreted by the genes *Thph1* and *Thph2* of *T. harzianum* favored maize root colonization and triggered the expression of defense genes associated with ISR. Apart from it, in maize, it was found that treatment with a distorted strain never expressed the accumulation of reactive oxygen species (ROS), whereas higher ROS production was documented in plants treated with a wild strain of *Trichoderma* expressing *Thph1* and *Thph2* genes. In addition to the transient expression of ROS, calcium influx and levels of calcium in maize leaves were also increased. Due to the consequence of ROS production, increased calcium influx, ISR, and defense gene accumulation in maize, the infection by leaf spot pathogen in maize caused by *Curvularia lunata* was significantly reduced. Further, mass spectrometry analysis indicated that 12 out of 40 proteins were upregulated and might have contributed to the reduction of *C. lunata* leaf blight in maize (Saravanakumar et al. 2016).

Thus, cellulase enzyme released during the interaction between maize roots and *Trichoderma* acts as an effector molecule to initiate MAMP-triggered immunity. Subsequently, DAMP-responsive effector molecule, cellobiose released from the roots of maize upon the action by the cellulase, induced the defense mechanisms in maize against leaf spot as a result of activation of C6 zinc finger protein-like elicitor *Thc6* that regulates two hydrolases *Thph1* and *Thph2* in *T. harzianum*. Thus, the transcription factors including MYC2 and ACO in ethylene pathway and PO involved in ROS signaling and molecular signals such as cellulase-like proteins and cellobiose triggered by the colonization of *T. harzianum* on to maize roots not only confer resistance to soilborne pathogens but also induce resistance against foliar pathogens.

4.8 Secondary Metabolites of *Trichoderma* and Their Diverse Mode of Action

Trichoderma is known to produce a huge diversity of secondary metabolites. The antibiotic activity of secondary metabolites produced by *Trichoderma* has been identified and described. Trichothecenes are sesquiterpenes produced by *Trichoderma*, inducing systemic resistance in plants. Trichodermin produced by *T. brevicompactum* and harzianum A produced by *T. arundinaceum* possess antifungal activity against *Fusarium*, *B. cinerea*, and *R. solani* (Tijerino et al. 2011; Malmierca et al. 2013). Peptaibols are secondary metabolites that produce antifungal and antibacterial activities. Trichokonins is a peptaibol detected in *T. pseudokoningii* that induces ROS accumulation locally and systemically. It utilizes SA, JA, and ET signalling genes (Luo et al. 2010). 6-Pentyl pyrones is a volatile metabolite of

Trichoderma and acts as an auxin-like compound and inhibits *Fusarium oxysporum* through the induction of PR1 genes (Rubio et al. 2009). The phytohormones are manipulated by *Trichoderma* to establish and to induce beneficial responses in plants. It promotes IAA synthesis and reduces ET production though ACCD activity (Contreras et al. 2009; Viterbo et al. 2010). *Trichoderma* produce the best known lactone, 6-pentyl-2H-pyran-2-one (6-PP). It promotes production of auxin and ET pathway and promotes plant growth, root architecture, and lateral shoots mediated through ARF7 and ARF19 transcription factors and EIN2 in root (Garnica-Vergara et al. 2016). *T. virens* interaction with cucumber produces peptaibol by activation of *Tex1* gene and suppresses *P. syringae* through the production of phytoalexins (Viterbo et al. 2007). Alamethicin peptaibols possess multiple activities in inducing defense responses including callose deposition, production of phenolic compounds, regulation of defense genes, and enhanced plant growth (Rippa et al. 2007). Polyketides produced by *Trichoderma* include macrolides, polyenes, and polyphenols. Mukherjee et al. (2012) observed the role of aspinolide C produced by *T. arundinaceum* in the induction of PR1b1 and PR-P2 genes in plants. Terpenes are highly diverse secondary metabolites produced by cytochrome p450 enzymes. It is involved in mycoparasitism and in plant interaction. The G3 gene of *T. hamatum* produces cytochrome p450 enzyme upon interaction with *Sclerotinia* and *Sclerotium* spp. (Carpenter et al. 2008). Malmierca et al. (2015) found that trichodiene (TD) is a VOC produced by *T. arundinaceum*, which served as a substrate for the synthesis of harzianum A. It is responsible for the expression of SA-related genes in tomato plants. Apart from the proteins and secondary metabolites, small RNAs of *Trichoderma* are considered as putative effectors. Ramirez-Valdespino et al. (2019) studied the interaction between *T. atroviride* and *Arabidopsis* through sRNA libraries in different time intervals. It revealed the effector activity of sRNA on symbiotic relationship establishment with *Arabidopsis* through chromatin modification and cell wall modifications. Further research is needed of the hour to dissect the molecular mechanism mediated by *Trichoderma* sRNAs (Fig. 4.1).

4.9 Interaction of *Trichoderma* Growth Promotional Molecules in the Rhizosphere

Communication through signal molecules between the host plant and *Trichoderma* spp. enables it to colonize both the external surface and internal regions of the plant system. The growth hormone auxin produced by *Trichoderma* colonizes roots and increases the surface area of colonization (Contreras et al. 2009). Plant root colonization is mediated by cysteine-rich hydrophobin-like protein TasHyd1 produced by *T. asperellum* and Qid74 produced by *T. harzianum* (Samolski et al. 2012; Viterbo and Chet 2006). Besides, Brotman et al. (2008) reported the involvement of an expansin-like protein called swollenin in plant root colonization. The cellulose binding module of the expansin-like protein binds with endopolygalacturonase to

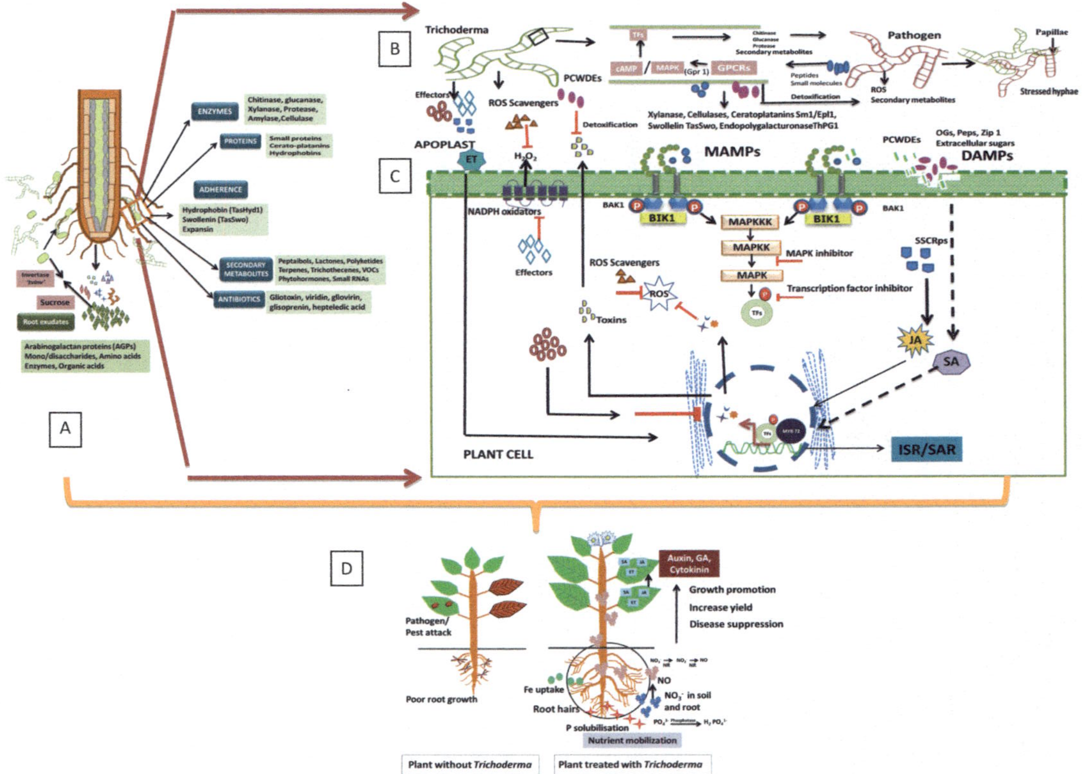

Fig. 4.1 Molecular interaction of *Trichoderma*–plant and *Trichoderma*–pathogen. (**a**) Sucrose in plant root exudates attracts *Trichoderma* which utilizes the carbon source using the invertase enzyme followed by adherence and colonization of root using hydrophobin, swollenin, and expansin. (**b**) MAMPs from *Trichoderma* are recognized by plant and stimulate defense response which is released by *Trichoderma* using effector molecules. (**c**) The peptides and small molecules from pathogen are recognized by *Trichoderma* through G-protein-coupled receptors, which activate the MAPK/cAMP pathway and regulate mycoparasitism or antibiosis against the pathogen. (**d**) *Trichoderma* is also involved in plant growth promotional activities like nutrient mobilization and phytohormone regulation and improves plant health

ensure penetration of *Trichoderma* into the root system (Brotman et al. 2008; Moran-Diez et al. 2009).

Different species of *Trichoderma* are effective colonizers on decaying and cellulosic materials and rhizosphere soil, which utilize the substrate by the production of their own secretion capacity for secondary metabolites, viz., CWDE, growth-promoting, and antibiotic nature. In addition, it also promoted the plant growth, reproduction, and biosynthesis of secondary metabolites (Schuster and Schmoll 2010). The application of *Trichoderma* enhanced crop production and productivity through PGPR activities. Zhang et al. (2014) reported that the use of *T. harzianum* (T-E5) as a biostimulant reduced the *Fusarium* wilt incidence and promoted plant growth in cucumber plants. Likewise, onion seeds treated with *T. harzianum* significantly promoted the seedling emergence and increased seedling length, seedling root length, number of leaves, and fresh weight of seedlings (Dabire et al. 2016). Field demonstration on wheat crop applied with different formulations of *T. harzianum* (Th3) increased plant growth parameters like rootlets and number of tillers (Sharma et al. 2012). Similarly, soil application of *T. viride* along with FYM and NPK increased the plant growth parameters, grain yield, and biomass of wheat crop (Mahato et al. 2018). Hajieghrari (2010) found that maize crop treatment with *T. hamatum* T614 isolate increased the leaf area, shoot weight, and fresh roots of seedling compared to the non-inoculated seedling. Similarly, bulb treatment with *T. harzianum* enhanced the plant growth of tuberose and resulted in increased flower production and quality (Nosir 2016). Root dipping and soil application of *T. asperellum*–based talc formulation effectively suppressed carnation wilt by increasing plant growth, number of shoots, stalk length, and flower yield (Vinodkumar et al. 2017).

The capacity of *Trichoderma* spp. to promote plant growth is attributed to the synthesis of secondary metabolites and production of enzymes to convert the unavailable form of nutrients to available form, auxin and gibberellin-based compounds (Parra et al. 2017). Seedlings of *Arabidopsis* treated with *T. virens* or *T. atroviride* produced auxin-related compounds such as indole-3-acetic acid, indole-3-acetaldehyde, and indole-3-ethanol which were responsible for increased biomass production (Contreras-Cornejo et al. 2014). Similarly, Harman et al. (2004) found that inoculation of *Trichoderma* in maize influenced the root system architecture, which resulted in increased crop yield, enhanced root biomass, and increased root hair development. The application of *Trichoderma* strains TH1 and T4 increased root weight, shoot weight, and dry root weight against *Pythium*. In addition, lateral root development and nodule formation were also found to be enhanced. Application of *Trichoderma* strains TH1, T4, and T12 significantly increased the soil fungal population and significantly reduced the cell-wall-degrading enzyme activities of *Pythium*. Besides, strains TH1, T4, and T12 reduced the alkaline phosphatase activity, and other strains, TH1, T4, and N47, reduced the urease activity for the suppression of *Pythium*. These enzymes directly influence the nutrient availability for *Pythium* in soil by avoiding the leakage of plant nutrients through damage made by *Pythium* spp. (Naseby et al. 2000).

Similarly, application of *Trichoderma* strains promoted significant growth of *Capsicum annuum* seedlings at 36 days after planting. In particular, *T. atroviride* increased plant height and dry shoot biomass and expressed the biocontrol activity against the nematode *Meloidogyne incognita*. Increased plant growth promotion and increased seedling emergence favored root colonization by *Trichoderma*, which is one of the critical characteristics of the species to utilize organic acids as a major trait to protect cucumber roots from infection of soilborne pathogens (Vasumathi et al. 2016). The production of organic acids alters the soil and nutrient uptake efficiency, which were useful for the metabolism of plants (Hermosa et al. 2013). Cucumber plants applied with *T. harzianum* favor the extensive secondary root hairs overexpressed in the cysteine-rich cell wall protein gene *qid74* and improved root architecture, thus increasing nutrient uptake through the increased total absorptive surface and increased plant biomass through efficient use of NPK and micronutrient translocation in the shoots (Samolski et al. 2012). Thus, research behind the molecules associated with plant growth promotion by *Trichoderma* spp. explains the involvement of secondary metabolites, CWDE, overexpression of cysteine-rich cell wall protein gene *qid74*, indole-3-acetaldehyde, indole-3-acetic acid, and indole-3-ethanol.

4.10 *Trichoderma* as Nutrient Mobilizers

Trichoderma spp. act as good nutrient mobilizers. The increased uptake of native mineral nitrogen (N) in the absence of N inputs has been reported in response to treatment with *T. virens* (GV41) in lettuce and rocket crops. In addition, the *Trichoderma* strains GV41 and T22 increased the uptake of N, under optimal N fertilization. Similarly, the higher concentration of K^+, Ca^{2+}, and PO_4^{3-} was influenced by the interaction during the application of *Trichoderma* and N fertilization (Fiorentino et al. 2018). Further analysis of leafy vegetable lettuce and rocket treated with bioformulations of *T. virens* (GV41) or *T. harzianum* (T22) along with different levels of nitrogen fertilizers revealed that overall fungal populations of rhizosphere soil of lettuce crop were increased in plots inoculated with *T. virens* (GV41) bioformulations and increased the N levels. In leafy rocket vegetable, treatment with *T. harzianum* (T22) maximized the fungal population in rhizosphere soil with optimum level of nitrogen (Fiorentino et al. 2018). *Trichoderma*, by boosting the availability of N, increased the total ascorbic acid content in leafy rocket. The total ascorbic acid in leafy vegetables was an important component of functional quality and constituent of vegetable crops. The role of *Trichoderma* on increased yield and plant growth in lettuce and rocket was due to biological and chemical processes in the rhizosphere soil accompanied with increased nutrient uptake and modulation of nutrient availability (Fiorentino et al. 2018).

Zhang et al. (2019) found that alfalfa seedlings treated with *Trichoderma* increased the N, P, and K contents of the soil. Interestingly, biomass of alfalfa was directly influenced by the abundance of *Trichoderma* and P available in soil. It indicated that the role of soil available P influenced alfalfa plant growth by *Trichoderma*. Further, in *Trichoderma*-treated soil, the soil microbial community identification through high-throughput sequencing of 16S rRNA and ITS region revealed the predominance of *Pseudomonas*, *Flavobacterium*, and *Arthrobacter* genera. The compounds 1-hexacosene and 1-heptadecanol were found in abundance only in the rhizosphere region of *Trichoderma*-treated plants (Zhang et al. 2019). Application of *T. harzianum* T22 enriched biofertilizers at a rate of 50% BioF/compost and 50% BioF/liquid with 50% N:P:K, effectively increasing the yield of tomato, improving the nutritional quality of the fruit, and reducing the external application of NPK fertilizers (Molla et al. 2012). *Trichoderma* spp. improved the nutrient uptake from the soil through enhanced root growth and promoted plant growth (Harman et al. 2004). A similar result has been reported by Altomare et al. (1999) who stated that the use of *T. harzianum* 1295-22 improved the uptake of N, Mn^{4+}, Fe^{3+}, and Cu^{2+} that led the healthier plant growth and development. Similarly, Li et al. (2015) demonstrated that in a pot culture study, *T. harzianum* SQR-T037 enhanced the biomass production and nutrient uptake of tomato seedlings. The *Trichoderma* on inoculation in liquid medium solubilized the Fe_2O_3, CuO, and metallic Zn and increased their concentration. Further, P-deficient hydroponic experiment (HE-P) revealed that tomato seedlings grown in the P-deficient nutrient solution inoculated with *T. harzianum* SQR-T037 influenced the uptake of nutrient root surface area, root volume, and the numbers of root tips.

Trichoderma consortium significantly increased the levels of N, P, K, Ca, and Mg in tomato seedlings. Apart from nutrient mobilization, increased root and shoot length and fresh and dry weights of shoot and root were also improved on treatment with a consortium of *Trichoderma* spp. The soil applied with *T. viride* in combination with *Azotobacter chroococcum* along with the recommended dosage of NPK fertilizers significantly enhanced biomass production of chickpea, root proteins, succinic and fumaric acid exudation, and nutrients available in soil and increased the activity of antioxidant-related enzymes like PAL, PPO, PO, and SOD (Velmourougane et al. 2017). Lilliana et al. (2015) found that inoculation of *Trichoderma* in bean plants increased the foliar concentrations of Cu, Fe, Ca, Mg, and K. Besides, *Trichoderma* also increased concentrations of P and Cu in the foliage of bean plants. Similarly, Martinez-Medina et al. (2011) demonstrated that under greenhouse conditions, increased nitrogen and phosphorous contents of melon plants were found only in co-inoculation of *T. harzianum* and *Glomus mosseae* rather than treatment with AMF alone. Similarly, plants co-inoculated with *T. harzianum* and AMF showed the same level of phosphorus content as that of plants inoculated with the AMF alone. This demonstrated the role of *Trichoderma* in mobilizing the nutrient in melon plants and improved shoot fresh weight of melon plants. Thus, root colonization by *Trichoderma* increased the efficiency of nutrient mobilization and plant growth by the interaction of *Trichoderma* with plant.

4.11 Molecular Signals of *Trichoderma* in the Rhizosphere Complements Plant and Soil Health

Plant and microorganism partnership has an evolutionary relationship and dates back to the origin of plants. The diverse relationship between plant and microorganisms exists due to molecular signals. Microorganisms reside as endophytic, epiphytic, and rhizospheric organisms of plant system. The root exudates play a vital role in the recruitment of specific organisms. The photosynthates of the plant system released as root exudates comprises of low-molecular-weight compounds like carbohydrates (arabinose, sucrose, glucose, galactose, raffinose, sterols, amino acid) glutamine, glutamic acid, phenylalanine, arginine, aspartate, leucine, lysine, enzymes (amylase, invertase, phenolase), lignins, and glucosinolates and high-molecular-weight compounds like organic acids, acetic acid, fumaric acid, lactic, and pyruvic acid (Dennis et al. 2010; Badri and Vivanco 2009; Hayat et al. 2017). The quality of root exudates is influenced by plant species, age of the plant, external environment, and biotic and abiotic stress conditions. The carbon-rich resources attract soil microorganisms and trigger root colonization. Similarly, cues from *Trichoderma* enable the plant to recognize and enable its colonization in root. The secretome of *Trichoderma* species is rich in carbohydrate-active enzymes (CAZymes) which are capable of modifying glycosidic bonds. The CAZymes are glycosidic hydrolases (GHs), glycosyl transferases (GTs), polysaccharide lyases (PLs), carbohydrate esterases (CEs), and carbohydrate-binding modules (CBM). Among these CAZymes, GHs are most predominantly produced to degrade polysaccharides like chitin, cellulose, and hemicellulose (di Cologna et al. 2018). Root colonization is the crucial step for successful plant–microbe interaction (Kamilova et al. 2005; Hayat et al. 2017). The colonization of root system includes attraction, adherence, colonization, and growth as rhizomicrobiome or as endophyte (Nihorimbere et al. 2010).

Vargas et al. (2009) reported that among the various compounds in root exudates, sucrose plays a vital role in the recruitment of opportunistic symbiont *Trichoderma*. This sucrose is produced in high concentration near root tips and are hydrolyzed by glycosidic hydrolase (GH), an invertase enzyme produced by *tvinv* gene of *T. virens* and acts as food for fungal cells. The earlier reports of Djonovic et al. (2006) and Viterbo et al. (2007) revealed that sucrose stimulates the expression of *tex1* and *sm1* genes in *Trichoderma* which induces ISR response. Thus, sucrose was reported as a nutrient for *Trichoderma* as well as a molecular signal for root colonization (Vargas et al. 2009). Apart from sucrose involved in the attraction of *Trichoderma* towards the root system, hydrophobins encoded by *TasHyd1* gene in *T. asperellum* and *qid*74 gene of *T. harzianum* encoding a cysteine-rich cell wall protein are involved in the attachment of fungi to the root system (Samolski et al. 2012). Further, root colonization is mediated by cell wall-degrading enzymes like endopolygalacturonase (ThPG1), swollenin, and expansin (Viterbo and Chet 2006; Brotman et al. 2008; Moran-Diez et al. 2009; Druzhinina et al. 2011). Thus, the plant root exudate can make up the microbial community in the root zone which in turn provides resistance against insect pests and diseases and is also involved in growth promotional activities. The recognition of *Trichoderma* by plants is facilitated through MAMPs, DAMPs, and elicitors released by *Trichoderma*, and the perception of these molecules by NBS-LRR proteins leads to elicitation of defense response; however, *Trichoderma* minimizes the stimulation of immunity by production of subsequent effectors or elicitor molecule. This phenomenon leads to a mutualistic relationship between *Trichoderma* and plant (Zamioudis and Pieterse 2012) and thus increases the plant health and soil health (Fig. 4.1).

4.12 Conclusion

The communication network between *Trichoderma* and plants is mediated through various signal molecules. Initially, *Trichoderma* recognizes the roots of crop plants by sensing the nutrients secreted in the rhizosphere. Subsequently, it interacts with the host through the production of various MAMP and DAMP molecules. The communication through MAMP and DAMP molecules induces an immune response and thereby confers resistance to both soilborne and foliar pathogens via SAR and ISR mechanisms. Apart from the same, *Trichoderma* produces effector molecules to overcome the MAMP- and DAMP-triggered immunity. It helps in colonization of the host both in the rhizosphere and endosphere. Subsequently after the colonization, *Trichoderma* produces growth hormones, enzymes, and antibiotics to overcome biotic and abiotic stress, which aids in improving plant health and soil health. However, only a very few molecules have been characterized for understanding their role in recognition, attachment, and colonization of root tissues. Till date there are no detailed explanations behind the involvement of receptors to recognize the secreted proteins of *Trichoderma* by the plant or how these proteins migrate into the plant. Though the proteins of elicitor nature suppress plant defense to favor colonization by *Trichoderma* in the root apoplast, understanding of the heritable nature of priming over different generations is only in the stage of infancy. A better understanding of the molecular communications between *Trichoderma* and plants will clearly explain how *Trichoderma* could differ from the mycorrhiza and pathogenic *Fusarium* leading to symbiotic relationship. To understand the dialogue between *Trichoderma* and plant roots, transcriptome profiling under ditrophic interactions will help to dissect the secret beneath the communication, leading to transcriptome reprogramming and epigenetic modification, which in turn help to upregulate the genes of interest to improve both soil health and plant health.

References

Abdel-Lateif K, Bogusz D, Hocher V (2012) The role of flavonoids in the establishment of plant roots endosymbioses with arbuscular mycorrhiza fungi, rhizobia and Frankia bacteria. Plant Signal Behav 7(6):636–641

Altomare C, Norvell WA, Bjorkman T, Harman GB (1999) Solubilization of phosphates and micronutrients by the plant growth- promoting and biocontrol fungus *Trichoderma harzianum* Rifai 1295-22. Appl Environ Microbiol 65:2926–2933

Avni A, Bailey BA, Mattoo AK, Anderson JD (1994) Induction of ethylene biosynthesis in *Nicotiana tabacum* by a *Trichoderma viride* xylanase is correlated to the accumulation of 1-aminocyclopropane-1-carboxylic acid (ACC) synthase and ACC oxidase transcripts. Plant Physiol 106:1049–1055

Badri DV, Vivanco JM (2009) Regulation and function of root exudates. Plant Cell Environ 32:666–681

Baroncelli R, Matarese F, Moncini L, Vannacci G, Vergara M (2016) Two endopolygalacturonase genes in *Trichoderma virens*: in silico characterization and expression during interaction with plants. J Phytopathol 164:18–28

Brotman Y, Briff E, Viterbo A, Chet I (2008) Role of swollenin, an expansin-like protein from *Trichoderma*, in plant root colonization. Plant Physiol 147:779–789

Carpenter MA, Ridgway HJ, Stringer AM, Hay AJ, Stewart A (2008) Characterization of a *Trichoderma hamatum* monooxygenase gene involved in antagonistic activity against fungal plant pathogens. Curr Genet 53:193–205

Chacon M, Rodriuez-Galan O, Benitez T, Sousa S, Rey M, Llobell A, Delgado-Jarana J (2007) Microscopic and transcriptome analysis of early colonization of tomato roots by *Trichoderma harzianum*. Int Microbiol 10:19–27

Chaparro JM, Badri DV, Vivanco JM (2014) Rhizosphere microbiome assemblage is affected by plant development. ISME J 8:790–803

Contreras CHA, Macias-Rodriguez L, Cortes-Penagos C, Lopez-Bucio J (2009) *Trichoderma* virens, a plant beneficial fungus, enhances biomass production and promotes lateral root growth through an auxin-dependent mechanism in *Arabidopsis*. Plant Physiol 149:1579–1592

Contreras-Cornejo HA, Macias-Rodriguez LI, Alfaro-Cuevas R (2014) *Trichoderma* improves growth of *Arabidopsis* seedlings under salt stress through enhanced root development, osmolite production and Na+ elimination through root exudates. Mol Plant Microbe Interact 27:503–514

Dabire TG, Bonzi S, Somda I, Legreve A (2016) Evaluation of the potential of *Trichoderma harzianum* as a plant growth promoter and biocontrol agent against *Fusarium* damping-off in onion in Burkina Faso. Asian J Plant Pathol 10(4):49–60

De Lorenzo G, Brutus A, Savatin DV, Sicilia F, Cervone F (2011) Engineering plant resistance by constructing chimeric receptors that recognize damage-associated molecular patterns (DAMPs). FEBS Lett 585:1521–1528

De Oliveira AL, Gallo M, Pazzagli L, Benedetti CE, Cappugi G, Scala A, Pantera B, Spisni A, Pertinhez TA, Cicero DO (2011) The structure of the elicitor Cerato-platanin (CP), the first member of the CP fungal protein family, reveals a double ψβ-barrel fold and carbohydrate binding. J Biol Chem 286:17560–17568

Dennis PG, Miller AJ, Hirsch PR (2010) Are root exudates more important than other sources of rhizodeposits in structuring rhizosphere bacterial communities. FEMS Microbiol Ecol 72:313–327

Di Cologna NDM, Gomez-Mendoza DP, Zanoelo FF, Giannesi GC, de Alencar Guimaraes NC, de Souza Moreira LR, Ferreira Filho EX, Ricart CAO (2018) Exploring *Trichoderma* and *Aspergillus* secretomes: proteomics approaches for the identification of enzymes of biotechnological interest. Enzyme Microb Technol 109:1–10

Djonovic S, Pozo MJ, Dangott LJ, Howell CR, Kenerley CM (2006) Sm1, a proteinaceous elicitor secreted by the biocontrol fungus *Trichoderma virens* induces plant defense responses and systemic resistance. Mol Plant Microbe Interact 19:838–853

Djonovic S, Vargas WA, Kolomiets MV, Horndeski M, Wiest A, Kenerley CM (2007) A proteinaceous elicitor Sm1 from the beneficial fungus *Trichoderma virens* is required for induced systemic resistance in maize. Plant Physiol 145:875–889

Druzhinina IS, Seidl-Seiboth V, Herrera-Estrella A, Horwitz BA, Kenerley CM, Monte E, Mukherjee PK, Zeilinger S, Grigoriev IV, Kubicek CP (2011) *Trichoderma*: the genomics of opportunistic success. Nat Rev Microbiol 16:749–759

Fiorentino N, Ventorino V, Woo SL, Pepe O, De Rosa A, Gioia L, Romano I, Lombardi N, Napolitano M, Colla G, Rouphael Y (2018) *Trichoderma*-based biostimulants modulate rhizosphere microbial populations and improve N uptake efficiency, yield, and nutritional quality of leafy vegetables. Front Plant Sci 9:743

Garnica-Vergara A, Barrera-Ortiz S, Munoz-Parra E, Raya-Gonzalez J, Mendez-Bravo A, Macias-Rodriguez L, Ruiz-Herrera LF, Lopez-Bucio J (2016) The volatile 6-pentyl- 2H-pyran-2-one from *Trichoderma atroviride* regulates *Arabidopsis thaliana* root morphogenesis via auxin signaling and ethylene insensitive 2 functioning. New Phytol 209:1496–1512

Hajieghrari B (2010) Effects of some Iranian *Trichoderma* isolates on maize seed germination and seedling vigor. Afr J Biotechnol 9(28):4342–4347

Harman GE, Howell CR, Viterbo A, Chet I, Lorito M (2004) Trichoderma species-opportunistic, avirulent plant symbionts. Nat Rev Microbiol 2:43–56

Hayat S, Faraz A, Faizan M (2017) Root exudates: composition and impact on plant–microbe interaction. In: Ahmad I, Husain FM (eds) Biofilms plant soil health. Wiley Publications, New York, NY, pp 179–193

Hermosa R, Viterbo A, Chet I, Monte E (2012) Plant-beneficial effects of *Trichoderma* and of its genes. Microbiology 158:17–25

Hermosa R, Rubio MB, Cardoza RE, Nicolas C, Monte E, Gutierrez S (2013) The contribution of *Trichoderma* to balancing the cost of plant growth and defense. Int Microbiol 16:69–80

Kamilova F, Validov S, Azarova T, Mulders I, Lugtenberg B (2005) Enrichment for enhanced competitive root tip colonizers selects for a new class of biocontrol bacteria. Environ Microbiol 7:1809–1817

Kou Y, Tan YH, Ramanujam R, Naqvi NI (2017) Structure-function analyses of the Pth11 receptor reveal an important role for CFEM motif and redox regulation in rice blast. New Phytol 214:330–342

Lamdan NL, Shalaby S, Ziv T, Kenerley CM, Horwitz BA (2015) Secretome of Trichoderma interacting with maize roots: role in induced systemic resistance. Mol Cell Proteomics 14:1054–1063

Li R-X, Cai F, Pang G, Shen Q-R, Li R, Chen W (2015) Solubilisation of phosphate and micronutrients by *Trichoderma harzianum* and its relationship with the promotion of tomato plant growth. PLoS One 10(6):e0130081

Li WC, Huang CH, Chen CL, Chuang YC, Tung SY, Wang TF (2017) *Trichoderma reesei* complete genome sequence, repeat-induced point mutation, and partitioning of CAZyme gene clusters. Biotechnol Biofuels 10:170

Lilliana HC, Andres C, Walter O, Sergio O (2015) The effect of various isolates of *Trichoderma* spp. on nutrient uptake in beans (*Phaseolus vulgaris*) in two soil types. Rev Colomb Cienc Hortic 9(2):268–278

Luo Y, Zhang DD, Dong XW, Zhao PB, Chen LL, Song XY, Wang XJ, Chen XL, Shi M, Zhang YZ (2010) Antimicrobial peptaibols induce defense responses and systemic resistance in tobacco against *Tobacco mosaic virus*. FEMS Microbiol Lett 313:120–126

Mahato S, Bhuju S, Shrestha J (2018) Effect of *Trichoderma viride* as biofertilizer on growth and yield of wheat. Malaysian J Sustain Agric 2(2):01–05

Malmierca MG, Cardoza RE, Alexander NJ, McCormick SP, Collado IG, Hermosa R, Monte E, Gutierrez S (2013) Relevance of trichothecenes in fungal physiology: disruption of *tri5* in *Trichoderma arundinaceum*. Fungal Genet Biol 53:22–33

Malmierca MG, McCormick SP, Cardoza RE, Alexander NJ, Monte E, Gutierrez S (2015) Production of trichodiene by *Trichoderma harzianum* alters the perception of this biocontrol strain by plants and antagonized fungi. Environ Microbiol 17:2628–2646

Marra R, Ambrosino P, Carbone V, Vinale F, Woo SL, Ruocco M, Ciliento R, Lanzuise S, Ferraioli S, Soriente I, Gigante S, Turra D, Fogliano V, Scala F, Lorito M (2006) Study of the three-way interaction between *Trichoderma atroviride*, plant and fungal pathogens by using a proteomic approach. Curr Genet 50:307–321

Martinez C, Blanc F, Le Claire E, Besnard O, Nicole M, Baccou CJ (2001) Salicylic acid and ethylene pathways are differentially activated in melon cotyledons by active or heat-denatured cellulase from *Trichoderma longibrachiatum*. Plant Physiol 127:334–344

Martinez-Medina A, Antonio Roldan A, Pascual JA (2011) Interaction between arbuscular mycorrhizal fungi and Trichoderma harzianum under conventional and low input fertilization field condition in melon crops: growth response and Fusarium wilt biocontrol. Appl Soil Ecol 47:98–10

Molla AH, Haque MM, Haque MA, Ilias GNM (2012) *Trichoderma*-enriched biofertilizer enhances production and nutritional quality of tomato (*Lycopersicon esculentum* Mill.) and minimizes NPK fertilizer use. Agric Res 1(3):265–272

Moran-Diez E, Hermosa R, Ambrosino P, Cardoza RE, Gutierrez S, Lorito M, Monte E (2009) The ThPG1 endopolygalacturonase is required for the *Trichoderma harzianum*-plant beneficial interaction. Mol Plant Microbe Interact 22:1021–1031

Moran-Diez ME, Trushina N, Lamdan NL, Rosenfelder L, Mukherjee PK, Kenerley CM, Horwitz BA (2015) Host-specific transcriptomic pattern of *Trichoderma virens* during interaction with maize or tomato roots. BMC Genomics 16(1):8

Mukherjee PK, Buensanteai N, Moran-Diez ME, Druzhinina IS (2012) Functional analysis of non-ribosomal peptide synthetases (NRPSs) in *Trichoderma virens* reveals a polyketide synthase (PKS)/NRPS hybrid enzyme involved in the induced systemic resistance response in maize. Microbiology 158:155–165

Naseby DC, Pascual JA, Lynch JM (2000) Effect of biocontrol strains of *Trichoderma* on plant growth, *Pythium ultimum* populations, soil microbial communities and soil enzyme activities. J Appl Microbiol 88:161–169

Nasser L, Weissman Z, Pinsky M, Amartely H, Dvir H, Kornitzer D (2016) Structural basis of haem-iron acquisition by fungal pathogens. Nat Microbiol 1:16156

Nihorimbere V, Ongena M, Cawoy H, Brostaux Y, Kakana P, Jourdan E, Thonart P (2010) Beneficial effects of Bacillus subtilis on field-grown tomato in Burundi: reduction of local Fusarium disease and growth promotion. Afr J Microbiol Res 4(11):1135–1142

Nosir WS (2016) *Trichoderma harzianum* as a growth promoter and bio-control agent against *Fusarium oxysporum* f. sp. *tuberosi*. Adv Crop Sci Technol 4(2):217

Parra EH, Alejo JC, Zapata JAR (2017) *Trichoderma* strains as growth promoters in *Capsicum annuum* and as biocontrol agents in *Meloidogyne incognita*. Chilean J Agric Res 77(4):318–324

Ramirez-Valdespino CA, Casas-Flores S, Olmedo-Monfil V (2019) *Trichoderma* as a model to study effector-like molecules. Front Microbiol 10:1030

Reddy VS, Shlykov MA, Castillo R, Sun EI, Saier MH Jr (2012) The major facilitator superfamily (MFS) revisited. FEBS J 279(11):2022–2035

Rippa S, Adenier H, Derbaly M, Beven L (2007) The Peptaibol Alamethicin induces an rRNA-cleavage-associated death in *Arabidopsis thaliana*. Chem Biodivers 4:1360–1373

Ron M, Avni A (2004) The receptor for the fungal elicitor ethylene inducing xylanase is a member of a resistance-like gene family in tomato. Plant Cell 16:1604–1615

Rotblat B, Enshell-Seijffers D, Gershoni JM, Schuster S, Avni A (2002) Identification of an essential component of the elicitation active site of the EIX protein elicitor. Plant J 32:1049–1055

Rubio MB, Hermosa R, Reino JL, Collado I, Monte E (2009) *Thctf1* transcription factor of *Trichoderma harzianum* is involved in 6-pentyl 2H-pyran-2-one production and antifungal activity. Fungal Genet Biol 46:17–27

Rubio MB, Domínguez S, Monte E, Hermosa R (2012) Comparative study of *Trichoderma* gene expression in interactions with tomato plants using high-density oligonucleotide microarrays. Microbiology 158:119–128

Rubio MB, Quijada NM, Perez E, Dominguez S, Monte E, Hermosa R (2014) Identifying beneficial qualities of *Trichoderma parareesei* for plants. Appl Environ Microbiol 80(6):1864–1873

Sabnam N, Roy Barman S (2017) WISH, a novel CFEM GPCR is indispensable for surface sensing, asexual and pathogenic differentiation in rice blast fungus. Fungal Genet Biol 105:37–51

Salas-Marina MA, Silva-Flores MA, Uresti-Rivera EE, Castro-Longoria E, Herrera-Estrella A, Casas-Flores S (2011) Colonization of *Arabidopsis* roots by *Trichoderma atroviride* promotes growth and enhances systemic disease resistance through jasmonic acid/ethylene and salicylic acid pathways. Eur J Plant Pathol 131:15–26

Samolski I, de Luis A, Vizcaino JA, Monte E, Suarez MB (2009) Gene expression analysis of the biocontrol fungus *Trichoderma harzianum* in the presence of tomato plants, chitin, or glucose using a high-density oligonucleotide microarray. BMC Microbiol 9:217

Samolski I, Rincon A, Pinzon LM, Viterbo A, Monte E (2012) The *qid74* gene from *Trichoderma harzianum* has a role in root architecture and plant biofertilization. Microbiology 158:129–138

Saravanakumar K, Fan L, Fu K, Yu C, Wang M, Xia H, Sun J, Li Y, Chen J (2016) Cellulase from *Trichoderma harzianum* interacts with roots and triggers induced systemic resistance to foliar disease in maize. Sci Rep 6:35543

Sarrocco S, Matarese F, Baroncelli R, Vannacci G, Seidl-Seiboth V, Kubicek CP, Vergara M (2017) The constitutive endopolygalacturonase TvPG2 regulates the induction of plant systemic resistance by *Trichoderma virens*. Phytopathology 107:537–544

Schuster A, Schmoll M (2010) Biology and biotechnology of *Trichoderma*. Appl Microbiol Biotechnol 87:787–799

Segarra G, Casanova E, Bellido D, Odena MA, Oliveira E, Trillas I (2007) Proteome, salicylic acid, and jasmonic acid changes in cucumber plants inoculated with *Trichoderma asperellum* strain T34. Proteomics 7:3943–3952

Seidl V, Marchetti M, Schandl R, Kubicek CP, Allmaier G, Kubicek CP (2006) Epl1, the major secreted protein of *Hypocrea atroviridis* on glucose, is a member of a strongly conserved protein family comprising plant defense response elicitors. FEBS J 273:4346–4359

Sharma P, Patel AN, Saini MK, Deep S (2012) Field demonstration of *Trichoderma harzianum* as a plant growth promoter in wheat (*Triticum aestivum* L). J Agric Sci 4:8

Shoresh M, Gal-On A, Leibman D, Chet I (2006) Characterization of a mitogen-activated protein kinase gene from cucumber required for *Trichoderma*-conferred plant resistance. Plant Physiol 142:1169–1179

Shoresh M, Harman GE, Mastouri F (2010) Induced systemic resistance and plant responses to fungal biocontrol agents. Annu Rev Phytopathol 48:21–43

Stergiopoulos I, De Wit PJGM (2009) Fungal effector proteins. Annu Rev Phytopathol 47:233–263

Takanashi K, Shitan N, Yazaki K (2014) The multidrug and toxic compound extrusion (MATE) family in plants. Plant Biotechnol 31:417–430

Taniguchi N, Honke K, Fukuda M (eds) (2011) Handbook of glycosyltransferases and related genes. Springer Science & Business Media, New York, NY

Tijerino A, Cardoza RE, Moraga J, Malmierca MG, Vicente F, Aleu J, Collado IG, Gutierrez S, Monte E, Hermosa R (2011) Overexpression of the trichodiene synthase gene *tri5* increases trichodermin production and antimicrobial activity in *Trichoderma brevicompactum*. Fungal Genet Biol 48:285–296

Vargas WA, Mandawe JC, Kenerley CM (2009) Plant-derived sucrose is a key element in the symbiotic association between *Trichoderma virens* and maize plants. Plant Physiol 151 (2):792–808

Vasumathi S, Eraivan AAK, Nakkeeran S (2016) Exploring the diversity of *Trichoderma* spp. and the synergistic activity of delivery systems for the management of *Fusarium* wilt of cucumber under protected cultivation. J Mycol Plant Pathol 46(4):311

Velmourougane K, Prasanna R, Singh S, Chawla G, Kumar A, Saxena AK (2017) Modulating rhizosphere colonisation, plant growth, soil nutrient availability and plant defense enzyme activity through *Trichoderma viride-Azotobacter chroococcum* biofilm inoculation in chickpea. Plant and Soil 421:157. https://doi.org/10.1007/s11104-017-3445-0

Vinodkumar S, Indumathi T, Nakkeeran S (2017) *Trichoderma asperellum* (NVTA2) as a potential antagonist for the management of stem rot in carnation under protected cultivation. Biol Control 113:58–64

Viterbo A, Chet I (2006) *TasHyd1*, a new hydrophobin gene from the biocontrol agent *Trichoderma asperellum*, is involved in plant root colonization. Mol Plant Pathol 7:249–258

Viterbo A, Harel M, Horwitz BA, Chet I, Mukherjee PK (2005) *Trichoderma* mitogen-activated protein kinase signaling is involved in induction of plant systemic resistance. Appl Environ Microbiol 71:6241–6246

Viterbo A, Wiest A, Brotman Y, Chet I, Kenerley C (2007) The 18 mer peptaibols from *Trichoderma virens* elicit plant defense responses. Mol Plant Pathol 8:737–746

Viterbo A, Landau U, Kim S, Chernin L, Chet I (2010) Characterization of ACC deaminase from the biocontrol and plant growth-promoting agent *Trichoderma asperellum* T203. FEMS Microbiol Lett 305:42–48

Woo SL, Scala F, Ruocco M, Lorito M (2006) The molecular biology of the interactions between *Trichoderma* spp., phytopathogenic fungi, and plants. Phytopathology 96:181–185

Yoshioka Y, Ichikawa H, Naznin HA, Kogure A, Hyakumachi M (2012) Systemic resistance induced in *Arabidopsis thaliana* by *Trichoderma asperellum* SKT-1, a microbial pesticide of seedborne diseases of rice. Pest Manag Sci 68:60–66

Zamioudis C, Pieterse CMJ (2012) Modulation of host immunity by beneficial microbes. Mol Plant Microbe Interact 25:139–150

Zhang F, Yang X, Ran W, Shen Q (2014) *Fusarium oxysporum* induces the production of proteins and volatile organic compounds by *Trichoderma harzianum* T-E5. FEMS Microbiol Lett 359:116–123

Zhang F, Xu X, Huo Y, Xiao Y (2019) *Trichoderma*-inoculation and mowing synergistically altered soil available nutrients, rhizosphere chemical compounds and soil microbial community, potentially driving alfalfa growth. Front Microbiol 9:3241

Chapter 5
Plant Immunity, Priming, and Systemic Resistance as Mechanisms for *Trichoderma* spp. Biocontrol

Rupali Gupta and Maya Bar

Abstract During plant–pathogen interactions, plants mount sophisticated defenses to ward off their enemies. Plant defenses can be triggered by biocontrol agents. In recent years, various biocontrol agents have been defined as systemic resistance inducers of plant immunity. *Trichoderma* spp. are opportunistic plant colonizers that trigger host-induced systemic resistance (ISR), typically eliciting rapid defense responses that include massive transcriptional reprogramming. Root and leaf colonization by *Trichoderma* can prime plant defense, enabling robust plant responses to subsequent pathogen challenges. The onset of host immune responses can also trigger systemic acquired resistance (SAR) in tissues distal to the infection site. *Trichoderma* spp. are potent inducers of plant ISR. Systemic priming leading to rapid defense responses upon pathogen attack is common to both SAR and ISR. In this chapter, we will outline the current knowledge of plant immunity, emphasizing mechanisms underlying induced systemic resistance triggered by *Trichoderma*.

Keywords *Trichoderma* · Biocontrol · Plant immunity · Systemic resistance

5.1 Introduction

The dynamic interaction between plants and beneficial microorganisms triggers plant immunity and has received increased attention in the past few years. Notably, these interactions lead to both physiological and biochemical reprogramming of plant defense (Perazzolli et al. 2011; Pieterse et al. 2014; Mauch-Mani et al. 2017). Recognition of microbial elicitors by plant receptors triggers various changes in the plant, leading to faster and/or stronger responses against phytopathogen invasion (Harman et al. 2004a; Contreras-Cornejo et al. 2011; Mathys et al. 2012; Nawrocka and Małolepsza 2013). Consequently, when the host plant is colonized by biocontrol

R. Gupta · M. Bar (✉)
Department of Plant Pathology and Weed Research, ARO, The Volcani Center, Rishon LeZion, Israel
e-mail: mayabar@agri.gov.il

A. K. Sharma, P. Sharma (eds.), *Trichoderma*, Rhizosphere Biology,
https://doi.org/10.1007/978-981-15-3321-1_5

agents, the interaction is potentiated, providing resistance against phytopathogens, and the robustness of the root system is promoted, enhancing nutrient uptake (Contreras-Cornejo et al. 2013; Crutcher et al. 2015; Macías-Rodríguez et al. 2018). This resistance persists for several months, and once terminated, re-induction is needed (Hermosa et al. 2012; Jogaiah et al. 2018). There is a large array of findings highlighting the potential of these biocontrol agents in suppression of plant diseases by direct management of soilborne and foliar pathogens via mycoparasitism, antibiosis, and competition as well as by augmentation of systemic plant resistance (Korolev et al. 2008; Yoshioka et al. 2012; Mehari et al. 2015; Carrero-Carrón et al. 2016; Cheng et al. 2018).

Trichoderma species have been the subject of extensive research. *Trichoderma* is considered an opportunistic plant symbiont (Harman et al. 2012) and, in addition to biocontrol activity, can also improve nutrient use efficacy and enhance plant growth (Salas-Marina et al. 2011; Nesler et al. 2015; Gupta et al. 2016). The success of *Trichoderma* colonization is owing to their rapid growth, intensive production of spores under different environmental conditions, efficiency in the utilization of nutrients, and capacity to communicate with plants through chemical signals (Badri et al. 2009; López-Bucio et al. 2015). *Trichoderma* spp. can exert activation of the plant immune system (induced systemic resistance, ISR) and pre-activation (priming) of molecular mechanisms of defense against a broad range of pathogens. Depending on the strain, the positive effects of *Trichoderma* on plants may have additional benefits, including amelioration of internal plant phytohormone homeostasis, improvement of photosynthetic efficiency, and alleviation of abiotic stresses (Shoresh et al. 2010; Contreras-Cornejo et al. 2014a; Rubio et al. 2017; De Palma et al. 2019).

For many years, it was thought that induced resistance for plant defense is based on direct activation of defense responses by the inducing biocontrol agent. Accumulation of pathogenesis-related (PR) proteins, for example, can occur directly upon activation of systemic acquired resistance (SAR). Nevertheless, the contribution of pathogenesis-related proteins to plant resistance is unresolved and may be insufficient to fully explain the wide array of defenses resulting from SAR (Van Wees et al. 2008). *Trichoderma*-mediated activation of ISR responses in the host is regulated by various defense pathways. Furthermore, in some cases, it has been reported that *Trichoderma*-activated ISR may not correlate with direct defense gene activation (Martínez-Medina et al. 2016). Interestingly, plants displaying SAR and/or ISR exhibit a rapid and stronger activation of specific defense responses after they have been infected by a pathogen. This capacity for augmented defense is called priming (Conrath et al. 2015). Priming accelerates the plant's immune system, activating the defenses to best cope with a wide range of biotic and abiotic stresses. Recent studies suggest that ISR and SAR may require specific cellular signaling components (Conrath et al. 2006), signifying exclusive "priming" regulatory mechanisms. In *Arabidopsis*, Brotman et al. (2012) demonstrated that *T. asperellum*-inoculated plants that suffered subsequent infection by *Pseudomonas syringae* (*Pst*) showed major differences in defense gene (related to ethylene (ET)/jasmonic acid (JA) signaling) expression patterns as compared with non-primed plants. Mathys et al. (2012) demonstrated that the establishment of ISR during the *T. hamatum* T382 priming phase is not dependent on JA/Et signaling pathways, although *PR1* is accumulated during this priming phase. In addition, the *Botrytis cinerea* post-challenge primed state is characterized by an increased response to JA and wounding pathways, which is a typical hallmark of ISR. In this respect, priming could represent an ecological adaptation toward environmental stress toleration. The relationship between *Trichoderma* spp. biocontrol potential and the differential induction of defense responses in the host is being investigated in several ongoing studies. This chapter focuses on the recent knowledge regarding *Trichoderma*–plant dialogue and the mechanisms by which plant immunity can be induced.

5.2 Induction of ISR in the Host

Both pathogenic and beneficial microorganisms in contact with the plant produce several molecules that prime defense mechanisms in plants and consequentially trigger an innate immune response. By eliciting early defense responses, signals from these microbes may be intensified through the production of various plant hormones such as salicylic acid (SA), jasmonic acid (JA), and ethylene (ET). Production of SA is generally associated with biotroph pathogen resistance, while JA and ET are related with resistance to necrotrophs (Glazebrook 2005; Ramirez-Prado et al. 2018). Resistance against several pathogens that is induced by local attack can often be extended systematically in different plant parts that are not suffering any injury. Van Bel and Gaupels (2004) defined the vasculature as offering an essential channel for translocation of diffusible signals that are related to root–shoot communication, thus allowing immunity in distal plant organs. Two main mechanisms of induced resistance have been recognized in plants on the basis of differences in their stimulation process: SAR and ISR. SAR is usually triggered by pathogens causing local infection. SAR involves an early increase in the synthesis of SA and results in the long-term systemic resistance to subsequent pathogen occurrence, the activation of PR genes, and possibly the accumulation of phytoalexin-type compounds (Shoresh et al. 2010). In contrast to the SAR response, ISR is primed by beneficial microbes, leading to enhanced defense and resistance to subsequent pathogens. ISR requires components of the JA and the ET signaling pathway (Van Wees et al. 2008; Pieterse et al. 2014).

Microbial-induced resistance was initially reported for *T. harzianum* T39 in cucumber gray mold and powdery mildew pathosystems (Elad et al. 1998), and it was described as systemically active (Bigirimana et al. 1997; De Meyer et al. 1998). The ability of *Trichoderma* spp. to systemically trigger plant resistance has been confirmed by several findings. *Trichoderma* spp. were shown to positively modify the response of numerous plants, including Solanaceae, Graminaceae, and Cucurbitaceae, against various biotroph and necrotroph pathogens (Harman et al. 2004b; Vinale et al. 2008a; Mauch-Mani et al. 2017). *Trichoderma asperellum* T203 induces resistance to *P. syringae* pv. *lachrymans* (*Psl*) in cucumber leaves. Though

Psl infection increased SA concentration, the *Trichoderma*–host interaction caused no further fluctuations in SA content when compared to control plants (Shoresh et al. 2005). The biocontrol activity of T203 was strongly reduced by inhibitors of JA and ET production, diethyldithiocarbamic acid and silver thiosulfate, respectively. These results demonstrate that both JA and ET are required for *T. asperellum* T203 biocontrol. ET content did not change following *Trichoderma* treatment, indicating that the requirement for ethylene signaling is independent of total ethylene levels (Shoresh et al. 2005).

In response to *T. asperellum* treatment, cucumber pathogenesis-related genes known to be involved in the SA signaling were not induced, suggesting that SAR is not the primary mechanism. However, upon subsequent treatment with the pathogen *Psl*, the expression of *PR* genes was higher than with pathogen or *Trichoderma* alone (Yedidia et al. 2003). The increased pathogenesis gene expression level was also associated with an accumulation of phenolic glycosides that serve as a strong antibiotic to numerous microbes. Concurrently, *Psl* colonization in leaves from plants treated with *Trichoderma* was significantly reduced when compared with colonization in *Trichoderma*-untreated plants. Thus, *T. asperellum* activates systemic resistance, without turning on the SA pathway in its entirety. Upon pathogen attack, the plant will respond more rapidly and strongly, increasing resistance. Thus, upstream regulatory genes are induced, delivering a more rapid defense response than would occur without *Trichoderma*. Interestingly, in maize plants treated with *T. harzianum* T22, a number of pathogenesis genes were constitutively switched on in the absence of pathogen challenge (Shoresh and Harman 2008). It would thus seem that there are different mechanisms at play in plant–*Trichoderma* interactions. Priming mechanisms may occur, or not, depending on the host plant–*Trichoderma* spp. combination. Interestingly, it was found that in the presence of *T. harzianum* T22, wild-type (WT) *Arabidopsis* plants were effectively protected from *Pythium* seedling blight, while no protection was observed in *npr1* mutants (Mastouri et al. 2010). Alfano et al. (2007) observed in the transcriptional analysis of the *T. hamatum*–plant interaction that only one classical SAR gene, PR5, was upregulated. According to Gallou et al. (2009), the defense response of *T. harzianum*-treated potato plants to *Rhizoctonia solani* required JA/ET and SA. These results support the notion that different *Trichoderma* species use different mechanisms to induce plant resistance.

Some findings suggest that *Trichoderma* strains are able to elicit long-term upregulation of SA-dependent genes in plants unchallenged by pathogens. However, following *B. cinerea* infection, *Trichoderma* treatment effected changes in SA-dependent gene expression. Following infection, JA signaling was activated, causing ISR to increase (Tucci et al. 2011). *Trichoderma asperellum Arabidopsis* root colonization activates ISR by triggering SA signaling. Furthermore, SA and JA/ET signaling are both involved in the ISR elicited by *Trichoderma* culture filtrates (CF) (Yoshioka et al. 2012). Salas-Marina et al. (2011) provided further insight into the mechanism underlying *Trichoderma*–plant interactions, where *T. atroviride* colonization of *Arabidopsis* roots induced expression of defense-related genes associated with both the SA and JA/ET pathways following subsequent phytopathogen challenge. Recent gene expression studies have provided many evidences pointing to JA/ET involvement in the transmission of *Trichoderma*-derived signals. Several genes associated with JA/ET signaling pathways were found to be upregulated after *Trichoderma* inoculation, demonstrating the key role of JA/ET in systemic defense (Gallou et al. 2009; Perazzolli et al. 2011, 2012; Meller-Harel et al. 2014; Okon Levy et al. 2015).

The role of JA and ET in mediating signals from *Trichoderma*-treated roots to plant leaves is established (Segarra et al. 2007; Korolev et al. 2008; Okon Levy et al. 2015). Gene expression studies indicate that plants inoculated with *T. asperellum* in the roots possessed upregulated lipoxygenase (*Lox1*), a gene that is involved in the synthesis of jasmonate. *Lox1* is upregulated in the roots very early following *Trichoderma* inoculation, possibly due to activation of JA synthesis. Phenylalanine ammonia lyase (PAL) *Pal1*, which is also upregulated by *Trichoderma* inoculation, leads to the production of phytoalexins and additional phenolic compounds (Shoresh et al. 2005; Shoresh and Harman 2008). Transient *Pal1* activation could indicate that *Trichoderma*-induced ISR relies on the accumulation of phytoalexins in some cases.

Ethylene induction is regarded as downstream of JA response in biocontrol-mediated ISR. In the absence of ethylene, ethylene receptor 1 (ETR1) and constitutive triple response 1 (CTR1) proteins work together to negatively control the ethylene signaling pathway (Hua and Meyerowitz 1998; Wang et al. 2002). Following *Trichoderma* inoculation, an increase and subsequent strong decrease in *ETR1* gene expression in leaves were observed, while the expression of *CTR1* was downregulated following *Trichoderma* treatment (Shoresh et al. 2005). Downregulation of both genes indicates that enhanced ethylene sensitivity leads to a stronger response upon subsequent pathogen challenge. In plants pretreated with *Trichoderma* and challenged with *Psl*, expression of *ETR1* and *CTR1* is reduced, suggesting that the increased ethylene sensitivity may be required for activation of this pathway. Remarkably, both genes are induced by *Trichoderma* in roots, suggesting downregulation of the ethylene response. This may indicate that *Trichoderma*-induced plant protection relies on JA/ET signaling and requires localized silencing of defense responses in favor of *Trichoderma* colonization, in contrast to SAR induced by pathogens. This idea is also supported by the downregulation of the pathogenesis-related (PR) protein, maize seed in roots inoculated with *Trichoderma*, such that local silencing of defense genes occurs in order to permit fungal colonization of the host roots (Chen et al. 2005; see below).

In various plant–pathogen systems, induction of plant systemic resistance by *Trichoderma* is well established, in no small part due to spatial separation between *Trichoderma* and pathogen. *Trichoderma* spp. can manage a wide range of phytopathogens, including fungi, oomycetes, bacteria, and nematodes (Shoresh et al. 2010; Hermosa et al. 2012; López-Bucio et al. 2015; Martínez-Medina et al. 2017; Sarrocco et al. 2017; Jogaiah et al. 2018). These works have demonstrated that *Trichoderma* induces plant immunity and that this induced immunity can serve in defending the plant against a broad range of pathogens and pests, primarily but not exclusively via the ET/JA pathway.

5.3 *Trichoderma* Priming for Enhanced Defense

In the modern concept of the plant immune system, plant resistance (R) proteins have evolved to recognize microbial compounds defined as pathogen-associated molecular patterns (PAMPs), such as flagellin, chitin, ergosterol, glycoproteins, and lipopolysaccharides (Zipfel 2009; Pieterse et al. 2012). Plants can also respond to several endogenous signals arising from damage caused by pathogen invasion, known as damage-associated molecular patterns (DAMPs) (Dodds and Rathjen 2010). Interaction between PAMPs and corresponding host receptors induces defense responses in the plant, referred to as PAMP-triggered immunity (PTI) (Pel and Pieterse 2012). Additionally, plant resistance receptor proteins (nucleotide-binding–leucine-rich repeat, NB-LRR) mediate recognition of pathogen-derived effector molecules, called effector-triggered immunity (ETI) (Dodds and Rathjen 2010). The onset of PTI and ETI often activates induced resistance in plant tissues distal to the infection site and involves several long-lasting signals, transmitting an improved defensive capacity through different plant organs (Shah and Zeier 2013; Conrath et al. 2015). Initial responses in plant defense include alterations in cytosolic calcium (Ca^{2+}) levels; generation of reactive oxygen (ROS) and nitrogen (RNS) species such as superoxide (O^{-2}), hydrogen peroxide (H_2O_2), and nitric oxide (NO); protein phosphorylation; and activation of phospholipases and GTP-binding proteins (Koornneef and Pieterse 2008; Naoumkina et al. 2010; Zhao 2015; Costa et al. 2018). One of the initial cellular responses following microbe recognition is the generation of ROS. Current challenges involve deciphering the local and systemic effects of ROS produced in different plant parts during *Trichoderma* interaction and how ROS are generated following the perception of microbe-assisted molecular patterns (MAMPs) to elicit immune responses. It would be interesting to identify which genes are responsible for ROS signaling and how resistance varies from defense responses in different plant tissues.

Mitogen-activated protein kinase (MAPK) signal transduction cascades mediate several aspects of plant stress responses and developmental biology (Xu and Zhang 2015; Komis et al. 2018). In plant immunity, induction of MAPK signaling cascades is recognized to activate/stimulate regulatory proteins (Segarra et al. 2007; Lorito et al. 2010; Druzhinina et al. 2011). Proteomic studies have demonstrated that the *Trichoderma*–plant interaction induces NBS/LRR resistance proteins (Marra et al. 2006; Shoresh and Harman 2008; Keswani et al. 2016). These are recognized as specificity determinants of plant immunity and, upon activation, elicit a cascade of signal transduction that results in defense responses.

As with PAMPs, several MAMPs from *Trichoderma* have been connected to an induced resistance mechanism (Hoitink et al. 2006; Vinale et al. 2008b; Hermosa et al. 2012; Vos et al. 2015) (Table 5.1). In addition to their enzymatic activity, several *Trichoderma* spp.-produced enzymes like cellulose and xylanase are known to stimulate resistance in plants, independent of their enzymatic activity. Heat-denatured cellulase from *T. longibrachiatum* stimulates melon defenses by inducing SA and ET signaling (Martinez et al. 2001). The first discovered *Trichoderma*

Table 5.1 *Trichoderma* spp. elicitors/MAMPs involved in induction of resistance in different plants against various phytopathogens

Trichoderma spp.	Elicitor(s)/MAMP	Host	Pathogen	Mechanism	References
	Proteins				
T. atroviride	Chitinase (endochitinase)	Apple	*Venturia inaequalis*	Induce defense mechanism	Faize et al. (2003)
T. virens	Chitinase (endochitinase)	Cotton, rice	*Rhizoctonia solani*	Induction of ROS, expression of several defense-related genes, and activation of some PR enzymes and the terpenoid pathway	Kumar et al. (2009); Shah et al. (2009)
T. virens	Protease (serine protease)	Cotton seedlings	*R. solani*	Plays a role in the biocontrol process	Pozo et al. (2004)
T. asperellum	Protease (aspartyl protease)	Cucumber seedlings	*R. solani*	Proteases play a role in *Trichoderma* both as a mycoparasite and as a plant opportunistic symbiont	Martinez et al. (2001); Viterbo et al. (2004)
T. longibrachiatum	Cellulase (endoglucanases, exoglucanases, and β-glucosidases)	Melon, maize	*Sphaerotheca fuliginea*, *Botrytis cinerea*	Activated and heat-denatured cellulases stimulate plant defenses through the induction of the SA and ET signaling pathways, respectively	Martinez et al. (2001); Saravanakumar et al. (2016)
T. viride		Tobacco, lima bean, corn, tobacco cell cultures	*R. solani*		Aidemark et al. (2010)
T. viride	Xylanase (Xyn2/Eix)–ethylene	Tomato and tobacco plants	*B. cinerea*	A xylanase that stimulates ET biosynthesis and hypersensitive response in tobacco leaf tissues	Rotblat et al. (2002)

(continued)

Table 5.1 (continued)

Trichoderma spp.	Elicitor(s)/MAMP	Host	Pathogen	Mechanism	References
T. virens, *T. atroviride*	Cerato-platanins Sm1/Epl1, small extracellular cysteine-rich proteins	Cotton, maize, tomato	*R. solani*, *Alternaria solani*, *B. cinerea*, and *P. syringae* pv. *tomato* (*Pst* DC3000)	Hydrophobin-like SSCP orthologues that can activate expression of defense responses in cotton and maize	Djonović et al. (2006); Seidl et al. (2006); Salas-Marina et al. (2015)
T. virens	Cerato-platanin small protein 2 (Sm2)	Maize	*Cochliobolus heterostrophus*	Plant protection conferred by the fungi–root interaction of *Trichoderma* with maize	Gaderer et al. (2015)
T. harzianum, *T. Formosa*, *T. asperellum*	Cerato-platanin protein (Epl-1)	Tomato, *Nicotiana benthamiana*	*B. cinerea*, *Tomato mosaic virus*	Involved in mycoparasitism, resistance induction, and self cell wall protection	Gomes et al. (2017); Cheng et al. (2018)
T. atroviride	Tanshinone I and tanshinone IIA	*Salvia miltiorrhiza*		Promoted *S. miltiorrhiza* hairy root growth and tanshinone biosynthesis	Ming et al. (2013)
T. asperellum	Hydrophobins (TasHyd1)	Cucumber, maize	*P. syringae*	Involved in plant root colonization and contributes to *Trichoderma*–plant interaction	Viterbo and Chet (2006)
T. longibrachiatum, *T. virens*	Type II hydrophobin	Tomato and *Arabidopsis thaliana*	*B. cinerea*, *R. solani*	Established a mutually beneficial interaction with the colonized plant	Ruocco et al. (2015); Guzmán-Guzmán et al. (2017)
T. asperelloides	Expansin-like proteins (swollenin)	Cucumber roots and leaves	*B. cinerea*	Expansin-like protein with a cellulose-binding domain capable of inducing local defense responses in cucumber roots and leaves against pathogens	Brotman et al. (2008)

T. harzianum (T34)	Endopolygalacturonase (ThPG1)	*A. thaliana*	*Pythium ultimum*	Involved in active colonization of tomato root and ISR-like defense in *Arabidopsis*	Morán-Diez et al. (2009), Sarrocco et al. (2017)
T. virens	Transferase (4-phosphopantetheinyl transferase)	*A. thaliana*	*R. solani*	Improve accumulation of salicylic acid (SA), jasmonic acid, camalexin and resistance to *B. cinerea*	Velázquez-Robledo et al. (2011); Contreras-Cornejo et al. (2014b)
	Indole compounds				
T. virens, *T. atroviride*	Idole-3-carboxyaldehyde (ICAld)–camalexin precursor	*A. thaliana*	*B. cinerea*	Induce growth and defense responses via canonical defense hormones SA and JA as well as camalexin	Contreras-Cornejo et al. (2011)
T. harzianum (T10)	Saccharides/derivatives, polysaccharide, oligosaccharide, fungal cell wall fragments, cell wall glucans	Red pepper seedlings	*Phytophthora capsici*	Induce systemic resistance against *P. capsici*	Sriram et al. (2009)
T. hamatum (T382)	Chitin-related compounds	*A. thaliana*	*B. cinerea*	Accelerate activation of the defense response against *B. cinerea* during ISR boost and a subsequent moderation of the *B. cinerea*-induced defense response	Mathys et al. (2012)
T. viride	Components of *Trichoderma* cells Glycosphingolipids (membrane)	Cultured rice cells	–	Elicits expression of the pathogen resistance-related genes PAL and PBZ1 genes	Uchiyama et al. (2009)
	Secondary metabolites				
T. viride	Peptaibol alamethicin (20mer peptaibol)	Lima bean	–	Elicitation of JA and SA biosynthesis in lima bean	Engelberth et al. (2001); Maischak et al. (2010)

(continued)

Table 5.1 (continued)

Trichoderma spp.	Elicitor(s)/MAMP	Host	Pathogen	Mechanism	References
T. pseudokoningii	Trichokonin (20mer peptaibol)	Tobacco	*Tobacco mosaic virus*	Enhances the ROS production, the accumulation of phenolic compounds at the application site, and virus resistance in tobacco plants through multiple defense signaling pathways	Luo et al. (2010)
T. virens	Trichovirin II (18mer peptaibols)	Cucumber	*P. syringae*	Elicitation of cucumber systemic defenses against *P. syringae*	Viterbo et al. (2007)
T. arundinaceum (IBT)	Trichodermin (trichothecene)	Tomato	*B. cinerea*, *R. solani*	Induce expression of tomato plant defense-related genes belonging to the SA and JA pathways against *B. cinerea*	Malmierca et al. (2016a, b)
T. harzianum	Trichodiene	Tomato	*B. cinerea*	Induce systemic resistance in plants against biotic stress	Malmierca et al. (2015b)
T. virens, *T. atroviride*, *T. reesei*	Polyketides	Maize		PKS/NRPS hybrid enzyme is involved in *Trichoderma*–plant interactions resulting in activation of defense responses	Mukherjee et al. (2012); Baker et al. (2012)
T. virens, *T. atroviride*	Harzianolide	Pea, tomato, and rape seed	*Sclerotinia sclerotiorum*	Enhancement of plant defense and growth in pea, canola, and tomato	Cai et al. (2013)
T. harzianum	Harzianopyridone				Vinale et al. (2008a)
	Pyrones (6-pentyl-a-pyrone)				

MAMP is an ET-inducing xylanase (Xyn2/Eix), which triggers plant defense responses in tomato and tobacco cultivars irrespective of its enzymatic activity (Furman-Matarasso et al. 1999; Hanania et al. 1999; Rotblat et al. 2002). ET-inducing xylanase is recognized by the tomato pattern recognition receptors (PRRs) LeEix1 and LeEix2, which traverse the cell membrane and have extracellular LRR domains (Ron and Avni 2004; Bar et al. 2009). It induces 1-aminocyclopropane-1-carboxylic acid (ACC) synthase expression, causing ET biosynthesis and ISR (Matarasso et al. 2005). Some *Trichoderma* proteins involved in plant root colonization can also act as MAMPs. Djonović et al. (2007) demonstrated that the beneficial filamentous fungus *T. virens* secretes a hydrophobin-like elicitor, Sm1, which is required for systemic resistance mediated by *T. virens* against *Colletotrichum graminicola* in maize. Maize plants germinated with SM1-overexpressing or SM1-deleted strains showed decreased or enhanced levels of systemic resistance, respectively, confirming SM1 function in eliciting plant defense. Orthologues of the small secreted cysteine-rich protein (SSCP) cerato-platanin family Epl1 from *T. atroviride* and *T. harzianum* are accumulated in the hyphae during interaction with the plant host and act as MAMPs in cotton and tomato, respectively (Seidl et al. 2006; Gomes et al. 2017). Gaderer et al. (2015) confirmed the function of another cerato-platanin protein (CPP) Sm2, *finding that T. virens* was more effective at priming plant defense than *T. atroviride*. The outcome with both *Trichoderma* species revealed defense induction was strongly reduced in plants inoculated with the *sm2/epl2* knockout strains than with *sm1/epl1* knockout strains. Interestingly, the cerato-platanin genes *sm1/epl1* are more abundantly expressed than *sm2/epl2* during fungal growth, while Sm2/Epl2 appear to be more important than Sm1/Epl1 for the enhancement of plant defense conferred by *Trichoderma* strains in the maize–*C. heterostrophus* pathosystem and may be exclusive inducers of plant systemic resistance triggered by *Trichoderma*–root interaction. These findings highlight the diversified functions of CPPs in fungi. The *T. asperellum*-induced MAPK (TIPK) protein was found to be essential for signal transduction in cucumber (Shoresh et al. 2006). The gene encoding this protein is homologous to *WIPK*, *MPK3*, and *MPK3a* (from tobacco, *Arabidopsis*, and parsley, respectively). *TIPK* is activated by *Trichoderma* as well as pathogen challenge in the roots and leaves of cucumber, with higher expression observed when plants were inoculated with *Trichoderma* prior to pathogen infection. Plants overexpressing *TIPK* were strongly resistant to *Psl* when compared with control plants. Concurrently, plants with reduced *TIPK* expression displayed increased sensitivity to pathogen challenge and lack of defense priming when pre-inoculated with *T. asperellum*. Thus, the defense priming effect of *Trichoderma* relies on MAPK activation in the host plant. The documented effects of JA, SA, or inhibitors of JA and ethylene demonstrate that TIPK activations occur upstream of the JA/ethylene signaling pathways. Segarra et al. (2009, 2010) suggested that *MYB72* is a node of convergence in *Trichoderma*-induced ISR triggered by soilborne beneficial microorganisms, although the combined application of *T. harzianum* Tr6 and *Pseudomonas* sp. Ps14 did not result in enhanced disease resistance (Alizadeh et al. 2013). In addition, overexpression of *MYB72* does not confer improved resistance to foliar pathogens, suggesting that MYB72 regulates and/or acts in concert with another signaling mechanism (Segarra et al. 2010). Additional regulatory proteins that act as an initial link between *Trichoderma* elicitors and plant host receptors in the triggering of plant resistance mechanisms will likely be identified in the future.

Several elicitors from *Trichoderma* have been reported to trigger Ca^{2+} fluxes and augment the accumulation of ROS (Hanson and Howell 2004). The variation of cytosolic Ca^{2+} mediated via MAMPs plays a pivotal role in systemic resistance in plants. Calcineurin B-like proteins (CBLs) act as Ca^{2+} sensors to activate specific protein kinases, termed CBL-interacting protein kinases (CIPKs). Two CIPKs from rice, OsCIPK14 and OsCIPK15, were rapidly stimulated by chitooligosaccharides and xylanase from *T. viride* (Kurusu et al. 2010). Overexpression of OsCIPK15 resulted in enhancement of TvX/EIX-induced cell death in rice plants (Kurusu et al. 2010). These results suggest that MAMP-mediated OsCIPK14/15 expression plays a decisive role in plant protection in rice. According to Meng and Zhang (2013), the induction of *Arabidopsis* MPK3 and MPK6 by ectopic expression of constitutively active MAPKK led to camalexin biosynthesis and hypersensitive response (HR)-like responses. Moreover, phytoalexin production is compromised in *mpk3* and *mpk6 Arabidopsis* mutants (Ren et al. 2008; Couto and Zipfel 2016).

Various oligosaccharides like chitin can also elicit plant defense responses, and oligomer scavenging is essential for colonization of fungal pathogens (De Jonge et al. 2010). Plants likely secrete chitinases to release the active polymers from the invading fungal cell walls for perceiving chitin, thus eliciting resistance responses. The mycoenzymatic action of *Trichoderma* chitinases can also release chitooligosaccharides, directly promoting the activation of defense mechanisms. *Trichoderma* secondary metabolites show antifungal activity at high doses but can act as MAMPs at very low concentrations. Several compounds like 6-pentyl-a-pyrone, harzianolide, cremenolide, and harzianopyridone boost plant defense responses and improve plant growth in numerous crops such as pea, tomato, and canola, at very low concentrations (Vinale et al. 2008b, 2014), signifying that plants resistance and their developmental responses to *Trichoderma* share common mechanisms.

Some of the chemically diverse secondary metabolites produced by *Trichoderma* are a source of antimicrobial activity. Some authors reported that secondary metabolites of *Trichoderma* spp. could be potentially effective as plant growth and biocontrol agents, with better outcomes than live cell application under greenhouse and field conditions (Vinale et al. 2014; Singh et al. 2016; Gupta et al. 2018). Mechanisms underlying the biocontrol activity of *Trichoderma*-based conidia formulations include secretion of cell wall-degrading enzymes and antimicrobial secondary metabolites. There are many reports detailing the antimicrobial activities of *Trichoderma* secondary metabolites (Dunlop et al. 1989; Evidente et al. 2003; Degenkolb et al. 2006; Vinale et al. 2006, 2008a, 2009, 2011; Xiao-Yan et al. 2006). In addition to the antimicrobial potential, several *Trichoderma* secondary metabolites induce systemic resistance in plants. Vinale et al. (2008a) reported that major secondary metabolites, viz., harzianolide and 6-n-pentyl-6H-pyran-2-one

(6PP) produced by *T. harzianum* strains T22, T39, and A6 and *T. atroviride* strain P1, improved growth, induced the expression of PR proteins, and reduced the disease symptoms of *B. cinerea* or *L. maculans* in tomato and canola seedlings, respectively. Moreover, application of 6PP on maize seedlings distinctly improved the activities of β-1,3-glucanase, peroxidase, and polyphenol oxidase in both shoot and root tissues, demonstrating an induction of defense responses in plants (El-Hasan and Buchenauer 2009).

Peptaibols such as alamethicin from *T. viride* induce plant production of volatile compounds related to JA signaling and cause an increase in endogenous levels of JA and SA in lima bean (Engelberth et al. 2001). 18mer peptaibols from *T. virens* strain Gv29-8 elicit systemic defense in cucumber against the leaf pathogen *Psl* (Viterbo et al. 2007). In addition, antimicrobial peptaibols from *T. pseudokoningii* SMF2 known as trichokonins also trigger defense signaling pathways in tobacco plants, increasing resistance against tobacco mosaic virus (Luo et al. 2010). Antifungal peptaibols produced by the *T. pseudokoningii* strain SMF2 directly inhibit mycelial growth and sporulation and induce cell apoptosis in the pathogenic fungus *B. cinerea* (Zhao et al. 2018). Furthermore, *Trichoderma* produces terpenes and sterols that can affect gene expression in both *Trichoderma* and plant. Malmierca et al. (2015a) showed that the terpene trichodiene produced in a *T. harzianum* erg1-silenced strain can act as an elicitor in tomato, as well as induce *Botrytis* genes involved in the synthesis of botrydial.

Recent findings have revealed that oxygen heterocyclic compounds (OHC), including polyketides referred to as PKS/NRPS from *T. reesei*, *T. atroviride*, and *T. virens*, are potentially involved in triggering defense response (Mukherjee et al. 2013). Harzianolides, harzianopyridone, pyrones, and other OHC compounds were also described as effective inducers of plant growth and defense responses (Mukherjee et al. 2012). Most OHC compounds are primarily reported as having antimicrobial and plant growth-promoting properties, whereas their plant defense-inducing activities are scarcely documented. Future findings will decipher the biological properties of these molecules in a comprehensive manner, allowing for the effective application of OHC metabolites in the promotion of disease resistance in plants.

5.4 Root and Leaf Colonization by *Trichoderma* spp.

Trichoderma colonization leads to systemic modification of transcriptome and proteome in various plant parts (Shoresh et al. 2010; Hermosa et al. 2012). Initiation of ISR requires *Trichoderma* strains to efficiently colonize the host plant. *Trichoderma* species are able to colonize different plant organs, causing substantial alterations in plant metabolism by producing compounds that stimulate plant growth (Elad 2000; Harman et al. 2004a). It is well documented that *Trichoderma* strains colonize plant roots prior to induction of plant growth and defense against pathogens. Some strains promote plant fitness and productivity via enhancing nutrient availability as well as resistance against various pathogens (Howell 2003; Harman 2006). Root colonization involves the ability to diagnose and adhere to plant roots, enter the plant, and endure toxic metabolites produced by the host plants in response to *Trichoderma* invasion. Yedidia et al. (1999) revealed the physical interaction between *Trichoderma* and the plant to be limited to the primary cell layers of plant epidermis and root outer cortex. They reported that the hyphae of *Trichoderma* are able to colonize the root cortex but that callose-enriched wall appositions efficiently restricted fungal growth into the vascular stele. Thus, a very active and symbiotic interaction evolves between *Trichoderma* and host, whereby the *Trichoderma* occupies a nutritional habitat and the host plant is protected from phyto-diseases.

The molecular cross talk that directs the interaction between *Trichoderma* spp. and host plant is still poorly understood. Advances in molecular methods have offered insight into *Trichoderma* colonization. The cross talk between different species and the plant roots likely results from interaction of various chemical signals from both partners. *Trichoderma* spp. produce and alter hormonal signals in order to aid in root colonization. Similar to mycorrhizae, growth of *Trichoderma* near the root region is enabled by root-secreted chemicals. Growth of *Trichoderma* into the rhizosphere is encouraged by the highly hydrated polysaccharides and the mono- and disaccharides secreted by plant roots. In addition, plant-derived sucrose and sucrose-dependent signaling in *Trichoderma* are important for the establishment of a mutualistic association (Vargas et al. 2009; Macías-Rodríguez et al. 2018). Solute transporters involved in the acquisition of root exudates have been identified in *Trichoderma* spp. (Vizcaíno et al. 2006; Vargas et al. 2009). Some of *Trichoderma* spp. such as *T. virens* or *T. atroviride* induce auxin-related effects that endorse root growth, facilitating colonization by enhancing the available surface area in *Arabidopsis* seedlings (Contreras-Cornejo et al. 2009). However, the defined role of auxins produced by the fungus and the regulation of root architecture are still under debate (Nieto-Jacobo et al. 2017), since *T. atroviride* produces the volatile compound 6-pentyl-2H-pyran-2-one (6PP), which causes changes to root architecture via auxin signaling in *Arabidopsis* (Garnica-Vergara et al. 2016). Viterbo et al. (2010) demonstrated the role of 1-aminocyclopropane-1-carboxylic acid (ACC) deaminase in regulation of canola root growth by *T. asperellum*.

Attachment of *Trichoderma* to the plant cell wall is mediated by secreted hydrophobin-like proteins coating the fungal cell surface and expansin-like proteins related to cell wall development. Hydrophobins play essential roles in fungal morphogenesis, including the formation of structures of infection, hyphal aggregation, and attachment of hyphae to hydrophobic surfaces and adhesion as well as cell-to-cell communication (Benítez et al. 2004). TasHyd1 from *T. asperellum* and Qid74 of *T. harzianum* have been shown to facilitate root surface attachment and protection of hyphal tips from plant defense compounds (Viterbo and Chet 2006; Samolski et al. 2012). *Trichoderma* spp. secrete the swollenin TasSwo, an expansin-like protein with cellulose-binding modules and endopolygalacturonase (ThPG1) able to modify the plant cell wall architecture, aiding root penetration (Brotman et al. 2008; Morán-Diez et al. 2009; Sarrocco et al. 2017). Once it enters the roots, *Trichoderma* can grow intercellularly, with a strengthening of plant epidermal and cortical cell walls along with development of barriers by the callose deposition and infiltrations of cellulose (Ruano-Rosa et al. 2016).

Plants respond to fungal invasion by producing and accumulating antimicrobial compounds: phytoalexins, flavonoids and terpenoids, phenolic derivatives, and aglycones. The colonization of plant roots by *Trichoderma* depends strongly on the competence of each strain to tolerate these compounds. This resistance has been interconnected with the presence of ATP-binding cassette (ABC) transport systems in the fungi, which are required for plant colonization (Ruocco et al. 2009; Sun et al. 2019). *Trichoderma* colonization of plant roots would thus be favored by strains highly resistant to toxic compounds, which overexpress hydrophobins and ABC transporters. In *Lotus japonicus* roots, inhibition of phytoalexin production during colonization with *T. koningii* was observed (Masunaka et al. 2011). Furthermore, a small secreted cysteine-rich protein (SSCP) found in different species of *Trichoderma* turned out to be a homologue of the *Cladosporium fulvum* avirulence protein Avr4 (Harman et al. 2004a). It has been suggested that the binding of Avr4 to chitin could support *Trichoderma* colonization (Stergiopoulos and de Wit 2009).

Some *Trichoderma* spp. also colonize host plant leaves. There are several factors that may influence microbial colonization in the phyllosphere. The plant surfaces are exposed to water vapor pressure deficit and water persistence, fluctuating temperatures, gases, wind, radiation, light, and polluting compounds. According to Lo et al. (1998), the root or soil application of a transgenic mutant of *T. harzianum* strain 1295–22 that expresses β-glucuronidase did not result in leaf colonization. Nevertheless, after foliar spray of T-1295–22 conidia, the spores germinated, and numerous hyphae were observed in the phyllosphere. In consequence, some strains can colonize leaf surfaces under specific conditions. The resident organisms of plant surfaces actively use leaf resources in their growth (Elad and Kirshner 1993). However, biocontrol potential might not be dependent on the growth of *Trichoderma* on leaf surfaces, as the colonization of the organism can activate systemic resistance or adversely affect growth of plant pathogens (Elad and Kirshner 1993).

Studies suggest that *Trichoderma* suppresses the early plant defense response in order to allow efficient colonization (Harman et al. 2004a). There is no doubt that the priming from *Trichoderma* triggers the expression of plant defense genes and promotes plant growth.

5.5 Plant Diseases Resistance Conferred by *Trichoderma*

Trichoderma treatment has been documented to promote innate immunity by induced resistance mechanisms in many plant varieties, viz., tomato, lettuce, tobacco, cucumber, cotton, maize, potato, cacao, bean, chickpea, melon, and *Arabidopsis* (Yedidia et al. 1999; Djonović et al. 2006; Gallou et al. 2009; Morán-Diez et al. 2009; Salas-Marina et al. 2011, 2015; Yoshioka et al. 2012; Jogaiah et al. 2018) infected with a broad range of pathogens (Table 5.2). Systemic immunity

Table 5.2 *Trichoderma*-mediated induced protection in various plants

Trichoderma spp.	Plant system	Pathogen	References
T. harzianum T39	Tomato, lettuce, tobacco, bean, cucumber, grape-vine, pepper	*Botrytis cinerea, Sphaerotheca fusca, Sclerotinia sclerotiorum, Plasmopara viticola, Cladosporium fulvum*	De Meyer et al. (1998); Elad (2000); Perazzolli et al. (2011); Meller-Harel et al. (2014)
T. harzianum	Cucumber, pepper, sunflower	*Phytophthora capsici, Rhizoctonia solani, Fusarium oxysporum*	Yedidia et al. (1999, 2000); Ahmed et al. (2003); Ezziyyani et al. (2007); Chen et al. (2017)
T. harzianum T22	Maize, tomato, lettuce	*Colletotrichum graminicola, Pythium ultimum, Calonectria pauciramosa, F. oxysporum* f. sp. *lactucae*	Harman et al. (2004b); Mastouri et al. (2010); Vitale et al. (2012); Innocenti et al. (2015)
T. harzianum T39	*Arabidopsis*	*B. cinerea*	Korolev et al. (2008)
T. harzianum RU01	Bean	*Uromyces appendiculatus*	Abeysinghe (2009)
T. harzianum MUCL 29707	Potato	*R. solani*	Gallou et al. (2009)
T. harzianum TH12	Oilseed rape	*S. sclerotiorum, Erysiphe cruciferarum*	Alkooranee et al. (2015, 2017)
T. asperellum T-203	Cucumber	*Pseudomonas syringae* pv. *lachrymans*	Yedidia et al. (2003); Shoresh et al. (2005);
T. asperellum T-34	Cucumber	*P. syringae* pv. *lachrymans*	Segarra et al. (2007)
T. asperellum SKT-1	*Arabidopsis*	*P. syringae* pv. *tomato* DC3000, cucumber mosaic virus	Yoshioka et al. (2012); Elsharkawy et al. (2013)
T. asperellum T1	Lettuce	*Corynespora cassiicola, Curvularia aeria*	Baiyee et al. (2019)
T. virens	Maize, *Arabidopsis*, tomato, cotton	*C. graminicola, B. cinerea, R. solani, F. oxysporum* f. sp. *lycopersici*	Howell et al. (2000); Djonović et al. (2006); Contreras-Cornejo et al. (2011); Mukherjee et al. (2018); Jogaiah et al. (2018)
*T. hamatum*382	Cucumber, tomato	*P. capsici, Xanthomonas euvesicatoria*	Khan et al. (2004); Alfano et al. (2007)
T. hamatum UoM 13	Pearl millet	*Sclerospora graminicola*	Nawrocka et al. (2018)
T. hamatum GD12	Rice, lettuce	*Magnaporthe oryzae, S. sclerotiorum, R. solani*	Ryder et al. (2012); Studholme et al. (2013);
T. atroviride	Tomato, *Arabidopsis*	*B. cinerea* and *P. syringae* pv. *tomato DC 3000*	Brunner et al. (2005); Salas-Marina et al. (2011)

(continued)

Table 5.2 (continued)

Trichoderma spp.	Plant system	Pathogen	References
T. atroviride TRS25	Cucumber	*R. solani*	Nawrocka et al. (2018)
T. arundinaceum	Tomato	*B. cinerea*, *R. solani*	Malmierca et al. (2015a, 2016a)
Trichoderma spp.	Hot pepper, bean, tomato	*P. capsici*, *R. solani*, *X. euvesicatoria*, *Alternaria solani*	Bae et al. (2011); Fontenelle et al. (2011); Mayo et al. (2015)

modification in plants is frequently related to elevated levels of PR proteins and/or accumulation of phytoalexin compounds (Yedidia et al. 2003; Shoresh et al. 2005; Salas-Marina et al. 2011). Maize roots colonized by either *T. harzianum* T22 or *T. virens* inhibited the soilborne disease caused by *C. graminicola* (Harman et al. 2004b; Djonović et al. 2007). *Arabidopsis* seedlings inoculated with either *T. virens* or *T. atroviride* and then challenged with *B. cinerea* developed less leaf necrotic lesions compared with *Trichoderma*-untreated plants (Contreras-Cornejo et al. 2011). Segarra et al. (2007) observed that the multiplication of *P. syringae* in cotyledons was restricted when cucumber plants were pretreated with *T. asperellum* T34. *Trichoderma* spp. induced defense responses and significant changes in SA, JA, ET, and abscisic acid content in melon plants, where different species of *Trichoderma* mitigated SA-mediated reactions in response to *Fusarium oxysporum* (Martínez-Medina et al. 2010, 2014). These results suggest that *T. harzianum* can manage *F. oxysporum* wilt in plants by both the stimulation of plant basal resistance and the attenuation of the hormonal responses caused by *F. oxysporum*.

Trichoderma has also been shown to be effective in induced resistance to a viral pathogen. *Trichoderma harzianum* (T22) was shown to decrease cucumber *mosaic virus* (CMV) disease in *Solanum lycopersicum* var. *cerasiforme*. The study found that T22 pre-treatment led to systemic resistance to CMV infection via jasmonic acid/ethylene and salicylic acid signaling (Vitti et al. 2016). Interestingly, root-colonizing *T. harzianum* TH12 and its culture filtrate (CF) both trigger ISR and SAR against infection by the fungi *S. sclerotiorum* and *Erysiphe cruciferarum* in oilseed rape (Alkooranee et al. 2015, 2017). Recently, Martínez-Medina et al. (2017) also observed that root colonization by *Trichoderma* hindered nematode multiplication both locally and systemically. *Trichoderma*-primed SA-regulated defense restricted nematode root invasion and enhanced JA-induced defenses, antagonizing the deregulation of JA-dependent immunity by pathogen nematodes. These findings demonstrate that different classes of pathogens can activate the same immunity pathways in plants.

Recently, the ability of *T. virens* spores or cell-free culture filtrate to induce defense responses in tomato (*S. lycopersicum*) against *F. oxysporum* was investigated (Jogaiah et al. 2018). The authors observed that the JA-deficient plants were more susceptible to *Fusarium* than the wild type (WT) when treated with *T. virens*.

SA-deficient *NahG* plants treated with culture filtrate were also found to be more susceptible to *Fusarium* than WT-treated tomato plants. The authors further demonstrated that the JA pathway is induced in WT tomato plants treated with *T. virens*, whereas the SA pathway is induced in WT tomato plants treated with *T. virens* culture filtrate. These results suggest that *T. virens* and its filtrate differentially induce JA and SA signaling cascades for the elicitation of *F. oxysporum* resistance in tomato.

5.6 Concluding Remarks and Outlook

During the last two decades, the use of *Trichoderma* spp. in both agriculture and research has greatly increased, owing to their multifaceted properties and expected success as bioweapons toward the development of sustainable agriculture (López-Bucio et al. 2015; Guzmán-Guzmán et al. 2018). *Trichoderma* undergoes long-lasting interactions with a wide range of plants, thereby playing a significant role in plant growth and pathogen resistance in various ecosystems (Kashyap et al. 2017). As such, there is an ongoing necessity to understand the molecular mechanisms that govern the ability of *Trichoderma* to enhance resistance and manage plant diseases. Management of phyto-diseases occurs via mycoparasitism, antibiosis and competence in the rhizosphere, production of antibiotics, and induction of systemic resistance (Fig. 5.1), the focus of this chapter. Bioprimed plants colonized by *Trichoderma* spp. can better resist various stresses and have an enhanced capacity to protect themselves from certain phytopathogens. Intracellular signaling responses that interpret fungal perception into genetic, metabolic, and developmental processes are the focus of recent plant–microbe interaction research. *Trichoderma* structural and secreted proteins as well as secondary metabolites act as MAMPs, while various enzymes produced by *Trichoderma*, by acting against plant cell wall and other fungi, generate DAMPs. MAMPs, PAMPs, and DAMPs can be recognized by specific plant receptors that trigger signaling pathways including MAPK cascades, leading to resistance as well as activation of phytohormone networking that cross-talk defense response and other growth-regulating substances (Xu and Zhang 2015). It will be interesting to determine whether one or more factors upstream of MPK3 or MPK6 in the MPK cascade, such as members of the MEK and MEKK families of kinases and/or sensors for MAMPs, PAMPs, DAMPs, or microbial effectors, also play a role in priming of plants. Some findings have opened the possibility that in addition to secondary metabolites and microbial effectors, fungal-derived hormones may also be involved in the *Trichoderma*–plant interaction. These hormones could be important during interactions with plants and could also be involved in various processes such as growth, hyphae or haustoria development, or conidia germination. Future research will reveal additional plant adaptation mechanisms in response to *Trichoderma*. In-depth understanding of *Trichoderma*–pathogen/*Trichoderma*–plant–pathogen interactions requires investigations relating to all three components of this system, to better understand the molecular dialogue that occurs during these

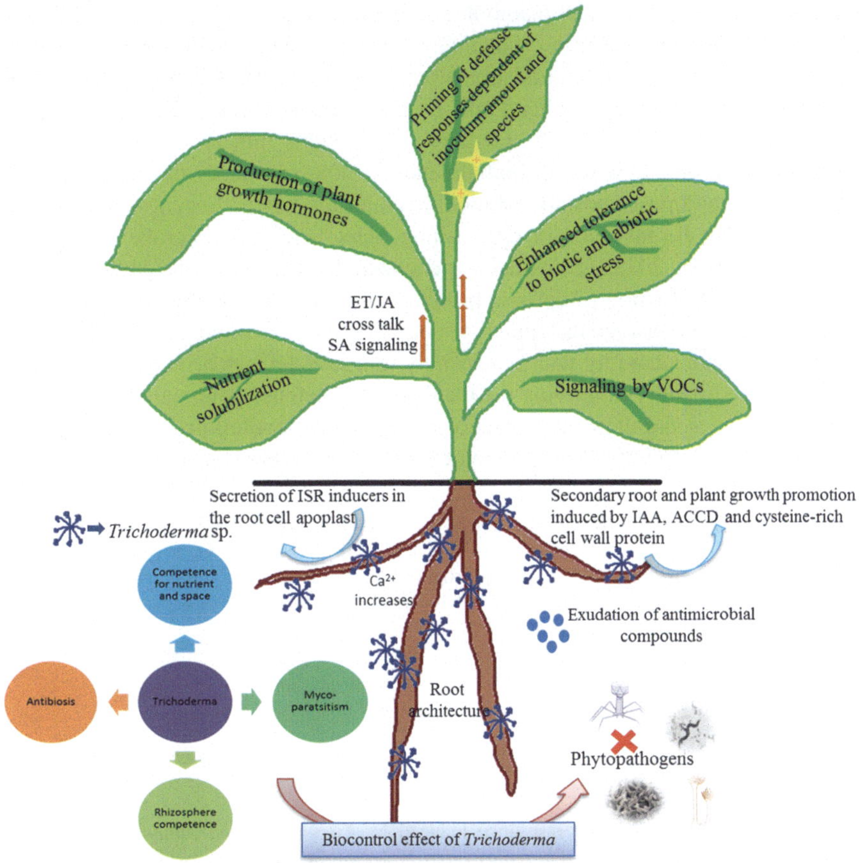

Fig. 5.1 Schematic illustration of *Trichoderma*–plant molecular signaling and mechanisms employed by *Trichoderma* to enhance plant growth and suppression of phytopathogens. *IAA* indole-3-acetic acid, *ACCD* 1-aminocyclopropane-1-carboxylate (ACC) deaminase, *ET* ethylene, *JA* jasmonic acid, *SA* salicylic acid, *ISR* induced systemic resistance

interactions. With the emergence of next-generation techniques for identification of novel candidate genes and proteins involved in host–pathogen interactions, *Trichoderma* has emerged as a favorite candidate for management of phytopathogens. The emerging knowledge will undoubtedly result in a great leap in our understanding of tripartite *Trichoderma*–plant–pathogen interactions.

Recent works have demonstrated that active defense-induced resistance is not an optimal approach for disease management, because the high associated costs can mitigate the benefits of plant protection. Conversely, induced resistance by means of priming offers an effective strategy to combat biotic and abiotic stresses with significant benefits in agricultural systems. Hence, knowledge regarding the molecular mechanisms underlying priming can be instrumental in the translation from the laboratory to the field for disease control. Such molecular mechanisms can include elevated levels of PRRs (cell membrane receptor recognizing MAMPs, PAMPs, or DAMPs), dormant cellular signaling enzymes (e.g., MPKs), transcription coactivators, and histone modifications (e.g., H3K4me3, H3K9ac) in defense gene promoters. Together with alterations in DNA methylation, these mechanisms together constitute a defensive "memory" of previous infections, maybe even in former generations. Few recent findings indicate that regulatory mechanisms, such as epigenetic and post-translational modifications, are key pathogenesis modulators and could also be involved in establishing plant–beneficial microbe interactions (De Palma et al. 2019). Knowledge generated by the scientific community on molecular mechanism(s) of defense priming will allow for the emergence of regulatory similarities between different types of plant-induced immunity. This emerging mechanistic understanding will increasingly translate defense priming into agricultural practice, thereby improving the plants natural defense potential against a wide range of stresses simultaneously. Hence, exploring the evolutionary and ecological contexts of priming will be an exciting challenge for future research.

Acknowledgments Research in the Bar group is supported by grants 20-02-0075 and 20-02-0113 from the Israeli Ministry of Agriculture and Rural Development. R. G. is supported by the Indo-China ARO Postdoctoral Fellowship Program. The authors thank Prof. Yigal Elad, Dalia Rav-David, and the members of the Elad and Bar groups for helpful discussions and continuous support (Publication No. 593/19 of the ARO).

References

Abeysinghe S (2009) Systemic resistance induced by *Trichoderma harzianum* RU01 against *Uromyces appendiculatus* on *Phaseolus vulgaris*. J. Natl Sci Found Sri Lanka 37:203–207

Ahmed AS, Ezziyyani M, Sánchez CP, Candela ME (2003) Effect of chitin on biological control activity of *Bacillus* spp. and *Trichoderma harzianum* against root rot disease in pepper (*Capsicum annuum*) plants. Eur J Plant Pathol 109:633–637

Aidemark M, Tjellström H, Sandelius AS, Stålbrand H, Andreasson E, Rasmusson AG, Widell S (2010) *Trichoderma viride* cellulase induces resistance to the antibiotic pore-forming peptide alamethicin associated with changes in the plasma membrane lipid composition of tobacco BY-2 cells. BMC Plant Biol 10:274

Alfano G, Ivey ML, Cakir C, Bos JIB, Miller SA, Madden LV, Kamoun S, Hoitink HAJ (2007) Systemic modulation of gene expression in tomato by *Trichoderma hamatum* 382. Phytopathology 97:429–437

Alizadeh H, Behboudi K, Ahmadzadeh M, Javan-Nikkhah M, Zamioudis C, Pieterse CM, Bakker PA (2013) Induced systemic resistance in cucumber and *Arabidopsis thaliana* by the combination of *Trichoderma harzianum* Tr6 and *Pseudomonas* sp. Ps14. Biol Control 65:14–23

Alkooranee JT, Yin Y, Aledan TR, Jiang Y, Lu G, Wu J, Li M (2015) Systemic resistance to powdery mildew in Brassica napus (AACC) and *Raphanus alboglabra* (RRCC) by *Trichoderma harzianum* TH12. PLoS One 10:e0142177

Alkooranee JT, Aledan TR, Ali AK, Lu G, Zhang X, Wu J, Fu C, Li M (2017) Detecting the hormonal pathways in oilseed rape behind induced systemic resistance by *Trichoderma harzianum* TH12 to *Sclerotinia sclerotiorum*. PLoS One 12:e0168850

Badri DV, Weir TL, van der Lelie D, Vivanco JM (2009) Rhizosphere chemical dialogues: plant–microbe interactions. Curr Opin Biotechnol 20:642–650

Bae H, Roberts DP, Lim HS, Strem MD, Park SC, Ryu CM, Melnick RL, Bailey BA (2011) Endophytic *Trichoderma* isolates from tropical environments delay disease onset and induce resistance against *Phytophthora capsici* in hot pepper using multiple mechanisms. Mol Plant-Microbe Interac 24:336–351

Baiyee B, Ito SI, Sunpapao A (2019) *Trichoderma asperellum* T1 mediated antifungal activity and induced defense response against leaf spot fungi in lettuce (*Lactuca sativa* L.). Physiol Mol Plant Pathol 106:96–101

Baker SE, Perrone G, Richardson NM, Gallo A, Kubicek CP (2012) Phylogenomic analysis of polyketide synthase-encoding genes in *Trichoderma*. Microbiology 158:147–154

Bar M, Sharfman M, Schuster S, Avni A (2009) The coiled-coil domain of EHD2 mediates inhibition of LeEix2 endocytosis and signaling. PLoS One 4:e7973

Benítez T, Rincón AM, Limón MC, Codon AC (2004) Biocontrol mechanisms of Trichoderma strains. Int Microbiol 7:249–260

Bigirimana J, De Meyer G, Poppe J, Elad Y, Höfte M (1997) Induction of systemic resistance on bean (*Phaseolus Vulgaris*) by *Trichoderma harzianum*. Mededelingen Van De Faculteit Landbouwkundige En Toegepaste Biologische Wetenschappen, Universiteit Gent 62:1001–1007

Brotman Y, Briff E, Viterbo A, Chet I (2008) Role of swollenin, an expansin-like protein from *Trichoderma*, in plant root colonization. Plant Physiol 147:779–789

Brotman Y, Lisec J, Méret M, Chet I, Willmitzer L, Viterbo A (2012) Transcript and metabolite analysis of the *Trichoderma*-induced systemic resistance response to *Pseudomonas syringae* in *Arabidopsis thaliana*. Microbiology 158:139–146

Brunner K, Zeilinger S, Ciliento R, Woo SL, Lorito M, Kubicek CP, Mach RL (2005) Improvement of the fungal biocontrol agent *Trichoderma atroviride* to enhance both antagonism and induction of plant systemic disease resistance. Appl Environ Microbiol 71:3959–3965

Cai F, Yu G, Wang P, Wei Z, Fu L, Shen Q, Chen W (2013) Harzianolide, a novel plant growth regulator and systemic resistance elicitor from *Trichoderma harzianum*. Plant Physiol Biochem 73:106–113

Carrero-Carrón I, Trapero-Casas JL, Olivares-García C, Monte E, Hermosa R, Jiménez-Díaz RM (2016) *Trichoderma asperellum* is effective for biocontrol of *Verticillium* wilt in olive caused by the defoliating pathotype of *Verticillium dahliae*. Crop Prot 88:45–52

Chen J, Harman GE, Comis A, Cheng GW (2005) Proteins related to the biocontrol of pythium damping-off in maize with *Trichoderma harzianum* Rifai. J Integr Plant Biol 47:988–997

Chen SC, Zhao HJ, Wang ZH, Zheng CX, Zhao PY, Guan ZH, Qin HY, Liu AR, Lin XM, Ahammed GJ (2017) *Trichoderma harzianum*-induced resistance against *Fusarium oxysporum* involves regulation of nuclear DNA content, cell viability and cell cycle-related genes expression in cucumber roots. Eur J Plant Pathol 147:43–53

Cheng CH, Shen BN, Shang QW, Liu LYD, Peng KC, Chen YH, Chen FF, Hu SF, Wang YT, Wang HC, Wu HY (2018) Gene-to-gene network analysis of the mediation of plant innate immunity by the eliciting plant response-like 1 (Epl1) Elicitor of *Trichoderma formosa*. Mol Plant-Microbe Interac 31:683–691

Conrath U, Beckers GJ, Flors V, García-Agustín P, Jakab G, Mauch F, Newman MA, Pieterse CM, Poinssot B, Pozo MJ, Pugin A (2006) Priming: getting ready for battle. Mol Plant-Microbe Interac 19:1062–1071

Conrath U, Beckers GJ, Langenbach CJ, Jaskiewicz MR (2015) Priming for enhanced defense. Annu Rev Phytopathol 53:97–119

Contreras-Cornejo HA, Macías-Rodríguez L, Cortés-Penagos C, López-Bucio J (2009) *Trichoderma virens*, a plant beneficial fungus, enhances biomass production and promotes lateral root growth through an auxin-dependent mechanism in *Arabidopsis*. Plant Physiol 149:1579–1592

Contreras-Cornejo HA, Macías-Rodríguez L, Beltrán-Peña E, Herrera-Estrella A, López-Bucio J (2011) *Trichoderma*-induced plant immunity likely involves both hormonal-and camalexin-dependent mechanisms in *Arabidopsis thaliana* and confers resistance against necrotrophic fungi *Botrytis cinerea*. Plant Signal Behav 6:1554–1563

Contreras-Cornejo HA, Ortiz-Castro R, López-Bucio J (2013) Promotion of plant growth and the induction of systemic defence by *Trichoderma*: physiology, genetics and gene expression. In: Trichoderma : biology and applications. CABI, Wallingford, pp 173–194

Contreras-Cornejo HA, Macías-Rodríguez L, Alfaro-Cuevas R, López-Bucio J (2014a) *Trichoderma* spp. improve growth of *Arabidopsis* seedlings under salt stress through enhanced root development, osmolite production, and Na^+ elimination through root exudates. Mol Plant-Microbe Interac 27:503–514

Contreras-Cornejo HA, Macías-Rodríguez L, Herrera-Estrella A, López-Bucio J (2014b) The 4-phosphopantetheinyl transferase of *Trichoderma virens* plays a role in plant protection against *Botrytis cinerea* through volatile organic compound emission. Plant and Soil 379:261–274

Costa A, Navazio L, Szabo I (2018) The contribution of organelles to plant intracellular calcium signalling. J Exp Bot 69:4175–4193

Couto D, Zipfel C (2016) Regulation of pattern recognition receptor signalling in plants. Nat Rev Immunol 16:537

Crutcher FK, Moran-Diez ME, Ding S, Liu J, Horwitz BA, Mukherjee PK, Kenerley CM (2015) A paralog of the proteinaceous elicitor *SM1* is involved in colonization of maize roots by *Trichoderma virens*. Fungal Biol 119:476–486

De Jonge R, Van Esse HP, Kombrink A, Shinya T, Desaki Y, Bours R, Van Der Krol S, Shibuya N, Joosten MH, Thomma BP (2010) Conserved fungal LysM effector *Ecp6* prevents chitin-triggered immunity in plants. Science 329:953–955

De Meyer G, Bigirimana J, Elad Y, Höfte M (1998) Induced systemic resistance in *Trichoderma harzianum* T39 biocontrol of *Botrytis cinerea*. Eur J Plant Pathol 104:279–286

De Palma M, Salzano M, Villano C, Aversano R, Lorito M, Ruocco M, Docimo T, Piccinelli AL, D'Agostino N, Tucci M (2019) Transcriptome reprogramming, epigenetic modifications and alternative splicing orchestrate the tomato root response to the beneficial fungus *Trichoderma harzianum*. Hortic Res 6:5

Degenkolb T, Gräfenhan T, Nirenberg HI, Gams W, Brückner H (2006) *Trichoderma brevicompactum* complex: rich source of novel and recurrent plant-protective polypeptide antibiotics (peptaibiotics). J Agric Chem 54:7047–7061

Djonović S, Pozo MJ, Dangott LJ, Howell CR, Kenerley CM (2006) Sm1, a proteinaceous elicitor secreted by the biocontrol fungus *Trichoderma virens* induces plant defense responses and systemic resistance. Mol Plant Microbe Interact 19:838–853

Djonović S, Vargas WA, Kolomiets MV, Horndeski M, Wiest A, Kenerley CM (2007) A proteinaceous elicitor Sm1 from the beneficial fungus *Trichoderma virens* is required for induced systemic resistance in maize. Plant Physiol 145:875–889

Dodds PN, Rathjen JP (2010) Plant immunity: towards an integrated view of plant–pathogen interactions. Nat Rev Genet 11:539

Druzhinina IS, Seidl-Seiboth V, Herrera-Estrella A, Horwitz BA, Kenerley CM, Monte E, Mukherjee PK, Zeilinger S, Grigoriev IV, Kubicek CP (2011) *Trichoderma*: the genomics of opportunistic success. Nat Rev Microbiol 9:749–759

Dunlop RW, Simon A, Sivasithamparam K, Ghisalberti EL (1989) An antibiotic from *Trichoderma koningii* active against soil borne plant pathogens. J Nat Prod 52:67–74

Elad Y (2000) *Trichoderma harzianum* T39 preparation for biocontrol of plant diseases-control of *Botrytis cinerea*, *Sclerotinia sclerotiorum* and *Cladosporium fulvum*. Biocontrol Sci Technol 10:499–507

Elad Y, Kirshner B (1993) Survival in the phylloplane of an introduced biocontrol agent (*Trichoderma harzianum*) and populations of the plant pathogen *Botrytis cinerea* as modified by abiotic conditions. Phytoparasitica 21:303–313

Elad Y, Kirshner B, Yehuda N, Sztjenberg A (1998) Management of powdery mildew and gray mold of cucumber by *Trichoderma harzianum* T39 and *Ampelomyces quisqualis* AQ10. BioControl 43:241–251

El-Hasan A, Buchenauer H (2009) Actions of 6-Pentyl-alpha-pyrone in controlling seedling blight incited by *Fusarium moniliforme* and inducing defence responses in maize. J Phytopathol 157:697–707

Elsharkawy MM, Shimizu M, Takahashi H, Ozaki K, Hyakumachi M (2013) Induction of systemic resistance against *Cucumber mosaic virus* in *Arabidopsis thaliana* by *Trichoderma asperellum* SKT-1. Plant Pathol J 29:193–200

Engelberth J, Koch T, Schüler G, Bachmann N, Rechtenbach J, Boland W (2001) Ion channel-forming alamethicin is a potent elicitor of volatile biosynthesis and tendril coiling. Cross talk between jasmonate and salicylate signaling in lima bean. Plant Physiol 125:369–377

Evidente A, Cabras A, Maddau L, Serra S, Andolfi A, Motta A (2003) Viridepyronone, a new antifungal 6-substituted 2 h-pyran-2-one produced by *Trichoderma viride*. J Agric Food Chem 51:6957–6960

Ezziyyani M, Requena ME, Egea-Gilabert C, Candela ME (2007) Biological control of *Phytophthora* root rot of pepper using *Trichoderma harzianum* and *Streptomyces rochei* in combination. J Phytopathol 155:342–349

Faize M, Malnoy M, Dupuis F, Chevalier M, Parisi L, Chevreau E (2003) Chitinases of *Trichoderma atroviride* induce scab resistance and some metabolic changes in two cultivars of apple. Phytopathology 93:1496–1504

Fontenelle ADB, Guzzo SD, Lucon CMM, Harakava R (2011) Growth promotion and induction of resistance in tomato plant against *Xanthomonas euvesicatoria* and *Alternaria solani* by *Trichoderma* spp. Crop Prot 30:1492–1500

Furman-Matarasso N, Cohen E, Du Q, Chejanovsky N, Hanania U, Avni A (1999) A point mutation in the ethylene-inducing xylanase elicitor inhibits the β-1-4-endoxylanase activity but not the elicitation activity. Plant Physiol 121:345–352

Gaderer R, Lamdan NL, Frischmann A, Sulyok M, Krska R, Horwitz BA, Seidl-Seiboth V (2015) Sm2, a paralog of the *Trichoderma* cerato-platanin elicitor Sm1, is also highly important for plant protection conferred by the fungal-root interaction of *Trichoderma* with maize. BMC Microbiol 15:2

Gallou A, Cranenbrouck S, Declerck S (2009) *Trichoderma harzianum* elicits defence response genes in roots of potato plantlets challenged by *Rhizoctonia solani*. Eur J Plant Pathol 124:219–230

Garnica-Vergara A, Barrera-Ortiz S, Muñoz-Parra E, Raya-González J, Méndez-Bravo A, Macías-Rodríguez L, Ruiz-Herrera LF, López-Bucio J (2016) The volatile 6-pentyl-2H-pyran-2-one from *Trichoderma atroviride* regulates *Arabidopsis thaliana* root morphogenesis via auxin signaling and ethylene insensitive 2 functioning. New Phytol 209:1496–1512

Glazebrook J (2005) Contrasting mechanisms of defense against biotrophic and necrotrophic pathogens. Annu Rev Phytopathol 43:205–227

Gomes EV, Ulhoa CJ, Cardoza RE, Silva RN, Gutiérrez S (2017) Involvement of *Trichoderma harzianum* Epl-1 protein in the regulation of *botrytis* virulence-and tomato defense-related genes. Front Plant Sci 8:880

Gupta R, Singh A, Gupta MM, Pandey R (2016) Cumulative role of bioinoculants on growth, antioxidant potential and artemisinin content in *Artemisia annua* L. under organic field conditions. World J Microbiol Biotechnol 32:167

Gupta R, Singh A, Kanaujia R, Kushwaha S, Pandey R (2018) Trichoderma harzianum ThU and its metabolites underscore alteration in essential oils of *Ocimum basilicum* and *Ocimum sanctum*. Proc Natl Acad Sci Ind Sec B Biol Sci 88:219–227

Guzmán-Guzmán P, Alemán-Duarte MI, Delaye L, Herrera-Estrella A, Olmedo-Monfil V (2017) Identification of effector-like proteins in *Trichoderma* spp. and role of a hydrophobin in the plant-fungus interaction and mycoparasitism. BMC Genet 18:16

Guzmán-Guzmán P, Porras-Troncoso MD, Olmedo-Monfil V, Herrera-Estrella A (2018) *Trichoderma* species: versatile plant symbionts. Phytopathology 109(1):6–16

Hanania U, Furman-Matarasso N, Ron M, Avni A (1999) Isolation of a novel SUMO protein from tomato that suppresses EIX-induced cell death. Plant J 19:533–541

Hanson LE, Howell CR (2004) Elicitors of plant defense responses from biocontrol strains of *Trichoderma virens*. Phytopathology 94:171–176

Harman GE (2006) Overview of mechanisms and uses of *Trichoderma* spp. Phytopathology 96:190–194

Harman GE, Howell CR, Viterbo A, Chet I, Lorito M (2004a) *Trichoderma* species—opportunistic, avirulent plant symbionts. Nat Rev Microbiol 2:43

Harman GE, Petzoldt R, Comis A, Chen J (2004b) Interactions between *Trichoderma harzianum* strain T22 and maize inbred line Mo17 and effects of these interactions on diseases caused by *Pythium ultimum* and *Colletotrichum graminicola*. Phytopathology 94:147–153

Harman GE, Herrera-Estrella AH, Horwitz BA, Lorito M (2012) Special issue: *Trichoderma*-from basic biology to biotechnology. Microbiol Read 158:1

Hermosa R, Viterbo A, Chet I, Monte E (2012) Plant-beneficial effects of *Trichoderma* and of its genes. Microbiology 158:17–25

Hoitink HAJ, Madden LV, Dorrance AE (2006) Systemic resistance induced by *Trichoderma* spp.: interactions between the host, the pathogen, the biocontrol agent, and soil organic matter quality. Phytopathology 96:186–189

Howell CR (2003) Mechanisms employed by *Trichoderma* species in the biological control of plant diseases: the history and evolution of current concepts. Plant Dis 87:4–10

Howell CR, Hanson LE, Stipanovic RD, Puckhaber LS (2000) Induction of terpenoid synthesis in cotton roots and control of *Rhizoctonia solani* by seed treatment with *Trichoderma virens*. Phytopathology 90:248–252

Hua J, Meyerowitz EM (1998) Ethylene responses are negatively regulated by a receptor gene family in *Arabidopsis thaliana*. Cell 94:261–271

Innocenti G, Roberti R, Piattoni F (2015) Biocontrol ability of *Trichoderma harzianum* strain T22 against *Fusarium* wilt disease on water-stressed lettuce plants. BioControl 60:573–581

Jogaiah S, Abdelrahman M, Tran LSP, Ito SI (2018) Different mechanisms of *Trichoderma virens*-mediated resistance in tomato against *Fusarium* wilt involve the jasmonic and salicylic acid pathways. Mol Plant Pathol 19:870–882

Kashyap PL, Rai P, Srivastava AK, Kumar S (2017) Trichoderma for climate resilient agriculture. World J Microbiol Biotechnol 33:155

Keswani C, Bisen K, Singh SP, Sarma BK, Singh HB (2016) A proteomic approach to understand the tripartite interactions between plant-*Trichoderma*-pathogen: investigating the potential for efficient biological control. In: Plant, soil and microbes. Springer, Cham, pp 79–93

Khan J, Ooka JJ, Miller SA, Madden LV, Hoitink HAJ (2004) Systemic resistance induced by *Trichoderma hamatum* 382 in cucumber against *Phytophthora* crown rot and leaf blight. Plant Dis 88:280–286

Komis G, Šamajová O, Ovečka M, Šamaj J (2018) Cell and developmental biology of plant Mitogen-activated protein kinases. Annu Rev Plant Biol 69:237–265

Koornneef A, Pieterse CM (2008) Cross talk in defense signaling. Plant Physiol 146:839–844

Korolev N, Rav-David D, Elad Y (2008) The role of phytohormones in basal resistance and *Trichoderma*-induced systemic resistance to *Botrytis cinerea* in *Arabidopsis thaliana*. BioControl 53:667–683

Kumar V, Parkhi V, Kenerley CM, Rathore KS (2009) Defense-related gene expression and enzyme activities in transgenic cotton plants expressing an endochitinase gene from *Trichoderma virens* in response to interaction with *Rhizoctonia solani*. Planta 230:277–291

Kurusu T, Hamada J, Nokajima H, Kitagawa Y, Kiyoduka M, Takahashi A, Hanamata S, Ohno R, Hayashi T, Okada K, Koga J (2010) Regulation of microbe-associated molecular pattern-induced hypersensitive cell death, phytoalexin production, and defense gene expression by calcineurin B-like protein-interacting protein kinases, OsCIPK14/15, in rice cultured cells. Plant Physiol 153:678–692

Lo CT, Nelson EB, Hayes CK, Harman GE (1998) Ecological studies of transformed *Trichoderma harzianum* strain 1295–22 in the rhizosphere and on the phylloplane of creeping bentgrass. Phytopathology 88:129–136

López-Bucio J, Pelagio-Flores R, Herrera-Estrella A (2015) *Trichoderma* as biostimulant: exploiting the multilevel properties of a plant beneficial fungus. Sci Hortic 196:109–123

Lorito M, Woo SL, Harman GE, Monte E (2010) Translational research on *Trichoderma*: from omics to the field. Annu Rev Phytopathol 48:395–417

Luo Y, Zhang DD, Dong XW, Zhao PB, Chen LL, Song XY, Wang XJ, Chen XL, Shi M, Zhang YZ (2010) Antimicrobial peptaibols induce defense responses and systemic resistance in tobacco against tobacco mosaic virus. FEMS Microbiol Lett 313:120–126

Macías-Rodríguez L, Guzmán-Gómez A, García-Juárez P, Contreras-Cornejo HA (2018) *Trichoderma atroviride* promotes tomato development and alters the root exudation of carbohydrates, which stimulates fungal growth and the biocontrol of the phytopathogen *Phytophthora cinnamomi* in a tripartite interaction system. FEMS Microbiol Ecol 94:137

Maischak H, Zimmermann MR, Felle HH, Boland W, Mithöfer A (2010) Alamethicin-induced electrical long distance signaling in plants. Plant Signal Behav 5:988–990

Malmierca MG, McCormick SP, Cardoza RE, Monte E, Alexander NJ, Gutiérrez S (2015a) Trichodiene production in a *Trichoderma harzianum* erg1-silenced strain provides evidence of the importance of the sterol biosynthetic pathway in inducing plant defense-related gene expression. Mol Plant-Microbe Interac 28:1181–1197

Malmierca MG, Barua J, McCormick SP, Izquierdo-Bueno I, Cardoza RE, Alexander NJ, Hermosa R, Collado IG, Monte E, Gutiérrez S (2015b) Novel aspinolide production by *Trichoderma arundinaceum* with a potential role in *Botrytis cinerea* antagonistic activity and plant defence priming. Environ Microbiol 17:1103–1118

Malmierca MG, Izquierdo-Bueno I, McCormick SP, Cardoza RE, Alexander NJ, Barua J, Lindo L, Casquero PA, Collado IG, Monte E, Gutiérrez S (2016a) Trichothecenes and aspinolides produced by *Trichoderma arundinaceum* regulate expression of *Botrytis cinerea* genes involved in virulence and growth. Environ Microbiol 18:3991–4004

Malmierca MG, Izquierdo-Bueno I, Mccormick SP, Cardoza RE, Alexander NJ, Moraga J, Gomes EV, Proctor RH, Collado IG, Monte E, Gutiérrez S (2016b) Botrydial and botcinins produced by *Botrytis cinerea* regulate the expression of *Trichoderma arundinaceum* genes involved in trichothecene biosynthesis. Mol Plant Pathol 17:1017–1031

Marra R, Ambrosino P, Carbone V, Vinale F, Woo SL, Ruocco M, Ciliento R, Lanzuise S, Ferraioli S, Soriente I, Gigante S (2006) Study of the three-way interaction between *Trichoderma atroviride*, plant and fungal pathogens by using a proteomic approach. Curr Genet 50:307–321

Martinez C, Blanc F, Le Claire E, Besnard O, Nicole M, Baccou JC (2001) Salicylic acid and ethylene pathways are differentially activated in melon cotyledons by active or heat-denatured cellulase from *Trichoderma longibrachiatum*. Plant Physiol 127:334–344

Martínez-Medina A, Pascual JA, Pérez-Alfocea F, Albacete A, Roldán A (2010) *Trichoderma harzianum* and *Glomus intraradices* modify the hormone disruption induced by *Fusarium oxysporum* infection in melon plants. Phytopathology 100:682–688

Martínez-Medina A, Alguacil MDM, Pascual JA, Van Wees SC (2014) Phytohormone profiles induced by *Trichoderma* isolates correspond with their biocontrol and plant growth-promoting activity on melon plants. J Chem Ecol 40:804–815

Martínez-Medina A, Flors V, Heil M, Mauch-Mani B, Pieterse CM, Pozo MJ, Ton J, van Dam NM, Conrath U (2016) Recognizing plant defense priming. Trends Plant Sci 21:818–822

Martínez-Medina A, Fernandez I, Lok GB, Pozo MJ, Pieterse CM, Van Wees S (2017) Shifting from priming of salicylic acid-to jasmonic acid-regulated defences by *Trichoderma* protects tomato against the root knot nematode *Meloidogyne incognita*. New Phytol 213:1363–1377

Mastouri F, Björkman T, Harman GE (2010) Seed treatment with *Trichoderma harzianum* alleviates biotic, abiotic, and physiological stresses in germinating seeds and seedlings. Phytopathology 100:1213–1221

Masunaka A, Hyakumachi M, Takenaka S (2011) Plant growth-promoting fungus, *Trichoderma koningii* suppresses isoflavonoid phytoalexin vestitol production for colonization on/in the roots of *Lotus japonicus*. Microb Environ 26:128–134

Matarasso N, Schuster S, Avni A (2005) A novel plant cysteine protease has a dual function as a regulator of 1-aminocyclopropane-1-carboxylic acid synthase gene expression. Plant Cell 17:1205–1216

Mathys J, De Cremer K, Timmermans P, Van Kerkhove S, Lievens B, Vanhaecke M, Cammue B, De Coninck B (2012) Genome-wide characterization of ISR induced in *Arabidopsis thaliana* by *Trichoderma hamatum* T382 against *Botrytis cinerea* infection. Front Plant Sci 3:108

Mauch-Mani B, Baccelli I, Luna E, Flors V (2017) Defense priming: an adaptive part of induced resistance. Annu Rev Plant Biol 68:485–512

Mayo S, Gutierrez S, Malmierca MG, Lorenzana A, Campelo MP, Hermosa R, Casquero PA (2015) Influence of *Rhizoctonia solani* and *Trichoderma* spp. in growth of bean (*Phaseolus vulgaris* L.) and in the induction of plant defense-related genes. Front Plant Sci 6:685

Mehari ZH, Elad Y, Rav-David D, Graber ER, Meller-Harel Y (2015) Induced systemic resistance in tomato (*Solanum lycopersicum*) against *Botrytis cinerea* by biochar amendment involves jasmonic acid signaling. Plant and Soil 395:31–44

Meller-Harel Y, Mehari ZH, Rav-David D, Elad Y (2014) Systemic resistance to gray mold induced in tomato by benzothiadiazole and *Trichoderma harzianum* T39. Phytopathology 104:150–157

Meng X, Zhang S (2013) MAPK cascades in plant disease resistance signaling. Annu Rev Phytopathol 51:245–266

Ming Q, Su C, Zheng C, Jia M, Zhang Q, Zhang H, Rahman K, Han T, Qin L (2013) Elicitors from the endophytic fungus *Trichoderma atroviride* promote Salvia miltiorrhiza hairy root growth and tanshinone biosynthesis. J Exp Bot 64:5687–5694

Morán-Diez E, Hermosa R, Ambrosino P, Cardoza RE, Gutiérrez S, Lorito M, Monte E (2009) The ThPG1 endopolygalacturonase is required for the *Trichoderma harzianum* – plant beneficial interaction. Mol Plant-Microbe Interac 22:1021–1031

Mukherjee PK, Buensanteai N, Moran-Diez ME, Druzhinina IS, Kenerley CM (2012) Functional analysis of non-ribosomal peptide synthetases (NRPSs) in *Trichoderma virens* reveals a polyketide synthase (PKS)/NRPS hybrid enzyme involved in the induced systemic resistance response in maize. Microbiology 158:155–165

Mukherjee PK, Horwitz BA, Herrera-Estrella A, Schmoll M, Kenerley CM (2013) *Trichoderma* research in the genome era. Annu Rev Phytopathol 51:105–129

Mukherjee PK, Hurley JF, Taylor JT, Puckhaber L, Lehner S, Druzhinina I, Schumacher R, Kenerley CM (2018) Ferricrocin, the intracellular siderophore of *Trichoderma virens*, is involved in growth, conidiation, gliotoxin biosynthesis and induction of systemic resistance in maize. Biochem Biophys Res Commun 505:606–611

Naoumkina MA, Zhao Q, Gallego-Giraldo LINA, Dai X, Zhao PX, Dixon RA (2010) Genome-wide analysis of phenylpropanoid defence pathways. Mol Plant Pathol 11:829–846

Nawrocka J, Małolepsza U (2013) Diversity in plant systemic resistance induced by *Trichoderma*. Biol Control 67:149–156

Nawrocka J, Małolepsza U, Szymczak K, Szczech M (2018) Involvement of metabolic components, volatile compounds, PR proteins, and mechanical strengthening in multilayer protection of cucumber plants against *Rhizoctonia solani* activated by *Trichoderma atroviride* TRS25. Protoplasma 255:359–373

Nesler A, Perazzolli M, Puopolo G, Giovannini O, Elad Y, Pertot I (2015) A complex protein derivative acts as biogenic elicitor of grapevine resistance against powdery mildew under field conditions. Front Plant Sci 6:715

Nieto-Jacobo MF, Steyaert JM, Salazar-Badillo FB, Nguyen DV, Rostás M, Braithwaite M, De Souza JT, Jimenez-Bremont JF, Ohkura M, Stewart A, Mendoza-Mendoza A (2017) Environmental growth conditions of *Trichoderma* spp. affects indole acetic acid derivatives, volatile organic compounds, and plant growth promotion. Front. Plant Sci 8:102

Okon Levy N, Meller Harel Y, Haile ZM, Elad Y, Rav-David D, Jurkevitch E, Katan J (2015) Induced resistance to foliar diseases by soil solarization and *Trichoderma harzianum*. Plant Pathol 64:365–374

Pel MJ, Pieterse CM (2012) Microbial recognition and evasion of host immunity. J Exp Bot 64:1237–1248

Perazzolli M, Roatti B, Bozza E, Pertot I (2011) *Trichoderma harzianum* T39 induces resistance against downy mildew by priming for defense without costs for grapevine. Biol Control 58:74–82

Perazzolli M, Moretto M, Fontana P, Ferrarini A, Velasco R, Moser C, Delledonne M, Pertot I (2012) Downy mildew resistance induced by *Trichoderma harzianum* T39 in susceptible grapevines partially mimics transcriptional changes of resistant genotypes. BMC Genomics 13:660

Pieterse CM, Van der Does D, Zamioudis C, Leon-Reyes A, Van Wees SC (2012) Hormonal modulation of plant immunity. Annu Rev Cell Dev Biol 28:489–521

Pieterse CM, Zamioudis C, Berendsen RL, Weller DM, Van Wees SC, Bakker PA (2014) Induced systemic resistance by beneficial microbes. Annu Rev Phytopathol 52:347

Pozo MJ, Baek JM, Garcıa JM, Kenerley CM (2004) Functional analysis of tvsp1, a serine protease-encoding gene in the biocontrol agent *Trichoderma virens*. Fungal Genet Biol 41:336–348

Ramirez-Prado JS, Abulfaraj AA, Rayapuram N, Benhamed M, Hirt H (2018) Plant immunity: from signaling to epigenetic control of defense. Trends Plant Sci 23:833–844

Ren D, Liu Y, Yang KY, Han L, Mao G, Glazebrook J, Zhang S (2008) A fungal-responsive MAPK cascade regulates phytoalexin biosynthesis in *Arabidopsis*. Proc Natl Acad Sci 105:5638–5643

Ron M, Avni A (2004) The receptor for the fungal elicitor ethylene-inducing xylanase is a member of a resistance-like gene family in tomato. Plant Cell 16:1604–1615

Rotblat B, Enshell-Seijffers D, Gershoni JM, Schuster S, Avni A (2002) Identification of an essential component of the elicitation active site of the EIX protein elicitor. Plant J 32:1049–1055

Ruano-Rosa D, Prieto P, Rincón AM, Gómez-Rodríguez MV, Valderrama R, Barroso JB, Mercado-Blanco J (2016) Fate of *Trichoderma harzianum* in the olive rhizosphere: time course of the root colonization process and interaction with the fungal pathogen *Verticillium dahliae*. BioControl 61:269–282

Rubio MB, Hermosa R, Vicente R, Gómez-Acosta FA, Morcuende R, Monte E, Bettiol W (2017) The combination of *Trichoderma harzianum* and chemical fertilization leads to the deregulation of phytohormone networking, preventing the adaptive responses of tomato plants to salt stress. Front Plant Sci 8:294

Ruocco M, Lanzuise S, Vinale F, Marra R, Turrà D, Woo SL, Lorito M (2009) Identification of a new biocontrol gene in *Trichoderma atroviride*: the role of an ABC transporter membrane pump in the interaction with different plant-pathogenic fungi. Mol Plant-Microbe Interac 22:291–301

Ruocco M, Lanzuise S, Lombardi N, Woo SL, Vinale F, Marra R, Varlese R, Manganiello G, Pascale A, Scala V, Turrà D (2015) Multiple roles and effects of a novel *Trichoderma* hydrophobin. Mol Plant-Microbe Interac 28:167–179

Ryder LS, Harris BD, Soanes DM, Kershaw MJ, Talbot NJ, Thornton CR (2012) Saprotrophic competitiveness and biocontrol fitness of a genetically modified strain of the plant-growth-promoting fungus *Trichoderma hamatum* GD12. Microbiology 158:84–97

Salas-Marina MA, Silva-Flores MA, Uresti-Rivera EE, Castro-Longoria E, Herrera-Estrella A, Casas-Flores S (2011) Colonization of *Arabidopsis* roots by *Trichoderma atroviride* promotes growth and enhances systemic disease resistance through jasmonic acid/ethylene and salicylic acid pathways. Eur J Plant Pathol 131:15–26

Salas-Marina MA, Isordia-Jasso MI, Islas-Osuna MA, Delgado-Sánchez P, Jiménez-Bremont JF, Rodríguez-Kessler M, Rosales-Saavedra MT, Herrera-Estrella A, Casas-Flores S (2015) The Epl1 and Sm1 proteins from *Trichoderma atroviride* and *Trichoderma virens* differentially modulate systemic disease resistance against different life style pathogens *Solanum lycopersicum*. Front Plant Sci 6:77

Samolski I, Rincón AM, Pinzón LM, Viterbo A, Monte E (2012) The *qid74* gene from *Trichoderma harzianum* has a role in root architecture and plant biofertilization. Microbiology 158:129–138

Saravanakumar K, Fan L, Fu K, Yu C, Wang M, Xia H, Sun J, Li Y, Chen J (2016) Cellulase from *Trichoderma harzianum* interacts with roots and triggers induced systemic resistance to foliar disease in maize. Sci Rep 6:35543

Sarrocco S, Matarese F, Baroncelli R, Vannacci G, Seidl-Seiboth V, Kubicek CP, Vergara M (2017) The constitutive endopolygalacturonase *TvPG2* regulates the induction of plant systemic resistance by *Trichoderma virens*. Phytopathology 107:537–544

Segarra G, Casanova E, Bellido D, Odena MA, Oliveira E, Trillas I (2007) Proteome, salicylic acid, and jasmonic acid changes in cucumber plants inoculated with *Trichoderma asperellum* strain T34. Proteomics 7:3943–3952

Segarra G, Van der Ent S, Trillas I, Pieterse CMJ (2009) MYB72, a node of convergence in induced systemic resistance triggered by a fungal and a bacterial beneficial microbe. Plant Biol 11:90–96

Segarra G, Casanova E, Avilés M, Trillas I (2010) *Trichoderma asperellum* strain T34 controls *Fusarium* wilt disease in tomato plants in soilless culture through competition for iron. Microb Ecol 59:141–149

Seidl V, Marchetti M, Schandl R, Allmaier G, Kubicek CP (2006) Epl1, the major secreted protein of *Hypocrea atroviridis* on glucose, is a member of a strongly conserved protein family comprising plant defense response elicitors. FEBS J 273:4346–4359

Shah J, Zeier J (2013) Long-distance communication and signal amplification in systemic acquired resistance. Front Plant Sci 4:30

Shah JM, Raghupathy V, Veluthambi K (2009) Enhanced sheath blight resistance in transgenic rice expressing an endochitinase gene from *Trichoderma virens*. Biotechnol Lett 31:239

Shoresh M, Harman GE (2008) The molecular basis of shoot responses of maize seedlings to *Trichoderma harzianum* T22 inoculation of the root: a proteomic approach. Plant Physiol 147:2147–2163

Shoresh M, Yedidia I, Chet I (2005) Involvement of jasmonic acid/ethylene signaling pathway in the systemic resistance induced in cucumber by *Trichoderma asperellum* T203. Phytopathology 95:76–84

Shoresh M, Gal-On A, Leibman D, Chet I (2006) Characterization of a mitogen-activated protein kinase gene from cucumber required for *Trichoderma*-conferred plant resistance. Plant Physiol 142:1169–1179

Shoresh M, Harman GE, Mastouri F (2010) Induced systemic resistance and plant responses to fungal biocontrol agents. Annu Rev Phytopathol 48:21–43

Singh A, Gupta R, Srivastava M, Gupta MM, Pandey R (2016) Microbial secondary metabolites ameliorate growth, *in planta* contents and lignification in *Withania somnifera* (L.) Dunal. Physiol Mol Biol Plants 22:253–260

Sriram S, Manasa SB, Savitha MJ (2009) Potential use of elicitors from *Trichoderma* in induced systemic resistance for the management of *Phytophthora capsici* in red pepper. J Biol Control 23:449–456

Stergiopoulos I, de Wit PJ (2009) Fungal effector proteins. Annu Rev Phytopathol 47:233–263

Studholme DJ, Harris B, Le Cocq K, Winsbury R, Perera V, Ryder L, Beale M, Ward J, Thornton CR, Grant M (2013) Investigating the beneficial traits of *Trichoderma hamatum* GD12 for sustainable agriculture insights from genomics. Front Plant Sci 4:258

Sun J, Zhang T, Li Y, Wang X, Chen J (2019) Functional characterization of the ABC transporter TaPdr2 in the tolerance of biocontrol the fungus *Trichoderma atroviride* T23 to dichlorvos stress. Biol Control 129:102–108

Tucci M, Ruocco M, De Masi L, De Palma M, Lorito M (2011) The beneficial effect of *Trichoderma* spp. on tomato is modulated by the plant genotype. Mol Plant Pathol 12:341–354

Uchiyama R, Aoki K, Sugimoto H, Taka N, Katayama T, Itonori S, Sugita M, Che FS, Kumagai H, Yamamoto K (2009) Phosphocholine-containing glycosyl inositol-phosphoceramides from *Trichoderma viride* induce defense responses in cultured rice cells. Biosci Biotechnol Biochem 73:74–78

Van Bel AJ, Gaupels F (2004) Pathogen-induced resistance and alarm signals in the phloem. Mol Plant Pathol 5:495–504

Van Wees SC, Van der Ent S, Pieterse CM (2008) Plant immune responses triggered by beneficial microbes. Curr Opin Plant Biol 11:443–448

Vargas WA, Mandawe JC, Kenerley CM (2009) Plant-derived sucrose is a key element in the symbiotic association between *Trichoderma virens* and maize plants. Plant Physiol 151:792–808

Velázquez-Robledo R, Contreras-Cornejo HA, Macías-Rodríguez L, Hernández-Morales A, Aguirre J, Casas-Flores S, López-Bucio J, Herrera-Estrella A (2011) Role of the 4-phosphopantetheinyl transferase of *Trichoderma virens* in secondary metabolism and induction of plant defense responses. Mol Plant-Microbe Interac 24:1459–1471

Vinale F, Marra R, Scala F, Ghisalberti EL, Lorito M, Sivasithamparam K (2006) Major secondary metabolites produced by two commercial *Trichoderma* strains active against different phytopathogens. Lett Appl Microbiol 43:143–148

Vinale F, Sivasithamparam K, Ghisalberti EL, Marra R, Barbetti MJ, Li H, Woo SL, Lorito M (2008a) A novel role for *Trichoderma* secondary metabolites in the interactions with plants. Physiol Mol Plant Pathol 72:80–86

Vinale F, Sivasithamparam K, Ghisalberti EL, Marra R, Woo SL, Lorito M (2008b) *Trichoderma* – plant–pathogen interactions. Soil Biol Biochem 40:1–10

Vinale F, Ghisalberti EL, Sivasithamparam K, Marra R, Ritieni A, Ferracane R, Woo S, Lorito M (2009) Factors affecting the production of *Trichoderma harzianum* secondary metabolites during the interaction with different plant pathogens. Lett Appl Microbiol 48:705–711

Vinale F, Girona IA, Nigro M, Mazzei P, Piccolo A, Ruocco M, Woo S, Rosa DR, Herrera CL, Lorito M (2011) Cerinolactone, a hydroxy-lactone derivative from *Trichoderma cerinum*. J Nat Prod 75:103–106

Vinale F, Sivasithamparam K, Ghisalberti EL, Woo SL, Nigro M, Marra R, Lombardi N, Pascale A, Ruocco M, Lanzuise S, Manganiello G (2014) *Trichoderma* secondary metabolites active on plants and fungal pathogens. Open Mycol J 8(1):127

Vitale A, Cirvilleri G, Castello I, Aiello D, Polizzi G (2012) Evaluation of *Trichoderma harzianum* strain T22 as biological control agent of *Calonectria pauciramosa*. BioControl 57:687–696

Viterbo ADA, Chet I (2006) TasHyd1, a new hydrophobin gene from the biocontrol agent *Trichoderma asperellum*, is involved in plant root colonization. Mol Plant Pathol 7:249–258

Viterbo A, Harel M, Chet I (2004) Isolation of two aspartyl proteases from *Trichoderma asperellum* expressed during colonization of cucumber roots. FEMS Microbiol Lett 238:151–158

Viterbo ADA, Wiest ARIC, Brotman Y, Chet I, Kenerley C (2007) The 18mer peptaibols from *Trichoderma virens* elicit plant defence responses. Mol Plant Pathol 8:737–746

Viterbo A, Landau U, Kim S, Chernin L, Chet I (2010) Characterization of ACC deaminase from the biocontrol and plant growth-promoting agent *Trichoderma asperellum* T203. FEMS Microbiol Lett 305:42–48

Vitti A, Pellegrini E, Nali C, Lovelli S, Sofo A, Valerio M, Scopa A, Nuzzaci M (2016) *Trichoderma harzianum* T-22 induces systemic resistance in tomato infected by *Cucumber mosaic virus*. Front Plant Sci 7:1520

Vizcaíno JA, Cardoza RE, Hauser M, Hermosa R, Rey M, Llobell A, Becker JM, Gutiérrez S, Monte E (2006) ThPTR2, a di/tri-peptide transporter gene from *Trichoderma harzianum*. Fungal Genet Biol 43:234–246

Vos IA, Moritz L, Pieterse CM, Van Wees S (2015) Impact of hormonal crosstalk on plant resistance and fitness under multi-attacker conditions. Front Plant Sci 6:639

Wang KLC, Li H, Ecker JR (2002) Ethylene biosynthesis and signaling networks. Plant Cell 14:131–151

Xiao-Yan S, Qing-Tao S, Shu-Tao X, Xiu-Lan C, Cai-Yun S, Yu-Zhong Z (2006) Broad-spectrum antimicrobial activity and high stability of Trichokonins from *Trichoderma koningii* SMF2 against plant pathogens. FEMS Microbiol Lett 260:119–125

Xu J, Zhang S (2015) Mitogen-activated protein kinase cascades in signaling plant growth and development. Trends Plant Sci 20:56–64

Yedidia I, Benhamou N, Chet I (1999) Induction of defense responses in cucumber plants (*Cucumis sativus* L.) by the biocontrol agent *Trichoderma harzianum*. Appl Environ Microbiol 65:1061–1070

Yedidia I, Benhamou N, Kapulnik Y, Chet I (2000) Induction and accumulation of PR proteins activity during early stages of root colonization by the mycoparasite *Trichoderma harzianum* strain T-203. Plant Physiol Biochem 38:863–873

Yedidia I, Shoresh M, Kerem Z, Benhamou N, Kapulnik Y, Chet I (2003) Concomitant induction of systemic resistance to *Pseudomonas syringae* pv. lachrymans in cucumber by *Trichoderma asperellum* (T-203) and accumulation of phytoalexins. Appl Environ Microbiol 69:7343–7353

Yoshioka Y, Ichikawa H, Naznin HA, Kogure A, Hyakumachi M (2012) Systemic resistance induced in *Arabidopsis thaliana* by *Trichoderma asperellum* SKT-1, a microbial pesticide of seedborne diseases of rice. Pest Manag Sci 68:60–66

Zhao J (2015) Phospholipase D and phosphatidic acid in plant defence response: from protein–protein and lipid–protein interactions to hormone signalling. J Exp Bot 66:1721–1736

Zhao P, Ren A, Dong P, Sheng Y, Li D (2018) Antimicrobial peptaibols, trichokonins, inhibit mycelial growth and sporulation and induce cell apoptosis in the pathogenic fungus *Botrytis cinerea*. Appl Biochem Microbiol 54:396–403

Zipfel C (2009) Early molecular events in PAMP-triggered immunity. Curr Opin Plant Biol 12:414–420

Chapter 6
Systemically Induced Resistance Against Maize Diseases by *Trichoderma* spp.

Jie Chen, Murugappan Vallikkannu, and Valliappan Karuppiah

Abstract Maize (*Zea mays*) is a vital crop for human food, animal feed, and biofuel generation. But production is altogether diminished because of several pathogens. *Trichoderma* species incorporates numerous imperative in farming and is used as biological control agents (BCAs). A group of elicitors produced by the *Trichoderma* hyphae are able to induce distinctive sorts of defensive signals inside the plants through activation of salicylic acid (SA), jasmonic acid (JA), brassinolide (BR), reactive oxygen species (ROS), and defense enzymes. Hyd1 and cellulase are the elicitors obtained from *Trichoderma harzianum* that were found to recognize root target proteins. Activation of these genes induces the plant to produce compounds associated with the reduction of pathogens invasion and improve the biochemical activities for the plant growth and development. The basic target of this chapter is to display the efficacy of *Trichoderma* on the systemically induced resistance against maize diseases.

Keywords *Trichoderma* · Plant pathogens · Maize · Induced resistance · Secondary metabolites · MAMP

6.1 Introduction

Maize is the major and important nutrient-rich staple food all over the world. Approximately 225 million tons/annum (He et al. 2018) of maize were produced in China. Maize has more benefits than other crops. It has also been used in the

J. Chen (✉) · M. Vallikkannu · V. Karuppiah
School of Agriculture and Biology, Shanghai Jiao Tong University, Shanghai, P. R. China

Key Laboratory of Urban Agriculture (South) Ministry of Agriculture, Shanghai Jiao Tong University, Shanghai, P. R. China

State Key Laboratory of Microbial Metabolism, Shanghai Jiao Tong University, Shanghai, P. R. China
e-mail: jiechen59@sjtu.edu.cn

A. K. Sharma, P. Sharma (eds.), *Trichoderma*, Rhizosphere Biology,
https://doi.org/10.1007/978-981-15-3321-1_6

production of fuel, oil, medicines, cosmetics, and livestock fodder. Especially in China, 70% of the maize is used as livestock fodder (Xing et al. 2017). The demand for corn is increasing year by year because of its enormous benefits. Nowadays, the production of maize is harder due to environmental conditions, serious disease and pest infection, lack of high resistance varieties, unhealthy soil, etc. (Saba et al. 2012; Chen et al. 2014).

During corn growth, it suffers multiple disease attack; however, some diseases occur in the late period of maize development such as northern leaf blight (*Exserohilum turcicum*), southern leaf blight (*Cochliobolus heterostrophus*), and ear rot (*Fusarium verticillioides*). Hence, it is difficult to practice biological control measures. On the other hand, it is very laborious for the application of a biocontrol agent. Hence, there is a need to discover a mutual, symbiotic biocontrol agent that helps to decrease the consumption of chemical fungicides and improve the yield of the product (Woo et al. 2006). Recently, several types of research on the expression of pathogen and bioagents are underway. For example, *Bacillus cereus*, *Trichoderma harzianum*, and *Trichoderma viride* are some of the biocontrol agents used against the maize leaf blight diseases (Huang et al. 2010).

Trichoderma spp., an environment-friendly fungus, play a significant part in biological control of crop infections and soil remediation. Most of the *Trichoderma* were accompanied with the soil and plant roots. Hence, it is very easy to communicate with the plant and root ecosystem (Martinez-Medina et al. 2016). *Trichoderma* species including *T. harzianum*, *T. virens*, *T. asperellum*, and *T. atroviride* were utilized as biocontrol operators of plant pathogens around the world. *Trichoderma* are sharp mycoparasites, are rhizosphere competent, and widely colonize the external root layers (Schuster and Schmoll 2010; Harman 2011). Further, *Trichoderma* increases plant growth under biotic and abiotic stresses (Hermosa et al. 2012; Contreras-Cornejo et al. 2013). The multi-facial role of *Trichoderma* against pathogens made it as an efficient biocontrol agent in the field. It has been used as both bio-fertilizer and biopesticide (Contreras-Cornejo et al. 2015). In plants, numerous changes were observed in transcriptomic regulations and the physiological process by the root colonization of *Trichoderma* (Alexandru et al. 2013). *Trichoderma* had the ability to maintain a good symbiotic relationship with plants as well as antagonistic relationship to disease pathogen. For example, the antagonistic relationship between *Trichoderma* and *Pythium ultimum* enhanced the endochitinase and beta-1,3-glucanase activity in the roots of maize and promoted growth (Harman et al. 2004; Chen et al. 2005). In addition to the mycoparasitism activity, *Trichoderma* induces both local and systemic resistance in plants. *Trichoderma* species produce biotechnologically and pharmaceutically important secondary metabolites like peptobials, non-ribosomal peptides, volatile and nonvolatile organic compounds, siderophores, and pyrones (Velázquez-Robledo et al. 2011; Müller et al. 2013).

This chapter presents the compilation of the induced resistance of *Trichoderma* species on maize plants against different plant pathogens and the mechanisms.

6.2 Induced Resistance: SAR and ISR

Plants have an enormous and dynamic defense scheme to protect themselves from biotic stresses like parasites and pathogen attack. It may be either a local response or a systemic response. The induced resistance of plants can be categorized into two types of mechanisms: systemic acquired resistance (SAR) and induced systemic resistance (ISR). The resistance mechanism has been differentiated by their regulatory pathways and produced elicitors. The major distinction between these two systems is what major defense signal pathway to be stimulated in plant. SAR is the salicylic acid-dependent pathway. Mostly, it upregulated pathogenesis-related proteins (17 PR genes). Salicylic acid or methyl salicylate acts as a signaling molecule at the time of pathogen attack, and the PR proteins are upregulated. It needs some time to produce the PR genes and salicylic acid throughout the plants after the invasion of virulent and non-virulent microorganisms. The NPR1 (salicylic acid receptor) is essential for the enhanced resistance and PR gene expression. Normally, in a plant system, NPR1 protein is in the form of the oligomer. Salicylic acid converts it into monomeric form by means of reducing the bonds of disulfide at the time of the SAR pathway induction. This monomeric form interacts with the PR genes and activates the TGA/OBF subclass of basic leucine-zipper transcription factors. ISR was not associated with the PR genes, and it is mostly contingent on the jasmonic acid and/or ethylene phytohormones to stimulate host plant defense responses. But NPR1 functional gene is also important to enhance the ISR resistance (Kinkema et al. 2000; Zhang et al. 2003; Vallad and Goodman 2004; Choudhary et al. 2007; Saldajeno et al. 2014). In most cases, biocontrol microbes present in rhizosphere soil mainly induce plant ISR against plant diseases.

6.3 *Trichoderma*-Associated Induced Resistance in Maize Plants

In addition to the mycoparasitic activity, *Trichoderma* species enable to trigger the induced resistance of maize plants against the pathogen in different ways including cross-communication, signaling molecules, and pathways. The plant pathogenic *Fusarium* species cause the major diseases of maize plants like corn stalk rot, ear rot, and root rot. The cysteine-rich *Fusarium graminearum* antifungal protein (FGAFP) protein of *Fusarium* had a higher antifungal activity, which made the toughest job to eradicate during initial stages (Patiño et al. 2018). Recently, loop-mediated isothermal amplification (LAMP) methodology was developed to find out the *Fusarium temperatum* at initial stages (Shan et al. 2019). A 6-pentyl-alpha-pyrone metabolite of *Trichoderma* strain T23 showed good antagonistic activity against *Fusarium* species (El-Hasan et al. 2018). *Trichoderma atroviride* AN240 suppressed more than 70% of mycotoxins including DON and ZEN (Błaszczyk et al. 2017). The detoxification of the DON through *Trichoderma* strains by means of a biological control-guided procedure reached the incredible consideration. As potential antagonists, *Trichoderma* strains inhibited the mycelial development and mycotoxin synthesis of *F. graminearum*. The conversion of DON to deoxynivalenol-3-glucoside (D3G) was considered as the end product of detoxification. It confirmed that *Trichoderma* strains retain a resistance mechanism to convert DON into D3G (Tian et al. 2016). Saravanakumar et al. (2018a) described that cellulase of *Trichoderma harzianum* strain Th22 induced the 2,4-dihydroxy-7-methoxy-2H-1, 4-benzoxazine-3(4H)-one (DIMBOA) and defense gene expression in corn root against deoxynivalenol. The expression of Thph1 and Thph2 genes encoding cellulase induced the maize JA, ET, SAR, DIMBOA, and PII-associated genes to prevent the infection of *Fusarium graminearum*. Pyrosequencing analysis showed that *T. harzianum* strain CCTCC-RW0024 increased the plant growth promoting bacteria and there by inhibited the *Fusarium graminearum* to improve the plant growth and development. Further, the metabolites of *T. harzianum* strain CCTCC-RW0024 was interacted with the pathogenicity related transcriptional cofactor FgSWi6 *Fusarium graminearum* to inhibit the *Fusarium graminearum* growth (Saravanakumar et al. 2017).

Anthracnose stalk rot and leaf blight are another severe diseases that can mostly be found on maize plants created by the fungalysin-encoding gene *Cgfl* of *Colletotrichum graminicola*. At the severe stage of this disease, it created early leaf senescence and decreased leaf area of photosynthesis (Jirak-Peterson and Esker 2011). The role of proteinaceous elicitor Sm1 of *T. virens* strain Gv29-8 was confirmed in the induced maize defense system against *Colletotrichum graminicola* (Djonović et al. 2007). A leaf spot caused by causal agent *Curvularia lunata* (Huang et al. 2004; Hua et al. 2006; Bisht et al. 2013) affect the adult stage leaves of maize using its toxins. To prevent maize plants from this disease, the mutant T66, a *T. harzianum* isolate, was constructed by the Fan et al. (2015) to stimulate systemic resistance against the foliar disease. A gene from the mutant was cloned and named as Thc6, which later is identified as a Zn-finger transcription factor upregulating the Thph1 and Thph2 cellulase genes in *T. harzianum*, which was found to devote to systemically induced resistance of maize against *C. lunata* leaf spot with *T. harzianum* colonization on maize root (Saravanakumar et al. 2016). In this study, the JA/ET signal transduction pathway is significantly activated in the leaf through the interaction of *Trichoderma* cellulases with maize root target like ZmATG3 or ZmGLP (Saravanakumar et al. 2018b). A platelet-activating factor acetylhydrolase-like gene (PAF-AH-like gene) of *Trichoderma harzianum* was cloned successfully. The PAF-AH-like gene was associated with antagonistic relationship against *Rhizoctonia solani*. It increased the secretion of chitinase and cellulase along with specific interaction with maize roots. Meanwhile, it proved that the PAF-AH-like gene in *Trichoderma* interacted with roots and regulated the hormone JA in maize leaf and JA pathway-related gene expression and induced maize resistance against *Curvularia lunata* (Yu et al. 2014). The seed treatment with selected *Trichoderma* strain restriction enzyme-mediated integration (REMI) mutant D9, C40, and H6 showed the disease reduction index against leaf spot (Sun et al. 2009).

6.4 Root Colonization Involved in Induced Resistance

Some of the bacteria and fungi species would be capable to colonize the plant roots and induce plant growth by suppressing the plant infections. It is widely demonstrated that root colonization with biocontrol microbes enables to stimulate the ISR pathway against plant disease control. The association of *Trichoderma* in the plant root ecosystem liberates the chemical signals, thereby communicating with the plants to induce the production of biocontrol-related hydrophobin, hydrolytic enzymes, antagonistic secondary metabolites, etc. (Djonović et al. 2007; Troian et al. 2014; Martínez-Medina et al. 2017). Those metabolites not only enable to degrade pathogen cell wall structure or inhibit pathogen growth but also induce plant resistance to reduce pathogen infections. Among them, secreted proteins like hydrophobin generated in the three layer of communication among the plant, *Trichoderma*, and pathogen induced the plant resistance and improve *Trichoderma* colonization on root. The higher activity of chitinolytic enzymes and phytohormones in plants by means of *Trichoderma* interaction increased the plant defense response (Shoresh and Harman 2008; Rawat et al. 2013).

The comparative genomic analysis of *T. atroviride* and *T. virens* provided in-depth knowledge of the mycoparasitism and colonization of the *Trichoderma* fungi (Kubicek et al. 2011; Hermosa et al. 2012). The root colonization of *Trichoderma* could be performed by some of the hydrophobin proteins such as hydrophobin-like elicitor SM1 in *T. virens* Gv29-8 and TasHyd1 elicitor from *T. asperellum*, which supports the root colonization and resistance mechanism (Viterbo and Chet 2006; Djonović et al. 2007). At the contact time of plant, *Trichoderma*, and pathogen, there was a chance of producing more number of secretome proteins, and this was commonly regulated by root colonization. To test this hypothesis, Lamdan et al. (2015) compared the secretomes of *T. virens* cocultured with maize root seedlings. The higher amount of Sm1 cerato-platanin family was found to be inducing the ISR activity. It was observed that the higher amount of Sm1 was present in the coculture of *T. virens* and maize seedlings, which resulted in the root colonization of *Trichoderma* with plant and thereby increased the growth and disease resistance. But the molecular mechanism between the *Trichoderma* and plant root colonization is unknown (Bisen et al. 2016). The recent development in molecular and genetic techniques revealed some of the mechanisms of attachment and colonization (Bais et al. 2006). Vargas et al. (2009) studied about the plant-derived sucrose interaction between *Trichoderma* and plant to induce the root colonization of *Trichoderma* in maize roots. The results showed that the intracellular invertase enzyme of *T. virens* (TvInv) was produced in the presence of plant-determined sucrose, which was not only used as a carbon source but also induced root colonization, chlamydospore production, and systemic induction in leaf photosynthesis. Similarly, a cell wall protein Qid74 played a vital part in the adherence and penetration of *T. harzianum* with plant root (Morán-Diez et al. 2009; Samolski et al. 2012). In addition, root colonization of *T. atroviride* in maize plants released a number of volatile terpenes like 6-pentyl-2H-pyran-2-one (6-PP), a signaling molecule from belowground, against the parasitism of *C. sonorensis* on *S. frugiperda* (Contreras-Cornejo et al. 2018).

After penetration of *Trichoderma* into the root system, it either compromise or fight against the plant defense system for invasion. Many of the *Trichoderma* species can tolerate the antimicrobial components of plants, chemicals, etc. The finding of ATP-binding cassette transporter of *T. atroviride* expressed the withstanding against plant-producing chemical factors and degradation of plant-secreting phenolic compounds (Ruocco et al. 2009; Chen et al. 2011). At the time of root colonization, the PRs in plant leaves and defense-associated enzymes such as β-1,3-glucanase, chitinase, and phenylalanine are continuously increasing (Shoresh et al. 2010; Tucci et al. 2011; Yoshioka et al. 2012; Saldajeno et al. 2014). Few kinds of cell wall-degrading enzymes, for example, cellulase and xylanase, could stimulate plant resistance. Fan et al. (2015) inferred a collection of elicitor-like substances possibly secreted from the ruined maize root cell wall by *Trichoderma*-based cellulase and xylanase, e.g., cellobiose may also be involved in stimulating the maize defense potential. The root may have turned out to be increasingly activated and thereby induce the ISR against foliar disease. It was inferred that the discharge of hydrolytic enzymes (e.g., pectinase, cellulase, and xylanase) during *Trichoderma* mycelia colonization on root tissue results in a limited degradation of plant root cell wall and releases some oligosaccharides, which subsequently work together with those hydrolytic enzymes to induce the plant defense response gene expression against pathogens.

6.5 Systemically Induced Resistance Mechanism

Recently, study on the mechanisms of biocontrol of plant diseases via *Trichoderma* has evolved to a new stage with the proposal of plant basic immunology theory with pathogen-associated molecular patterns/microbe-associated molecular patterns/damage-associated molecular pattern (PAMPs/MAMPs/DAMPs) as the core content and the application of omics technology. The three models are evidenced to support maize systemically induced resistance in our study. Systemic resistance generated in maize through seed or soil treatment with *Trichoderma* agents is significant for multiple plant diseases along development stage. SM1, SM2 small protein-like elicitors, and Epl-1 protein, a plant resistance elicitor induces plant defense responses. Djonović et al. (2007) mentioned that the overexpressed *T. atroviride* strain produced a higher amount of Sm1 cerato-platanin elicitor than wild type and increased the defense system of maize by inducing the JA-/ET-/ green leaf volatiles (GLV)-mediated signaling pathways. The soil treatment with *T. atroviride* spore suspension can induce maize silk defense gene expression against *F. verticillioides* infection through the silk channel. The transcriptional analysis suggested the set of five functional categories of genes especially signaling and hormone metabolism.

During the shift from vegetative to reproduction phase, these specific responses were observed and demonstrated that the salicylic acid and steroids are the key elements to control the immunity network of silk during ISR. The expression of hormone-subordinate qualities was in great concurrence with the ABA, JA, and SA levels (Agostini et al. 2018). The platelet-initiating factor acetylhydrolase (PAF-AH) is a member of phospholipase A2 (PLA2) protein super family. These proteins reduce ester bonds on Sn-2 site of glycerol phospholipids to discharge free unsaturated fats and lysophospholipids, which activate signal transduction pathways. Yu et al. (2014) mentioned the importance of *T. harzianum* platelet-initiating factor acetylhydrolase (PAF-AH) on the physiology, stress reaction, and mycoparasitism. The sterol and unsaturated fatty acids of *T. harzianum* generated by gas chromatography-mass spectrum showed the hydrolytic action of platelet-initiating factor acetylhydrolase and also demonstrated the role of platelet-initiating factor acetyl-hydrolase on the stress response related with an elevation of ergosterol and eicosanoid. Curiously, platelet-initiating factor acetylhydrolase-mediated thermotolerance in *T. harzianum* was related with the raise of PAF-AH transcript. It was speculated that PAF-AH may be related to proteins of cytoplasm to control the synthesis of ergosterol and thereby maintain the structure of the *Trichoderma* plasma membrane. The regulation of cytochrome c, HEX1, and Cu/Zn SOD gene was reduced in PAF-AH mutant (KO40), proposing that PAF-AH regulates the proteins required for stress resistance. The expression of SOD was reduced in the mutant KO40, while it was increased in the transformant containing overexpression of PAF-AH gene (OE1) when compared to the natural strain. Catalase activity of KO40 was increased compare to the wild type likely because of the effect of Cu/Zn SOD gene expression and increase in ROS by the diminished PAF-AH gene expression. Moreover, communications between *T. harzianum* and maize roots upregulated the expression of platelet-initiating factor acetylhydrolase due to the starvation of carbon and nitrogen sources in the rhizosphere. Likewise, the expression PAF-AH of *T. harzianum* was increased by the three-way interaction of *Trichoderma*, corn roots, and *C. lunata*. The genetic expression of PAF-AH was also regulating the expression of genes responsible for the hydrolysis of cellulose and chitin. But, mostly, the expression is associated with chitinase. It was evidenced by the lower expression in the PAF-AH knockout transformant compared to the natural strain. The association of *T. harzianum* on the maize roots increased the systemic resistance of plants challenged with *C. lunata* through the upregulation of the jasmonic acid pathway-associated genes. Besides, LC-MS analysis on the maize leaves also proved that jasmonic acid was increased by means of increase in jasmonic acid pathway-associated genes, proposing that platelet-initiating factor acetylhydrolase convince resistance against the leaf pathogen.

Trichoderma secrete a minimum of two cellobiohydrolases (CBHI and CBHII) and three endo-xylanases (XYNI–XYNIII). Regarding Thc6, it regulates the generation of cellulase and thereby induces the maize resistance. Thph gene induces hydrolysis of cellulose and helps protect against pathogens. The upregulation of MYC2 protein induces the plant JA pathway and thereby improved the plant defense. Plant peroxidase (POD) has been involved in the oxidation of hydrogen peroxide. The cellulase-like proteins Thph1 and Thph2 cellulases show vital action on the specificity and recognition to hold on the specific symbiotic fungi. The role of fungal is to digest the cellulose of plant cell walls; further, it was used to convert biomass into energy. But only a few studies have revealed the cellulose-based induction of plant defense mechanism. Thc6, Thph1, and Thph2 regulate network and corn resistance induction. Thph1 and Thph2 protein secretion was detected in the corn root epidermis colonized by *Trichoderma* with Western blot. Protein Thph1 and Thph2 treating corn root system can systematically induce corn defense reaction to *Curvularia* leaf spot. HPLC analysis found cellobiose released from epidermis tissues of corn root colonized by *Trichoderma*. qRT-PCR indicates that the defense-related genes, Opr7, and Erf1 expression were upregulated. Furthermore, the corn roots treated with purified protein also stimulate a stronger burst of active oxygen than wild strain. Quantitative proteomics analysis showed that the corn root system grown from the seeds treated with Thph1 knockout mutant would lead to the downregulation of MYC2 and ACO, the JA/ET pathway-related proteins in corn defense reaction, while the active oxygen scavenging-related protein POD expression also became downregulated, which further demonstrated that *Trichoderma* cellulase and cellobiose can be seen as the inducers of ISR signal from root to leaf through long-distance transduction system as elicitor, thus resulting in the defense reaction of corn leaves to *Curvularia* (Saravanakumar et al. 2016). For screening potential targets in maize root interactive with *T. harzianum* cellulase genes, the cDNA library and label-free proteomics were constructed from the maize roots treated with *T. harzianum* strains including knockout Thph1, Thph2, and wild-type and catalyst cellobiohydrolase 1 (CBH1). The protein expression was significantly varied between the treatments. Plants treated with the *T. harzianum* wild type (WT), Thph1, Thph2 mutants and enzyme CBH 1 showed highed protein expression compared to the control. The Y2H assay revealed that autophagocytosis-related protein (ZmATG3) and germin-like protein (ZmGLP) of the maize root interact with the Thph1 and Thph2, respectively. The cellular localization of ZmATG3 and ZmGLP in the maize root confirms the induction of induced systemic resistance (ISR) in maize (Saravanakumar et al. 2018a).

In order to clarify the MAMP model of the *Trichoderma*-based stimulation of maize systemic resistance against leaf spot pathogen, a class II hydrophobin gene Hyd1 was cloned from *T. harzianum*. The Hyd1 localized on the plasma membrane, and the Hyd1 secreted into the extracellular to promote *Trichoderma* colonization on maize roots, and it was also proved that the protein could be expressed on the cell membrane of the maize roots, which provided precondition that Hyd1 acted as MAMPs and interacted with maize roots to induce maize resistance against foliar spot pathogen. B73 seeds treated with the spore suspension of different Hyd1 transformants and grew in the greenhouse until at the four-leaf stage were inoculated with *C. lunata*. The results showed the maize leaf lesion area treated with Hyd1 mutant (KO16) was larger than wild type (WT). Notably, the lesion area of the Hyd1 overexpressed transformant (OE3 and OE5) was lowest. It suggested that

Hyd1 induce plant systemic resistance (Yu et al. 2019). Yeast two-hybrid systems were applied to screen the interaction protein in maize roots that were the response to or recognized by Hyd1. Maize root ubiquilin1-like protein that interacted with Hyd1 was screened by yeast two-hybrid method, and their direct interactions were verified by a variety of methods. The motif of the ubiquilin1-like protein that interacted with Hyd1 was identified in yeast cells, and it was also proved that the signal peptide sequence of the Hyd1 was essential to the interaction between Hyd1 and ubiquilin1-N. In addition, eight cysteine amino acids of the Hyd1 contributed in the above interaction. In order to further uncover that Hyd1 and ubiquilin1-like gene had a synergistic effect on increasing plant resistance, Hyd1 and ubiquilin1-like genes were co-expressed in *Arabidopsis thaliana* by a transgenic method, and it significantly promoted plant resistance to *Botrytis cinerea*, for which transgenic plant leaf lesion area was only 50% compared with the wild type. The Hyd1 and ubiquilin1-like genes were localized on the plasma membrane, and Hyd1 promoted the stability of the ubiquilin1-like protein in *Nicotiana benthamiana*. In the RNA-seq analysis of global gene expression in maize leaves at the 24 h time point after spraying *C. lunata*, the results showed that it was mainly associated with brassinolide (BR) pathway signaling transduction, in which some genes changed close to three times and partial genes were testified again by qPCR at 24 and 72 h. Hyd1 induced the plant resistance, which was probably regulated by BAK1 through its pathway signaling transduction and was preliminarily verified by the mutant BANK1-4. Furthermore, it indicated that the genes of the JA/ET pathway were also associated with Hyd1. In addition, some calmodulin (CaM) and WRKY genes also changed a bit. MAMP model of the *Trichoderma* systemic induced maize resistance against leaf spot pathogen was proposed as follows: *T. harzianum* secreted Hyd1 interact with receptor-like protein ubiquilin1-like in maize roots, and this ubiquilin1-like bind to the BR receptor BAK1; this protein complex could stimulate the JA, calmodulin (CAM) and WRKY gene expression to increase the plant resistance. It provided an important clue about the molecular interaction mechanism between *Trichoderma* MAMP hydrophobins and pseudo-response regulator (PRR)-like protein in the plant (Yu et al. 2014, 2019; Saravanakumar et al. 2018b).

Based on the microbial ecology, reprogrammed soil rhizosphere or endophytic microbiome also devotes to the systemic control of maize diseases by soil or seed treatment with *Trichoderma* agents, and it was inferred that those changed microbiome is also involved in the systemically induced resistance. Simulated inoculation of *Trichoderma asperellum* reduced the maize stalk rot and ear rot. High-throughput sequencing showed that *T. asperellum* considerably increased the endophytic microbial communities in the plant stem. In addition, the abundance of endophytic microbes reduced from the basal to apical stem of the *Trichoderma*-treated soil. It represents that *Trichoderma* encouraged the defense system in the basal part than the apical parts. Additionally, *T. asperellum* treatment reduced the accretion of deoxynivalenol (DON) and fumonisin B1 (FB1) in the stem and ear of maize.

6.6 Conclusion

Trichoderma has been widely applied to control maize diseases. However, the mechanism of systemically induced resistance against maize diseases by *Trichoderma* spp. is still unknown. Understanding of the plant defense mechanisms stimulated by *Trichoderma* along maize development from seedling to adult stage is still very limited, even though a set of evidences are obtained showing that seed or soil treatment with *Trichoderma* agent can trigger whole-plant immunity responses against soilborne and airborne diseases, for example, stalk rot, leaf spot, and ear rot, in seedling and adult stages. Advancement in exploration of maize immune responses elicited by *Trichoderma* displays the creation of mutualistic relations and synchronization among the elicitor or MAMPs and DAMPs. The model is consistent with the basic model of plant immunity mechanism, a zigzag model depicted as pathogen-associated molecular pattern (PAMP)/MAMPs by Jones and Dangl (2006). A group of MAMPs such as Sm1/EPI1, invertase (TvInv), hydrophobin1 (TasHyd1), polyketide, cellulases, exo-1,4-β-d-glucanase and endo-1,4-β-d-glucanase from *Trichoderma* species has been identified to be interact with the maize. Those elicitors systemically activate the pattern trigger immunity (PTI) in maize against maize diseases along plant development. Meanwhile, cellobiose based activation of PTI in maize evidenced that the *Trichoderma* stimulate DAMP based immunity *in maize plants*. Further, some specific interactions are identified between maize root and *Trichoderma*, such as interaction of *T. harzianum* cellulase Thph1 and Thph2 with maize root target protein ZmATG3 and ZmGLP and *T. harzianum* Hyd1 with maize root ubiquilin1-like protein; however, so far, it is still unsure whether those maize root targets are receptors because of the lack of strict confirmation test with gene editing in maize plant overexpressing or deleting those genes. In addition, the elicitation of defense mechanism and signaling pathways by the different metabolites of *Trichoderma* have to be studied. It is also required to trace the molecules and speculate the redundancy of MAMP recognition and defense activation.

Acknowledgment The review covers research achievements supported by the grant from the National Key Projects (2017YFD0200403; 2017YFD0200901), National Nature Science Foundation (31872015, 31672072), and Chinese Agriculture Research System CARS-02.

References

Agostini RB, Postigo A, Rius SP, Rech GE, Campos-Bermudez VA, Vargas WA (2018) Long-lasting primed state in maize plants: salicylic acid and steroid signaling pathways as key players in the early activation of immune responses in silks. Mol Plant Microbe Interact 32:95–106

Alexandru M, Lazăr D, Ene M, Şesan TE (2013) Influence of some *Trichoderma* species on photosynthesis intensity and pigments in tomatoes. Rom Biotech Lett 18:8499–8510

Bais HP, Weir TL, Perry LG, Gilroy S, Vivanco JM (2006) The role of root exudates in rhizosphere interactions with plants and other organisms. Annu Rev Plant Biol 57:233–266

Bisen K, Keswani C, Patel JS, Sarma BK, Singh HB (2016) *Trichoderma* spp.: efficient inducers of systemic resistance in plants. In: Microbial-mediated induced systemic resistance in plants. Springer, Singapore, pp 185–195

Bisht S, Kumar P, Srinivasanraghavan A, Purohit J (2013) In vitro management of Curvularia leaf spot of maize using botanicals, essential oils, and bio-control agents. Bioscan Suppl Med Plants 8:731–733

Błaszczyk L, Basińska-Barczak A, Ćwiek-Kupczyńska H, Gromadzka K, Popiel D, Stępień Ł (2017) Suppressive Effect of *Trichoderma* spp. on toxigenic *Fusarium* species. J Appl Microbiol 66:85–100

Chen J, Harman GE, Comis A, Cheng GW (2005) Proteins related to the biocontrol of *Pythium* damping-off in maize with *Trichoderma harzianum Rifai.* J Integr Plant Biol 47:988–997

Chen LL, Yang X, Raza W, Li J, Liu Y, Qiu M, Zhang F, Shen Q (2011) *Trichoderma harzianum* SQR-T037 rapidly degrades allelochemicals in rhizospheres of continuously cropped cucumbers. Appl Microbiol Biotechnol 89:1653–1663

Chen J, Dou K, Gao Y, Li Y (2014) Mechanism and application of *Trichoderma* spp in biological control of corn diseases. Mycosystema 33:1154–1167

Choudhary DK, Prakash A, Johri BN (2007) Induced systemic resistance (ISR) in plants: mechanism of action. Ind J Microbiol 47:289–297

Contreras-Cornejo H, Ortiz-Castro R, López-Bucio J (2013) Promotion of plant growth and the induction of systemic defence by *Trichoderma*: physiology, genetics, and gene expression. In: Mukherjee PK, Horwitz BA, Singh US, Mukherjee M, Schmoll M (eds) *Trichoderma*: biology and applications. CABI International, Wallingford, pp 175–196

Contreras-Cornejo HA, López-Bucio JS, Méndez-Bravo A, Macías-Rodríguez L, Ramos-Vega M, Guevara-García ÁA, López-Bucio J (2015) Mitogen-activated protein kinase 6 and ethylene and auxin signaling pathways are involved in Arabidopsis root-system architecture alterations by *Trichoderma* atroviride. Mol Plant Microbe Interact 28:701–710

Contreras-Cornejo HA, Del-Val E, Macías-Rodríguez L, Alarcón A, González-Esquivel CE, Larsen J (2018) *Trichoderma atroviride*, a maize root associated fungus, increases the parasitism rate of the fall armyworm *Spodoptera frugiperda* by its natural enemy Campoletis sonorensis. Soil Biol Biochem 122:196–202

Djonović S, Vargas WA, Kolomiets MV, Horndeski M, Wiest A, Kenerley CM (2007) A proteinaceous elicitor Sm1 from the beneficial fungus *Trichoderma virens* is required for induced systemic resistance in maize. Plant Physiol 145:875–889

El-Hasan A, Schoene J, Hoeglinger B, Walker F, Voegele RT (2018) Assessment of the antifungal activity of selected biocontrol agents and their secondary metabolites against *Fusarium graminearum.* Eur J Plant Pathol 150:91–103

Fan L, Fu K, Yu C, Li Y, Li Y, Chen J (2015) Thc6 protein, isolated from *Trichoderma harzianum*, can induce maize defense response against *Curvularia lunata*. J Basic Microbiol 55:591–600

Harman GE (2011) Multifunctional fungal symbionts: new tools to enhance plant growth and productivity. New Phytol 189:647–649

Harman GE, Petzoldt R, Comis A, Chen J (2004) Interactions between *Trichoderma* harzianum strain T22 and maize inbred line Mo17 and effects of these interactions on diseases caused by *Pythium ultimum* and Colletotrichum graminicola. Phytopathology 94:147–153

He A, Liu J, Wang XH, Zhang Q, Song W, Chen J (2018) Soil application of *Trichoderma* asperellum GDFS1009 granules promotes growth and resistance to *Fusarium graminearum* in maize. J Integr Agric 17:60345–60347

Hermosa R, Viterbo A, Chet I, Monte E (2012) Plant-beneficial effects of *Trichoderma* and of its genes. Microbiology 158:17–25

Hua L-F, Zhi Y-H, Tao W-Y, Xiong-Mian P (2006) Research progress for maize Curvularia leaf spot disease. J Maize Sci 12:97–101

Huang J, Zheng L, Hsiang T (2004) First report of leaf spot caused by *Curvularia verruculosa* on *Cynodon* sp. In Hubei, China. Plant Pathol 54(2):253

Huang CJ, Yang KH, Liu YH, Lin YJ, Chen CY (2010) Suppression of southern corn leaf blight by a plant growth-promoting rhizobacterium *Bacillus cereus* C1L. Ann Appl Biol 157:45–53

Jirak-Peterson JC, Esker PD (2011) Tillage, crop rotation, and hybrid effects on residue and corn anthracnose occurrence in Wisconsin. Plant Dis 95:601–610

Jones JDG, Dangl JL (2006) The plant immune system. Nature 144:323–329

Kinkema M, Fan W, Dong X (2000) Nuclear localization of NPR1 is required for activation of PR gene expression. Plant Cell 12:2339–2350

Kubicek CP, Herrera-Estrella A, Seidl-Seiboth V, Martinez DA, Druzhinina IS, Thon M, Zeilinger S, Casas-Flores S, Horwitz BA, Mukherjee PK, Mukherjee M (2011) Comparative genome sequence analysis underscores mycoparasitism as the ancestral life style of *Trichoderma*. Genome Biol 12:40

Lamdan NL, Shalaby S, Ziv T, Kenerley CM, Horwitz BA (2015) Secretome of *Trichoderma* interacting with maize roots: role in induced systemic resistance. Mol Cell Proteomics 14:1054–1063

Martinez-Medina A, Pozo MJ, Cammue BP, Vos CM (2016) Belowground defence strategies in plants: the plant–*Trichoderma* dialogue below ground defence strategies in plants. Springer, New York, NY, pp 301–327

Martínez-Medina A, Fernandez I, Lok GB, Pozo MJ, Pieterse CM, Van Wees S (2017) Shifting from priming of salicylic acid-to jasmonic acid-regulated defences by *Trichoderma* protects tomato against the root knot nematode *Meloidogyne incognita*. New Phytol 213:1363–1377

Morán-Diez E, Hermosa R, Ambrosino P, Cardoza RE, Gutiérrez S, Lorito M, Monte E (2009) The ThPG1 endopolygalacturonase is required for the *Trichoderma* harzianum–plant beneficial interaction. Mol Plant Microbe Interact 22:1021–1031

Müller A, Faubert P, Hagen M, zu Castell W, Polle A, Schnitzler J-P, Rosenkranz M (2013) Volatile profiles of fungi–chemotyping of species and ecological functions. Fungal Genet Biol 54:25–33

Patiño B, Vázquez C, Manning JM, Roncero MIG, Córdoba-Cañero D, Di Pietro A, Martínez-del-Pozo Á (2018) Characterization of a novel cysteine-rich antifungal protein from *Fusarium graminearum* with activity against maize fungal pathogens. Int J Food Microbiol 283:45–51

Rawat L, Singh Y, Shukla N, Kumar J (2013) Salinity tolerant *Trichoderma harzianum* reinforces NaCl tolerance and reduces population dynamics of *Fusarium oxysporum* f. sp ciceri in chickpea (Cicer arietinum L) under salt stress conditions. Archiv Phytopathol Plant Protect 46:1442–1467

Ruocco M, Lanzuise S, Vinale F, Marra R, Turrà T, Woo SL, Lorito M (2009) Identification of a new biocontrol gene in *Trichoderma atroviride*: the role of an ABC transporter membrane pump in the interaction with different plant-pathogenic fungi. Mol Plant Microbe Interact 22:291–301

Saba H, Vibhash D, Manisha M, Prashant K, Farhan H, Tauseef A (2012) *Trichoderma*–a promising plant growth stimulator and biocontrol agent. Mycosphere 3:524–531

Saldajeno MGB, Naznin HA, Elsharkawy MM, Shimizu M, Hyakumachi M (2014) Enhanced resistance of plants to disease using *Trichoderma* spp. In: Biotechnology and biology of *Trichoderma*. Elsevier, Amsterdam, pp 477–493

Samolski I, Rincón AM, Pinzón LM, Viterbo A, Monte E (2012) The qid74 gene from *Trichoderma* harzianum has a role in root architecture and plant biofertilization. Microbiology 158:129–138

Saravanakumar K, Fan L, Fu K, Yu C, Wang M, Xia H, Sun J, Li Y, Chen J (2016) Cellulase from *Trichoderma* harzianum interacts with roots and triggers induced systemic resistance to foliar disease in maize. Sci Rep 6:35543

Saravanakumar K, Li Y, Yu C, Wang Q-q, Wang M, Sun J, Gao J-x, Chen J (2017) Effect of *Trichoderma harzianum* on maize rhizosphere microbiome and biocontrol of *Fusarium* Stalk rot. Sci Rep 7:1771

Saravanakumar K, Dou K, Lu Z, Wang X, Li Y, Chen J (2018a) Enhanced biocontrol activity of cellulase from *Trichoderma harzianum* against *Fusarium graminearum* through activation of defense-related genes in maize. Physiol Mol Plant Pathol 103:130–136

Saravanakumar K, Wang S, Dou K, Lu Z, Chen J (2018b) Yeast two-hybrid and label-free proteomics based screening of maize root receptor to cellulase of *Trichoderma harzianum*. Physiol Mol Plant Pathol 104:86–94

Schuster A, Schmoll M (2010) Biology and biotechnology of *Trichoderma*. Appl Microbiol Biotechnol 87:787–799

Shan L, Hafiz AH, Jun Z, Dandan Z, Daniel PJ, Xiaofeng D, Wei G (2019) A loop-mediated isothermal amplification (LAMP) assay for the rapid detection of toxigenic *Fusarium temperatum* in maize stalks and kernels. Int J Food Microbiol 291:72–78

Shoresh M, Harman GE (2008) Genome-wide identification, expression and chromosomal location of the genes encoding chitinolytic enzymes in Zea mays. Mol Gen Genet 280:173

Shoresh M, Harman GE, Mastouri F (2010) Induced systemic resistance and plant responses to fungal biocontrol agents. Annu Rev Phytopathol 48:21–43

Sun HQ, He ZD, Gao YF, Liu R (2009) The mechanisms of *Trichoderma* spp in inhibition to maize foliar pathogens. Jiangsu Agric Sci 3:105–107. (in Chinese)

Tian Y, Tan Y, Liu N, Yan Z, Liao Y, Chen J, De Saeger S, Yang H, Zhang Q, Wu A (2016) Detoxification of deoxynivalenol via glycosylation represents novel insights on antagonistic activities of *Trichoderma* when confronted with *Fusarium graminearum*. Toxins 8:335

Troian RF, Steindorff AS, Ramada MHS, Arruda W, Ulhoa CJ (2014) Mycoparasitism studies of *Trichoderma harzianum* against *Sclerotinia sclerotiorum*: evaluation of antagonism and expression of cell wall-degrading enzymes genes. Biotechnol Lett 36:2095–2101

Tucci M, Ruocco M, De Masi L, De Palma M, Lorito M (2011) The beneficial effect of Trichoderma spp. on tomato is modulated by the plant genotype. Mol Plant Pathol 12:341–354

Vallad GE, Goodman RM (2004) Systemic acquired resistance and induced systemic resistance in conventional agriculture. Crop Sci 44:1920–1934

Vargas WA, Mandawe JC, Kenerley CM (2009) Plant-derived sucrose is a key element in the symbiotic association between *Trichoderma* virens and maize plants. Plant Physiol 151:792–808

Velázquez-Robledo R, Contreras-Cornejo H, Macías-Rodríguez L, Hernández-Morales A, Aguirre J, Casas-Flores S, López-Bucio J, Herrera-Estrella A (2011) Role of the 4-phosphopantetheinyl transferase of *Trichoderma virens* in secondary metabolism and induction of plant defense responses. Mol Plant Microbe Interact 24:1459–1471

Viterbo ADA, Chet I (2006) TasHyd1, a new hydrophobin gene from the biocontrol agent *Trichoderma asperellum*, is involved in plant root colonization. Physiol Mol Plant Pathol 7:249–258

Woo S, Scala F, Ruocco M, Lorito M (2006) The molecular biology of the interactions between *Trichoderma* spp., phytopathogenic fungi, and plants. Phytopathology 96:181–185

Xing F, Liu X, Wang L, Selvaraj JN, Jin N, Wang Y, Zhao Y, Liu Y (2017) Distribution and variation of fungi and major mycotoxins in pre-and post-nature drying maize in North China Plain. Food Control 80:244–251

Yoshioka Y, Ichikawa H, Naznin HA, Kogure A, Hyakumachi M (2012) Systemic resistance induced in Arabidopsis thaliana by *Trichoderma asperellum* SKT-1, a microbial pesticide of seed-borne diseases of rice. Pest Manag Sci 68:60–66

Yu C, Fan L, Wu Q, Fu K, Gao S, Wang M, Gao J, Li Y, Chen J (2014) Biological role of *Trichoderma harzianum*-derived Platelet-Activating Factor Acetylhydrolase (PAF-AH) on stress response and antagonism. PLoS One 9:e100367

Yu C, Dou K, Wang S, Wu Q, Ni M, Zhang T, Lu Z, Tang J, Chen J (2019) Elicitor hydrophobin Hyd1 interacts with Ubiquilin1-like to induce maize systemic resistance. J Integr Plant Biol. https://doi.org/10.1111/jipb.12796

Zhang Y, Tessaro MJ, Lassner M, Li X (2003) Knockout analysis of Arabidopsis transcription factors TGA2, TGA5, and TGA6 reveals their redundant and essential roles in systemic acquired resistance. Plant Cell 15:2647–2653

Chapter 7
Induced Immunity Developed by *Trichoderma* Species in Plants

B. N. Chakraborty, U. Chakraborty, and K. Sunar

Abstract One of the most interesting and important microorganisms in nature is *Trichoderma*, which, from being a mycoparasitic biocontrol agent (BCA), has now emerged as one with multiple traits such as antagonism to pathogen, competition with pathogens for nutrients, induction of systemic resistance in the host, overall plant growth promotion and also alleviation of abiotic stresses. Besides, interestingly, though they were reported earlier as soil and root colonizers only, it is now evident that several species of *Trichoderma* are endophytic. Interactions between plant and *Trichoderma* involve recognition, penetration, attachment and colonization and, finally, nutrient transfer from the root. Appressoria-like structures have been reported to be formed by *Trichoderma* which helps in root penetration, and two proteins TasHyd1 and *qid*74 were found to mediate the attachment of appressoria to the roots. *Trichoderma* produces a large number of secondary metabolites such as xylanases, cellulases, polygalacturonases, cerato-platanins, swollenins, peptaibols, 6-pentyl-α-pyrones and trichothecenes. These secondary metabolites help *Trichoderma* in various capacities, such as cell wall-degrading enzymes, elicitors and antimicrobial compounds. *Trichoderma* can trigger plant resistance towards pathogen attack by inducing plant immune response. *Trichoderma viride*, *T. harzianum*, *T. virens*, *T. aureoviride* and *T. asperellum* are being used as microbial inducers of plant immunity. An immunity-inducing protein (Sm1/Ep11) of the cerato-platanin family and elicitor produced by *Trichoderma* increase the expression of genes involved in defence, which in turn induces immunity. *Trichoderma* spp. release microbe-associated molecular patterns (MAMPs) required for molecular recognition leading to a signal cascade within the plant involving signalling molecules such as salicylic acid (SA), jasmonate (JA) and ethylene (ET). The defence responses triggered by *Trichoderma*, both locally and systemically,

B. N. Chakraborty (✉)
Department of Biological Sciences, Aliah University, Kolkata, India

U. Chakraborty
Department of Botany, University of North Bengal, Siliguri, India

K. Sunar
Department of Botany, Balurghat Mahila Mahavidyalaya, Balurghat, India

A. K. Sharma, P. Sharma (eds.), *Trichoderma*, Rhizosphere Biology,
https://doi.org/10.1007/978-981-15-3321-1_7

include enhanced accumulation of PR proteins, phytoalexins and terpenoids and enhanced activities of several enzymes such as phenylalanine ammonia lyase, peroxidase, polyphenol oxidase and lipoxygenase. Activation of such defence responses finally leads to crop protection through induced resistance.

Keywords Biocontrol · Secondary metabolites · Plant immunity · Induced resistance · Defence enzymes

7.1 Introduction

Trichoderma can be considered as nature's 'boon' to human beings because of its multiple beneficial traits, which if utilized suitably can lead to sustainable eco-friendly agriculture, replacing or supplementing the use of chemicals (Chakraborty et al. 2011). *Trichoderma* has been exploited for biocontrol against phytopathogens since they have properties such as mycoparasitism and antagonism (Chowdhury et al. 2017). Almost all the different species of *Trichoderma* are fast-growing soil-inhabiting fungi commonly occurring in different types of soil and may constitute up to as much as 3% of the total population. Many of them form endophytic associations, interacting with other rhizosphere microorganisms which in turn influence growth of plant, yield and disease development. It is now well documented that colonization of root by selected *Trichoderma* isolates increases the resistance of plants to different pathogens (Harman et al. 2004). *Trichoderma* acts through different mechanisms, which may have a direct effect on pathogens (Vinale et al. 2008a, b), and through indirect mechanisms such as improving nutritional status of plants (Shoresh and Harman 2008a, b) or partially activating plant immunity (Shoresh et al. 2010). Some of the efficient strains of *Trichoderma* can also induce physiological changes in the plants and activate defence mechanisms but do not cause any damage to the plant (Alfano et al. 2007; Chacón et al. 2007; Brotman et al. 2012; Mathys et al. 2012). The changes in the plant may be a regulatory strategy to stop the colonization by the microorganisms (Zamioudis and Pieterse 2012).

The first trait of *Trichoderma* which was found to be linked to its biocontrol potential was mycoparasitism, by which it could directly antagonize the pathogen. Analysis of genome sequence of *Trichoderma atroviride* and *T. virens* has made it possible to have an understanding of the role of mycoparasitism as a lifestyle in a common ancestor of *Trichoderma* (Kubicek et al. 2011). The evolution of positive interactions of plants with *Trichoderma* was most probably facilitated by the availability of root-derived nutrients and easily available prey which led to the rhizosphere colonization by ancestors of *Trichoderma* (Druzhinina et al. 2011). Some of the proposed mechanisms of action attributed to *Trichoderma* include production of toxic metabolites, competition and induction of systemic resistance (Harman et al. 2004). Root colonization by *Trichoderma* is enhanced by root secretions such as polysaccharides, monosaccharides and disaccharides and also the mucigel layer. This enhanced endosymbiosis in turn activates plant defences (Vargas et al. 2009).

Recent studies have revealed the molecular basis of interaction between *Trichoderma* and plant. Molecular recognition of the two partners occurs through a signalling network which involves plant hormones jasmonate, salicylic acid and ethylene. Two signalling pathways have been reported involving these hormones. The first one, which is induced by elicitors and/or pathogens, leading to systemic acquired resistance (SAR), is mediated by salicylic acid, whereas the second pathway, mediated by jasmonate and ethylene, is induced by beneficial microorganisms and leads to induced systemic resistance (ISR). In ISR, priming with the beneficial microorganism(s), there is no activation of defence responses directly, but the plant is kept in a state of readiness for attack by pathogens (Van Wees et al. 2008).

Though the recognition process of plant and *Trichoderma* is not clearly understood, some of the plant immunity-inducing elicitors have been reported in *Trichoderma*. These include xylanase (which induces ethylene), the elicitor Sm1 (proteinaceous but non-enzymatic) and peptaibols (Hanson and Howell 2004; Djonovic et al. 2006, 2007; Viterbo et al. 2007). Some researchers (Ron and Avni 2004; Petutschnig et al. 2010) have been successful in characterizing a few receptors which recognize the MAMPs associated with *Trichoderma*. Among the MAMPs associated with *Trichoderma*, the xylanase with the ability to induce ethylene (ET) (Xyn2/Eix) was the first to be recognized as such, and this was reported to have the ability of inducing defence responses in certain tobacco and tomato cultivars (Rotblat et al. 2002). It was further revealed that *Trichoderma* could induce a delayed expression of the defence genes of the salicylic acid and jasmonate/ethylene pathways both locally and systemically (Salas-Marina et al. 2011). Besides, it also triggers a long-lasting upregulation of SA-dependent gene in plants through the JA signal transduction pathway leading to ISR (Tucci et al. 2011; Yoshioka et al. 2012). Thus, involvement of defence hormones such as salicylate, jasmonate and camalexin has been shown to be important factors in increasing immunity of plants.

During the endophytic colonization of the roots by *Trichoderma*, without any outward symptom, initially its growth is checked by the plant's enhanced defence mechanisms, which include reinforcement of cell walls, reactive oxygen species (ROS) generation and increased accumulation of antimicrobial compounds (Chacón et al. 2007; Contreras-Cornejo et al. 2011; Salas-Marina et al. 2011). However, after the penetration becomes established only in some of the outer layers of the root cortex, the enhanced defence machinery gets back to normal levels (Yedidia et al. 2003; Masunaka et al. 2011).

In the current global scenario, several abiotic stresses affect crop production detrimentally, and many of these also act as predisposing factors for disease augmentation. More recent studies on *Trichoderma*-plant interaction have also shown that it can induce tolerance against abiotic stresses and improve growth (Yasmeen and Siddiqui 2017; Hidangmayum and Dwivedi 2018) in several vegetable and other crops (Bae et al. 2009; Brotman et al. 2013; Contreras-Cornejo et al. 2016; Zeilinger et al. 2016; Kashyap et al. 2017). Tolerance to abiotic stresses and better growth in plants have been observed when *Trichoderma* colonization occurs. It has been suggested that this may be due to increased production of metabolites such as antioxidants including phenols, phytoalexins and hormones, as well as enhanced activities of enzymes (Brotman et al. 2013; López-Bucio et al. 2015).

7.2 Root Colonization and Interactions

Growth of *Trichoderma* in the rhizosphere is encouraged by the root secretions such as polysaccharides, monosaccharides and disaccharides and also the mucigel layer. Besides, the sucrose from plants is also important in root colonization by *Trichoderma*, as well as for defence mechanism coordination and increase in photosynthetic rate (Vargas et al. 2009). Transporters such as di/tripeptide ones and permeases are necessary for the acquisition of root exudates by *Trichoderma* (Vizcaíno et al. 2006; Vargas et al. 2009). Plant and *Trichoderma* interactions involve recognition, penetration, attachment and colonization, by nutrient transfer from the root (Mukherjee et al. 2012). For root penetration, hyphae of *Trichoderma* form appressoria-like structures, and it has been shown that two proteins TasHyd1 (from *T. asperelloides*) and *qid*74 (from *T. harzianum*) mediate the attachment of appressoria to the roots.

Hydrophobins, small hydrophobic proteins present in the outer cell wall layer coating the fungal cell surface of *Trichoderma* and expansin-like proteins involved in cell wall development, mediate adherence to the root surface. A cysteine-rich cell wall protein and another protein swollenin TasHyd1 which has a binding domain for cellulose are encoded by the gene *qid*74. Cellulose is recognized by these proteins which in turn modify plant cell wall architecture by secreting cellulolytic and proteolytic enzymes for root colonization facilitation (Viterbo and Chet 2006; Brotman et al. 2008; Samolski et al. 2012). Further, *T. harzianum* also produces enzymes like endopolygalacturonase which degrade plant cell walls and thus facilitate root colonization (Morán-Diez et al. 2009).

Reactions of the host to *Trichoderma* penetration occur at sites far away from the actual site of penetration. The growth of *Trichoderma* into the epidermal and cortical intercellular spaces is restricted by callose-enriched wall appositions and also prevents the fungal entry into the vascular tissues. Since the hosts accumulate antimicrobial compounds, the efficiency of different *Trichoderma* strains in root colonization is dependent to a great degree on how much they can tolerate the toxic compounds. *Trichoderma virens* has been reported to produce growth regulators such as auxins which can enhance root growth and thus facilitate colonization of the root (Contreras-Cornejo et al. 2009).

Initial responses of plants to microbes, whether beneficial or pathogenic, are to consider them as pathogens, leading to the triggering of immune responses through MAMP perception by receptors (Jones and Dangl 2006; Boller and Felix 2009). However, *Trichoderma* reprograms their transcriptome and proteome and manipulates the plant immune response (Marra et al. 2006; Alfano et al. 2007; Segarra et al. 2007). The general mechanism involved in colonization of root and interaction is the exchanges of molecular signals and deposition of fungal elicitors in root apoplast (Shoresh and Harman 2008a, b; Contreras-Cornejo et al. 2016).

7.3 Metabolites from *Trichoderma* Inducing Plant Responses

During the course of evolution, different species of *Trichoderma* have developed mechanisms through which they establish contact with other microorganisms in the soil as well as with the host. It is the opportunistic capacity of different species of *Trichoderma* such as *T. hamatum* which makes it possible for them to live in different environmental conditions, within the host as endophytes and as soil microorganisms in the rhizosphere. These capacities for opportunism have enabled such species to live both as common components of the soil and rhizosphere and also as endophytes. To achieve this, they have a well-developed signalling network involving several metabolites released by them, enabling them to enter the host or break down cell walls of potential pathogens. Some of the commonly occurring metabolites released by *Trichoderma* sp. include xylanases, cellulases, polygalacturonases, cerato-platanins, swollenins, peptaibols, 6-pentyl-α-pyrones and trichothecenes (Hermosa et al. 2013). Some of the important metabolites/enzymes are enumerated below.

Xylanase Eix/Xyn2 has been reported to be a MAMP, which, over and above its ability to degrade xylan, is also a potent elicitor of hypersensitive response (HR) as well as defence mechanisms in several hosts such as tomato and tobacco. The expression of 1-aminocyclopropane-1-carboxylic acid (ACC) synthase is induced by this xylanase leading to the biosynthesis of ethylene and then ISR (Matarasso et al. 2005). In tobacco, on the other hand, *T. reesei* produces another kind of xylanase which induces death in the cells of leaves, followed by typical enhanced defence including an oxidative burst and activation of a specific cellular signal transduction cascade.

Trichoderma spp. can produce cellulases required for degrading the cell walls, and it has also been reported that when *Trichoderma* is applied on cut petioles of tobacco and corn, they raised the levels of endogenous JA, followed by the transient emission of ethylene (Piel et al. 1997). Infiltration of cellulase from *T. longibrachiatum* to cotyledons of melon activated early defence strategies and also enhanced oxidative burst (Martinez et al. 2001). These were found to be associated with the signalling pathways of ethylene and salicylic acid, which are responsible for enhancement of chitinase and peroxidase activities.

A plant cell wall-degrading hydrolytic enzyme, polygalacturonase which produces oligogalacturonides by breaking down plant pectin (De Lorenzo et al. 2011), has been found to be essential for successful root colonization by *T. harzianum* (Morán-Diez et al. 2009). In tomato plant following inoculation with *Rhizoctonia solani* and challenge inoculation with *T. harzianum*, *Thpg1* gene was detected through transcriptome and proteome analyses which has also been linked to cell wall modification, root architecture and activation of systemic acquired resistance. An inhibitor from *Arabidopsis* had no effect on this enzyme. However, higher concentration of bean polygalacturonase inhibitor protein was found to be affected.

During root colonization by *Trichoderma* sp., certain small secreted cysteine-rich proteins (SSCPs) such as Sm1 and Epl1 from *T. virens* and *T. atroviride*, respectively, have been found to accumulate in the hyphae. These are hydrophobin-like proteins or cerato-platanins and have been reported in cotton and maize to act as elicitors of jasmonate-induced ISR, but not SAR (Hermosa et al. 2012). The abundance of SSCPs in the genomes of *T. virens* and *T. atroviride* has been related to their rhizosphere competence (Djonovic et al. 2007; Kubicek et al. 2011). The wide diversity among the three species of sequenced *Trichoderma* has been correlated with rapid evolution of SSCPs. Swollenin protein in *T. reesei*, encoded by *Swo1*, has two functional domains at the N-terminal and C-terminals, the carbohydrate-binding domain with a cellulose-binding function and expansin-like domain, respectively. The expansin-like domain was reported to have sequence similarity to plant expansins. Though the carbohydrate-binding domain of swollenin probably has no role in ISR, experiments in different crops have revealed that it has MAMP activity (Brotman et al. 2008).

A secondary metabolite, peptaibol, composed of 5–20 amino acids long peptide having antibiotic activities against phytopathogens has been found to be produced not only by *Trichoderma* which induces the biosynthesis of VOCs mainly through jasmonate signalling pathway but also to some extent by upregulating salicylic acid biosynthesis in lima bean (Hermosa et al. 2012). Production of such JA-dependent VOCs is reduced by the increase in production of SA which modifies plant defence strategies against pathogens as well as controls above-ground herbivorous insects by attracting their natural enemies (parasitoids and predators) as well as pollinators (Pineda et al. 2010).

Many peptaibols (approximately 300) with antimicrobial activities have been found to be produced by different species of *Trichoderma*. It has been reported that *T. longibrachiatum* produces 11 such peptaibols; *T. brevicompactum* has been shown to produce a peptaibol polymer containing 20 peptaibol residues (Nielsen et al. 2005; Ruiz et al. 2007). In tobacco plants, it has been shown that a group of 20-mer peptaibols, trichokonins, produced by *T. pseudokoningii*, can induce production of reactive oxygen species and phenol accumulation systemically as well as locally at the site of application. They also have the ability to induce systemic resistance against tobacco mosaic virus, probably through upregulation of genes of defence signalling pathway (Luo et al. 2010). In another study using cucumber, it was shown that *T. virens*, which produces an 18-mer peptaibol, elicits systemic resistance responses against *Pseudomonas syringae*, a leaf pathogen.

It has been reported that 18-mer peptaibols of *T. virens* can elicit systemic defence responses in cucumber against the leaf pathogen *P. syringae*. The defence pathways (involving JA and SA) are induced only after 48 h. It has further been shown that fungal elicitor or peptaibol application directly elicits a higher expression of phenylalanine ammonia lyase (marker of SA gene) and hydroperoxide lyase (marker of JA gene), respectively (Viterbo et al. 2007).

A typical yellow pigmentation and coconut-like aroma found in certain strains of *Trichoderma* have been shown to be due to the production of a volatile 6-pentyl-2*H*-pyran-2-one (6PP), which also inhibits the growth of plant pathogens such as *Fusarium oxysporum*. Besides, this compound, at low concentrations, can improve plant growth both under- and above ground, as well as enhance seed germination and seed growth. These activities suggest this compound acts like an auxin compound or has the ability to induce auxin production (Rubio et al. 2009). The ability of *Trichoderma* to induce multiple systemic defence pathways through the production of such metabolites is proved by the observations that in canola, 6PP and harzianopyridone at 1 mg/L induce activation of the *PR-1* gene, which suggests the activation of SAR response dependent on SA, whereas 6PP, harzianopyridone or azaphilone can also induce a defence dependent on chitinase *PR-3* (Vinale et al. 2008a, b).

Trichoderma species of the clade *Brevicompactum*, such as *T. brevicompactum*, produce trichothecenes which are sesquiterpenes (C15 terpene compounds). This fungus produces trichodermin, which is an antifungal compound, and is used as a biocontrol agent (Tijerino et al. 2011). Harzianum A, produced by *T. arundinaceum*, is also a trichothecene with antifungal activity and has been shown to be active against *B. cinerea* and *R. solani* in tomato plants probably by eliciting systemic priming responses and defence (Malmierca et al. 2012, 2013).

Besides the ones listed above, a large number of other metabolites with varied functions from different species are now coming to light. A study by Wu et al. (2017), based on metabolomics, proteomics, genomics, transcriptomics and physiological and biochemical analyses, established that *T. asperellum* induces resistance through different mechanisms which included the biocontrol factors involved in antibiosis, mycoparasitism and induced systemic resistance. A novel strain of *Trichoderma asperellum*, GDFS1009, identified by the authors had a high growth and sporulation activity and could inhibit pathogens causing plant diseases such as corn stalk rot and *Fusarium* wilt of cucumber. This *T. asperellum* strain produced cell wall-degrading enzymes such as chitinase, protease and glucanase required for mycoparasitism.

This strain was reported to produce chitinase, glucanase and protease, which can degrade the cell walls of fungi and participate in the mycoparasitism. They also produced xylanase which was shown to induce plant resistance and enhance plant immunity. In addition, it produced a number of primary metabolites, antimicrobial compound precursors and different secondary metabolites with antimicrobial activity, such as alkanes and polyketides.

In a recent review by Li et al. (2019), 390 non-volatile compounds were listed which were produced by 20 known *Trichoderma* species and several spp. which included peptaibols, steroids, amides, terpenes, lactones, peptides, diketopiperazines, tetronic acid derivatives, polyketides, pyridines, cyclopentenones and derivatives of pyranone. The biological activities of these compounds were wide-ranging and included antiviral, antibacterial, antifungal, nematicidal, antitumour and anti-inflammatory. The activities further included cytotoxicity, plant growth enhancement, induction of systemic resistance, production of siderophores, inhibition of hyperplasia, inhibition of certain enzymes and DPPH radical scavenging. Therefore, it is clear that, in spite of the fact that a large body of information is available on different *Trichoderma* spp., it is still not yet complete. There are probably many more such metabolites in *Trichoderma*, which yet remain to be identified, but it is quite certain that, using more and more techniques, these will be identified in the near future.

7.4 Induction of Immunity and Resistance in Host

Plants have their own resistance mechanism which can be activated and used as a strategy for control of diseases, and this is known as induced resistance. During expression of immunity in plants, first of all, the microbial molecules are recognized by the plants, which are then transduced through different pathways leading to responses, which in turn results in activation of defence reactions whereby the microbes are inhibited. This immunity can be triggered in the plants by elicitors both chemical or biological which keeps the plant primed against future pathogen attack by pathogens and pests (Chakraborty 2012). This primed state is characterized by enhanced defence reactions which are expressed during pathogen attack. Plants recognize elicitors or pathogen-associated molecular patterns (PAMPs) which can be lipopolysaccharides, peptidoglycan, B-glucans, chitin or flagellin, and this leads to the typical plant defence responses such as strengthening of cell walls, enhanced accumulation of PR proteins and phytoalexin biosynthesis (Chakraborty 2005a). Plants also have other receptors such as pathogen recognition receptors (PRRs) and resistance proteins involved in pathogen detection. The R proteins are mostly NB-LRR (nucleotide binding leucine rich repeat) proteins which interact with pathogen effector proteins.

Induced systemic resistance (ISR) is expressed systemically in plant parts which confers increase level of induced immunity against a broad spectrum of attackers (Walters et al. 2013). The nature of interaction and the type of proteins involved often determine the particular defence pathway to be activated. Plant hormones regulate a network of interconnected signalling pathways in order to develop induced resistance.

Although many phytohormones are now well known to mediate induced defence signalling, SA and JA are considered to play major roles in regulating signalling during plant defence. They activate specific pathways involved in signalling, which can act individually, synergistically or antagonistically, depending on the pathogen involved. The biosynthesis of SA, a small phenolic compound, occurs via the shikimic acid pathway, which is dependent on metabolites derived from carbohydrate pathways such as pentose phosphate pathway and glycolysis (Chakraborty and Chakraborty 2008). Because of its exciting potential, research has focused on systemic acquired resistance to a great extent based on all advanced techniques. A key to using this technology for crop protection is that SAR can be activated by spraying with SA or its synthetic functional analogues as 1,2,3-benzothiadiazole-7-carbothioic acid S-methyl ester (BTH) or bioinoculants even in the absence of pathogen. Induction of PR proteins in plants is one of the main components of SAR (Chakraborty and Sharma 2008). Many PR proteins have antipathogen activity, including chitinases, glucanases and antifungal protein osmotin, which may degrade fungal cell walls and lysozymal activity, suggestive of antibacterial functions and nuclease activity, which is effective against viruses. Microarray studies show that when SAR is developed, hundreds of genes show increased or decreased expression. These altered gene expressions have significant effects such as cell wall strengthening and ROS production (Chakraborty 2005b). JA is also involved in systemic immunity against microbes. JA along with ethylene is the dominant regulator of ISR. ISR, which is different from SAR, is a form of broad-spectrum resistance induced in the above-ground tissues of plants in response to colonization by microorganisms which are non-pathogenic but can promote growth (Chakraborty et al. 2014). Similar to ISR, SAR is also induced in the systemic (noninfected) parts of the plant and is broad spectrum in nature. Hence, SA and JA are important regulators of both local and systemic induced defence responses. However, SA-derived and JA-derived signalling often overlap with signalling derived from other plant hormones such as gibberellins, ethylene, auxin and abscisic acid (ABA). The ability of microbes with multiple beneficial traits to promote growth and induce systemic resistance paves the way for their use in agriculture while reducing the use of harmful chemicals (Chakraborty 2016).

Microorganisms can trigger plant resistance towards pathogen attack by inducing plant immune response. *Trichoderma* spp. are beneficial microorganisms that have been used ever since for biocontrol of plant diseases; hence, they have been subjects of extensive research and have led to development of bioformulations and commercial products, and these are now marketed globally (Harman et al. 2004; Singh 2006; Perazzolli et al. 2011; Singh et al. 2012). *Trichoderma* strains and their products with ability to induce ISR have been registered for commercialization. Effector-triggered immunity is induced in plants by elicitors such as AVR proteins or other resistance gene products (Zhang and Zhou 2010). *T. viride*, *T. harzianum*, *T. virens*, *T. aureoviride* and *T. asperellum* are being used as microbial inducers of plant immunity. Sm1/Ep11, an immunity-inducing protein of the cerato-platanin family, is an elicitor produced by *Trichoderma* (Djonovic et al. 2006; Seidl et al. 2006), and this protein increases the expression of genes involved in defence, which in turn induces immunity (Table 7.1) (Djonovic et al. 2007). Sharma et al. (2011) carried out a detailed review of biocontrol genes of *Trichoderma* species. Gulijimila et al. (2012) cloned full-length cDNA and genomic DNA sequences of hyb2, a small hydrophobic protein obtained from *T. asperellum* T4. The chitinase activity of hyb2 has further been enhanced through gene recombination technology (Cong et al. 2012).

Table 7.1 Defence-related genes of *Trichoderma* spp.

Trichoderma species	Genes	Biotic/abiotic stress	Product/function	Reference
T. harzianum	Sm1	Biotic and abiotic stress	Elicitor for triggering plant defence	Freitas et al. (2014)
	tri 5	Soilborne pathogens	Trichothecene inhibitor of protein and DNA synthesis	Cardoza et al. (2011)
	erg 1		Squalene epoxidase for ergosterol synthesis	Godio et al. (2007)
	ThPG1	*Rhizoctonia solani*	Endopolygalacturonase for cell wall degradation	Morán-Diez et al. (2009)
	Thkel1	Osmotic and salt stress	Regulate glucosidase in *A. thaliana*	Hermosa et al. (2011)
	Th-Chit42	*Sclerotinia sclerotiorum*	Mycoparasitic activity	Steyaert et al. (2003)
	exc1, exc2, chit 42, chit3, bgn13.1, prb1	*Fusarium oxysporum*	Chitinases, glucanases, protease synthesis	Lopez Mondejar et al. (2011)
	qid74	*R. solani*	Surface adherence in mycoparasitism	Rosado et al. (2007)
T. virens	tac1	*Pythium ultimum*	Mycoparasitic activity	Abbas et al. (2017)
	Tvsp1	*R. solani*	Serine protease synthesis	Pozo et al. (2004)
	TgaA, TgaB	*Sclerotium rolfsii*	Antagonistic activity	Mukherjee et al. (2004)
T. longibrachiatum	egl1	*P. ultimum*	Biocontrol activity	Migheli et al. (1998)
T. brevicompactum	tri5	*Candida albicans*	Antifungal activity	Tijerino et al. (2011)
T. asperellum	chit42, chit3	*Fusarium graminearum*, *Fusarium oxysporum* f. sp. *cucumerinum*	Chitinases, glucanases, protease synthesis	Wu et al. (2017)

7.5 Signalling Involved in Induced Immunity and Biocontrol

The immune system of plants can detect the specific domains of microorganisms known as microbe-associated molecular patterns (MAMPs) (Hermosa et al. 2012). *Trichoderma* spp. release these MAMPs required for molecular recognition, and this probably contributes to a signal cascade within the plant involving signalling molecules such as SA, JA and ET. Though initially JA and/or ET was considered to be involved in signalling during ISR induced by *Trichoderma*, it has now been shown that SA is also involved. SA signalling has been detected in *Arabidopsis* during ISR induced by *Trichoderma* following colonization of root. Treatment of *Arabidopsis* with cell-free culture filtrates of *Trichoderma* induced ISR, and in this case, a combination of SA and JA/ET pathways was evident (Yoshioka et al. 2012). In the absence of any pathogen attack, the upregulation of SA gene markers triggered by *Trichoderma* persists over a long period of time. On the other hand, *Trichoderma* pretreatment when challenged with a pathogen and the upregulated gene expression is immediately modulated for induction of defence genes, mostly through JA/ET pathway, leading to the development of ISR (Tucci et al. 2011). This results in enhanced plant immunity against pathogen attack, leading to improved growth of plants. A characteristic feature of ISR induced by beneficial microorganisms is that it primes the plant to activate the final defence mechanisms when under pathogen attack or other forms of stress which makes the resistance more durable (Van Wees et al. 2008). The defence responses triggered by *Trichoderma*, both locally and systemically, include enhanced accumulation of several enzymes such as PAL, PO, PPO and LO; PR proteins; phytoalexins; terpenoids; and some antioxidants (Howell et al. 2000).

It is now established that the biocontrol mechanism of *Trichoderma* involves both direct mycoparasitism and induced systemic resistance in plants. Since the *Trichoderma*, pathogen and host interaction is specific, it is clear that signals from the fungal pathogen are recognized by *Trichoderma* and leads to transcription of mycoparasitism-related genes. Heterotrimeric G-proteins and mitogen-activated protein kinases (MAPKs) involved processes such as the production of hydrolytic enzymes and antifungal metabolites, and the formation of infection structures is involved in biocontrol.

Of the three subgroups of G-protein α subgroup I, Gα subunit Tga1 of *T. atroviride* is involved in both coiling and conidiation (Rocha-Ramirez et al. 2002), as also other processes such as hyphal growth and production of antifungal metabolites and chitinase enzyme which may have some role in *Trichoderma* biocontrol (Reithner et al. 2005). The Δ*tga1* mutant had much lower chitinase activities; transcription of the *nag1* (N-acetylglucosaminidase encoding) and *ech42* (endochitinase 42-encoding) genes was also decreased (Reithner et al. 2005). Although this mutant could produce infection structures, it had lost its ability of growing over the host fungi and lysing it, but it could still inhibit the growth of host fungi by overproduction and secretion of low-molecular-weight metabolites.

However, some of the secondary metabolites were produced in lesser amount in the mutant. Thus, it has been suggested that Tga1 has opposite roles in biosynthetic regulation of different antifungal metabolites (Zeilinger and Omann 2007).

Of the different MAPKs of *Trichoderma*, the most studied ones are YERK1, a family of fungal and yeast extracellular kinases. This family also contains other MAPKs, Pmk1, Fmk1, Bmp1 and Ubc3/Kpp2 from *Magnaporthe grisea*, *Fusarium oxysporum*, *B. cinerea* and *Ustilago maydis*, respectively. The roles reported for this MAPK from *T. virens* in the production of mycoparasitism-related enzymes are contradictory (Mendoza-Mendoza et al. 2003; Mukherjee et al. 2003), but other roles such as in the expression of genes coding for cell wall proteins during development, hydrophobicity and conidiation have also been reported (Mendoza-Mendoza et al. 2007). MAPK from *T. atroviride* (Tmk1) showed 98% identity with *T. virens* TmkA/Tvk1 (Reithner et al. 2007). Both the MAPKs affected the host specificity as Δ*tmk1* mutants still could parasitize *R. solani* but were unable to attack *B. cinerea* though they could attach and coil around the host. Upon germination of *T. virens* spores near cucumber roots, Δ*tmkA* mutants had the ability of root colonization, but could not induce systemic resistance against *Pseudomonas syringae* pv. *lachrymans*. This indicates the role of MAPK signalling for development of ISR in the plant (Viterbo et al. 2005).

A number of processes in fungi are dependent on cAMP signalling; the levels of cAMP are regulated by two enzymes, an adenylate cyclase for biosynthesis and a specific phosphodiesterase for degradation. The induction of endoglucanase in *T. reesei* was positively regulated by cAMP (Seidl et al. 2006). Studies have been made to determine whether conidiation in *T. viride* and *T. atroviride* was regulated by cAMP since survival of *Trichoderma* and its dispersal in the environment were to a great extent dependent on production of conidia (Nemcovic and Farkas 1998; Casas-Flores et al. 2006). An adenylate cyclase-encoding gene (*tac1*) was cloned from *T. virens*, and it was shown that Δ*tac1* mutants with reduced levels of intracellular cAMP also could not grow well on agar nor sporulate in darkness and were also unable to grow on host fungi. Besides, they also showed reduced secondary metabolite production (Mukherjee et al. 2007).

Cellulase induction in *T. reesei* has been correlated with signalling pathway involving cAMP-dependent protein kinase, known to be a secondary messenger which may also be involved in overproduction of this enzyme (Tisch et al. 2011; Sesták and Farkas 1993). It was found that in *T. reesei*, cellulase gene expression was regulated by cAMP which was carbon source dependent as was evident in the enhanced expression in presence of sophorose, which is formed during hydrolysis of cellulose by cellobiose transglycosylation (Nogueira et al. 2015). It is thus clear that other than signalling pathways induced in the host by different species of *Trichoderma*, they themselves have signalling mechanisms involving molecules such as G-proteins, MAPKs and cAMP, which are activated in response to stimuli from the host or environment. Besides being involved in mycoparasitism, these signalling mechanisms are also now indicated in induced resistance.

7.6 Crop Protection Through Induced Resistance

Harman et al. (2004) have documented induction of plant immunity following application of *Trichoderma* sp. It has also been found that colonization of root by *Trichoderma* strains has led to extensive changes in the plant genome and metabolism which is responsible for the accumulation of antimicrobial compounds. In tomato plants, Alfano et al. (2007) demonstrated that *Trichoderma hamatum*-induced genes were associated with abiotic and biotic stresses as well as protein metabolism. Such biocontrol agent could also induce gene which code for extensin-like proteins, but not thaumatin-like proteins (PR-5) that are considered as molecular markers of systemic acquired resistance.

Maize seeds primed with *T. harzianum* T22 prevented disease caused by *Colletotrichum graminicola* (Harman et al. 2004), whereas grapevine coated with *T. harzianum* T39 could induce immunity against downy mildew pathogen (Perazzolli et al. 2011). Induction of both systemic and local resistance in several crops such as cotton, rice, maize, tobacco, bean and lettuce by *T. harzianum* and in other pulses, cereals, plantation and horticultural crops by *T. asperellum* has been reported (Table 7.2). The ability of *T. asperellum* SKT-1 and its cell-free culture filtrate to elicit ISR against *Pseudomonas syringae* pv. *tomato* DC3000 causing bacterial leaf speck pathogen in *Arabidopsis thaliana* Col-0 plants has been reported

Table 7.2 *Trichoderma* species elicited induced resistance in crop plants towards pathogen

Trichoderma spp.	Strain	Plant	Pathogen	Reference
T. harzianum	T-39	Tomato	*Botrytis cinerea*	Meller et al. (2013)
		Cucumber	*B. cinerea*	Levy et al. (2015)
		Bean	*B. cinerea*	
		Strawberry	*Podosphaera xanthii*	
		Grapevine	*Plasmopara viticola*	Perazzolli et al. (2012)
	T-22	Bean	*B. cinerea*	Harman et al. (2004)
		Tomato	*Alternaria solani*	Seaman (2003)
	T-1	Pepper	*Phytophthora capsici*	Ahmed et al. (2000)
T. virens	G-6	Cotton	*Rhizoctonia solani*	Howell et al. (2000)
	G-6-5	Bean	*Colletotrichum lindemuthianum*	Bigirimana et al. (1997)
	G-11	Tobacco	*B. cinerea*	De Meyer et al. (1998)
T. atroviride	P-1	Tomato	*A. solani*	Salas-Marina et al. (2015)
T. asperellum	T-203	Cucumber	*Pseudomonas syringae* pv. *lachrymans*	Shoresh et al. (2005)
	GDFS1009	Corn	*Fusarium graminearum*	Wu et al. (2017)
		Cucumber	*Fusarium oxysporum* f. sp. *cucumerinum*	
	Ta-34	Mandarin	*Fusarium solani*	Chakraborty et al. (2019)

(Yoshioka et al. 2012). Studies with several mutants of *Arabidopsis* such as JA-resistant mutant *jar1*, ET-resistant mutant *etr1*, SA-signalling-impaired transgenic NahG and the NPR1-activity-impaired mutant *npr1* revealed the involvement of hormones when ISR was induced by *T. asperellum* SKT-1 and its cell-free culture filtrate. Leaves of Col-0 plants showed increased levels of *PR-1*, *PR-2* and *PR-5* genes which were SA inducible. The levels of JA/ET-induced genes such as *PDF1.2a*, *PR-3*, *PR-4* and *AtVsp1* were also enhanced but to a lesser degree than the SA-inducible ones.

In order to improve the soil health condition, enhanced fertilizer efficiency and productivity of crops, straw compost as an organic ameliorant can be used which is rich in carbon and also has the relative nutrient content such as potassium and silicon. Crop productivity can be enhanced by maintaining the level of carbon in the soil through organic matter management. By using consortia of *Trichoderma* sp., *Streptomyces* sp. and *Bacillus* sp., compost quality of rice straw can be improved to control the potential pathogen in straw. Simarmata et al. (2015) have demonstrated that rice straw which had been composted with consortia of *Trichoderma* sp. and biofertilizer could improve the soil health and accelerate the ISR against sheath blight, brown spot and bacterial leaf blight diseases of rice which enhance the crop productivity. Wu et al. (2017) have identified a novel strain of *T. asperellum* (GDFS1009) which exhibited a high growth rate and sporulation capacity as well as strong inhibitory effects against *Fusarium oxysporum* f. sp. *cucumerinum* causing cucumber *Fusarium* wilt and *Fusarium graminearum* that cause corn stalk rot disease. Cell wall-degrading enzymes such as protease, chitinase and glucanase are secreted by *T. asperellum* which play a role in mycoparasitism. On the other hand, secreted xylanases have been shown to induce resistance and enhance immunity in plants against pathogens. Based on RNA sequencing (RNA-seq) and gas chromatography-mass spectrometry (GC-MS), it was shown that *T. asperellum* produces primary metabolites which are precursors of antimicrobial compounds and also produces a number of antimicrobial secondary metabolites which include alkanes and polyketides.

In tea, the most devastating are the root diseases caused by various fungal pathogens of which sclerotial blight, brown root rot, red root rot, black root rot and charcoal stump rot are common. Management of such root diseases can be achieved by exogenous application of antagonistic microorganisms specially biocontrol agents (BCA) like *Trichoderma harzianum*, *T. asperellum* at 200 g per planting pit. One of the important stem diseases, collar canker, can be managed by avoiding planting of susceptible clones in gravelly soils and areas prone to drought as well as by improving the organic matter of such soils and by using plants having healthy root system. Disease can be cured by pruning the affected portions of healthy wood and by applying spore suspension of *Trichoderma* spp. to cut ends (Chakraborty and Chakraborty 2018). Induction of immunity in mandarin plants by joint inoculation of bioinoculants (*T. harzianum* and *Glomus mosseae*) against *Fusarium* root rot disease has been demonstrated (Allay and Chakraborty 2010, 2013). Further attempt for induction of resistance was made by colonizing mandarin roots with *G. gigantea* (AMF) and treating with *T. asperellum* (BCA) and

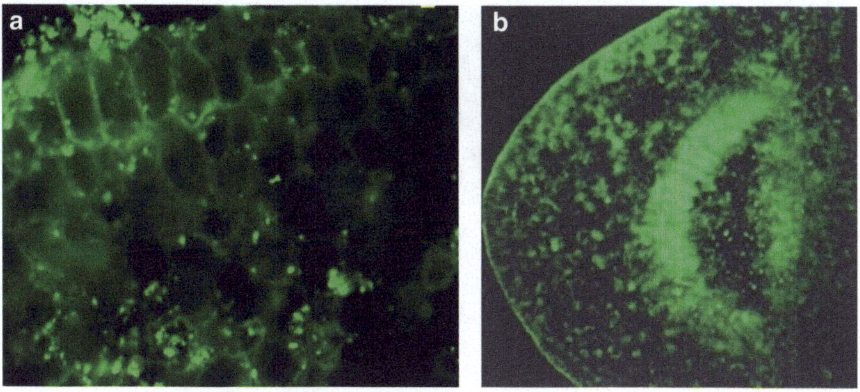

Fig. 7.1 Cellular localization of chitinase in mandarin root (**a**) and leaf tissue (**b**) following application of *Trichoderma asperellum* and challenge inoculation with *Fusarium solani* probed with PAb of chitinase and reacted with FITC conjugates

Pseudomonas poae (PGPR). Cellular localization of defence enzymes mainly chitinase was demonstrated using PAb of chitinase followed by labelling with FITC conjugates. In indirect immunofluorescence tests, strong bright apple green fluorescence was observed in the epidermal cells and homogenously in cortical tissues in roots. Enhancement of chitinase was revealed in both roots and leaves (epidermal cells and mesophyll tissues) following induction of resistance (Fig. 7.1).

Finally immunogold localization of chitinase in mandarin roots subsequent to colonization with bioinoculants and following inoculation with *F. solani* was evident as heavy deposition of gold particles near the cell wall of inoculated roots as observed by transmission electron microscopy. A few gold particles were also seen in the parts of cell walls that adhered firmly to intercellularly growing fungal hyphae. When plants were treated with *T. asperellum* (Ta-34) followed by challenge inoculation with *F. solani*, elicitation of chitinase was evident as intense gold labelling over both cell walls and wall appositions which activate the defence response in plants (Fig. 7.2). Improvement of health status in mandarin plants and suppression of disease incidence have been correlated with inhibition of the pathogen directly in the soil, immunity developed in the host or enriched availability of nutrients (Chakraborty et al. 2019).

7.7 Prospects

Responses of plants and pathogen during disease development can be fully understood using advanced molecular techniques, such as proteomics and transcriptomics, which in turn can prove useful while developing disease control strategies including biocontrol. Genome and transcriptome analyses can determine the expression of virulence genes which would allow the screening of virulence suppressive factors.

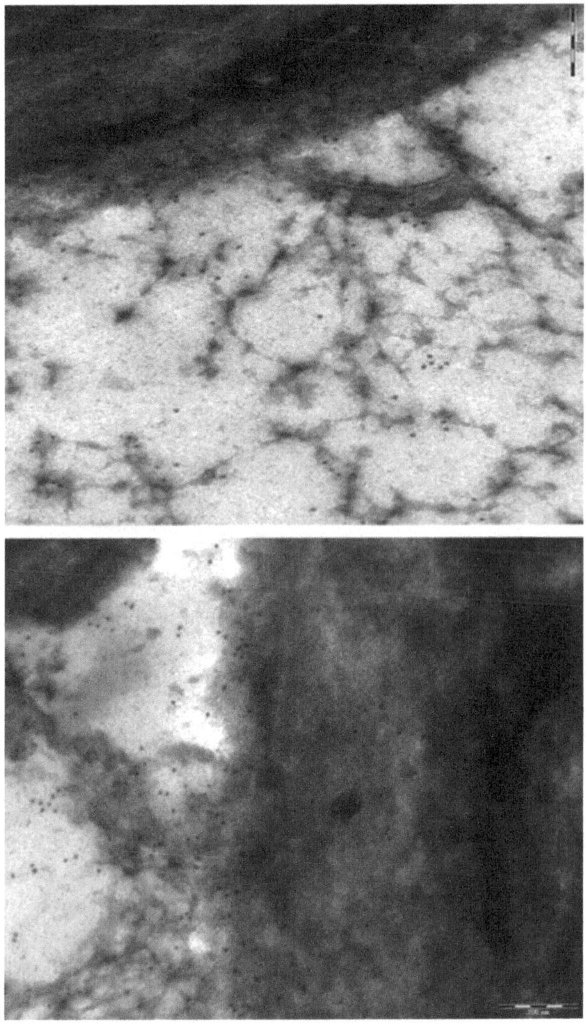

Fig. 7.2 Transmission electron micrographs of immunogold-labelled mandarin root following treatment with *Trichoderma asperellum* and challenge inoculated with *Fusarium solani* probed with PAb of chitinase

It is now possible to screen the genome library of certain beneficial species while searching for new PGP traits. It was proved difficult to study the expression of some *Trichoderma* genes as their activity may be connected solely with defence against other microbes or multicellular organisms. Some strategies can be initiated to activate silent gene clusters by growing fungi in conditions that stimulate competition and allow the usual biosynthetic pathways to operate.

Roles of *Trichoderma* gene products can be elucidated by detailed metabolomic-genomic study by which many new compounds may also be discovered and the role of these compounds may be unravelled, before their commercial use as biocontrol agents.

7.8 Conclusion

It is thus clear that *Trichoderma* possesses multiple beneficial traits and can be considered as a boon to humans and specially to agriculture. Earlier studies have now given way to more advanced ones using modern molecular techniques, such as proteomics and transcriptomics. The results from such studies can prove useful while developing disease control strategies including biocontrol. Genome and transcriptome analyses can determine the expression of virulence genes which would allow the screening of virulence suppressive factors. Though most of the activities of *Trichoderma* are now well documented, expression of some *Trichoderma* genes as their activity may be connected solely with defence against other microbes or multicellular organisms. Besides, research has also indicated that more secondary metabolites produced by *Trichoderma* are yet to be discovered and their roles elucidated. Commercial use of *Trichoderma* not only as a biocontrol agent but also as a growth promoter and immunity inducer should be taken up more vigorously as such microorganism offers eco-friendly sustainable solutions to different problems arising in the field.

References

Abbas A, Jiang D, Fu Y (2017) *Trichoderma* spp. as antagonist of *Rhizoctonia solani*. J Plant Pathol Microbiol 8:402

Ahmed AS, Sanchez CP, Candela ME (2000) Evaluation of induction of systemic resistance in pepper plants (*Capsicum annuum*) to *Phytophthora capsici* using *Trichoderma harzianum* and its relation with capsidiol accumulation. Eur J Plant Pathol 106:817–824

Alfano G, Ivey ML, Cakir C, Bos JIB, Miller SA, Madden LV, Kamoun S, Hoitnik HA (2007) Systemic modulation of gene expression in tomato by *Trichoderma hamatum* 382. Phytopathology 97:429–437

Allay S, Chakraborty BN (2010) Activation of defence response of mandarin plats against *Fusarium* root rot diseases using *Glomus mosseae* and *Trichoderma hamatum*. J Mycol Plant Pathol 40:499–511

Allay S, Chakraborty BN (2013) Induction of resistance in *Citrus reticulata* against *Fusarium solani* by dual application of AMF and *Trichoderma asperellum*. Int. J. Bioresour Stress Manag 4:588–592

Bae H, Sicher RC, Kim MS, Kim SH, Strem MD (2009) The beneficial endophyte *Trichoderma hamatum* isolate DIS219b promotes growth and delays the onset of the drought response in *Theobroma cacao*. J Exp Bot 60:3279–3295

Bigirimana J, de Meyer G, Poppe J, Elad Y, Hofte M (1997) Induction of systemic resistance on bean (*Phaseolus vulgaris*) by *Trichoderma harzianum*. Med Fac Landbouww Univ Gent 62:1001–1007

Boller T, Felix G (2009) A renaissance of elicitors: perception of microbe-associated molecular patterns and danger signals by pattern recognition receptors. Annu Rev Plant Biol 60:379–406

Brotman Y, Briff E, Viterbo A, Chet I (2008) Role of swollenin, an expansin-like protein from *Trichoderma*, in plant root colonization. Plant Physiol 147:779–789

Brotman Y, Lisec J, Méret M, Chet I, Willmitzer L, Viterbo A (2012) Transcript and metabolite analysis of the *Trichoderma*-induced systemic resistance response to *Pseudomonas syringae* in *Arabidopsis thaliana*. Microbiology 158:139–146

Brotman Y, Landau U, Cuadros-Inostroza A (2013) *Trichoderma* plant root colonization: escaping early plant defence responses and activation of the antioxidant machinery for saline stress tolerance. PLoS Pathog 9:e1003221

Cardoza RE, Malmierca MG, Hermosa MR, Alexander NJ, McCormick SP, Proctor RH, Tijerino AM, Rumbero A, Monte E, Gutiérrez S (2011) Identification of loci and functional characterization of trichothecene biosynthesis genes in filamentous fungi of the genus *Trichoderma*. Appl Environ Microbiol 77:4867–4877

Casas-Flores S, Rios-Momberg M, Rosales-Saavedra T, Martinez-Hernandez P, Olmedo-Monfil V, Herrera-Estrella A (2006) Cross talk between a fungal blue-light perception system and the cyclic AMP signaling pathway. Eukaryot Cell 5:499–506

Chacón MR, Rodriguez-Galan O, Benítez T, Sousa S, Rey M, Llobell A, Delgado-Jarana J (2007) Microscopic and transcriptome analyses of early colonization of tomato roots by *Trichoderma harzianum*. Int Microbiol 10:19–27

Chakraborty BN (2005a) Antimicrobial proteins in plant defence. In: Chakravorti SS (ed) New perspectives in the frontiers of chemical research. Royal Society of Chemistry (EIS), Kolkata, pp 470–483

Chakraborty U (2005b) Formation and scavenging of active oxygen species in plants. In: Chakravorti SS (ed) New perspectives in the frontiers of chemical research. Royal Society of Chemistry (EIS), Kolkata, pp 497–507

Chakraborty BN (2012) Recognition of fungi and activation of defence responses in plants. In: Sigh HP, Chowdappa P, Chakraborty BN, Podile AR (eds) Molecular approaches for plant fungal disease management. Westville Publishing House, New Delhi, pp 70–95

Chakraborty BN (2016) Scoping the potential uses of beneficial microorganisms for biopesticide industry and entrepreneurship development in crop protection. In: Chowdappa P, Sharma P, Singh D, Misra AK (eds) Perspectives of Plant Pathology in genomic era. Indian Phytopathological Society, IARI, New Delhi, pp 607–627

Chakraborty BN, Chakraborty U (2008) Involvement of salicylic acid in plant defence against stresses. In: Khan NA, Singh S (eds) Abiotic stresses in plants. International Publishers, New Delhi, pp 233–246

Chakraborty BN, Chakraborty U (2018) Tea diseases: early detection and their management strategies. In: Das S, Dutta S, Chakraborty BN, Singh D (eds) Recent approaches for management of plant diseases. Indian Phytopathological Society, IARI, New Delhi, pp 175–208

Chakraborty BN, Sharma M (2008) Pathogenesis related proteins in plant defence. In: Reddy SM, Gour HN (eds) Review of plant pathology, vol 4. Scientific Publishers, Jodhpur, pp 105–138

Chakraborty BN, Chakraborty U, Dey PL (2011) Potential application of *Trichoderma* as biocontrol agents, their molecular characterization and diversity analysis. In: Singh A (ed) Plant diseases and their biological control. Aavishkar Publishers, Jaipur, pp 186–216

Chakraborty BN, Chakraborty U, Sunar K, Dey PL (2014) Harnessing beneficial microbial resources for crop improvement. In: Sigh DP, Singh HB (eds) Trends in soil Microbial ecology. Studium Press LLC, Houston, TX, pp 175–201

Chakraborty BN, Chakraborty U, Allay S (2019) Wilt root rot complex in mandarin plants and activation of defence against pathogen. In: Bhattacharyya A, Chakraborty BN, Pandey RN, Singh D, Dubey SC (eds) Wilt diseases of crops. Indian Phytopathological Society, IARI, New Delhi, pp 293–321

Chowdhury AK, Debnath A, Roy A, Bhattacharya PM, Chattopadhyay C (2017) Biological control in 21st century: opportunities and challenges in subsistence farming system in India. In: Pandey RN, Chakraborty BN, Singh D, Sharma P (eds) Microbial antagonists: role in biological control of plant diseases. Indian Phytopathological Society, IARI, New Delhi, pp 39–66

Cong D, Li Y, Xian H (2012) Purification, renaturation and characterization of chitinase gene from *Trichoderma asperellum*. Chin Agric Sci Bull 28:34–38

Contreras-Cornejo HA, Macías-Rodríguez L, Cortés-Penagos C, López-Bucio J (2009) *Trichoderma virens*, a plant beneficial fungus, enhances biomass production and promotes lateral root growth through an auxin-dependent mechanism in *Arabidopsis*. Plant Physiol 149:1579–1592

Contreras-Cornejo HA, Macías-Rodríguez L, Beltrán-Peña E, Herrera-Estrella A, López-Bucio J (2011) *Trichoderma*-induced plant immunity likely involves both hormonal and camalexin dependent mechanisms in *Arabidopsis thaliana* and confers resistance against necrotrophic fungus *Botrytis cinerea*. Plant Signal Behav 6:1554–1563

Contreras-Cornejo HA, Macías-Rodríguez L, Del-Val E, Larsen J (2016) Ecological functions of *Trichoderma* spp. and their secondary metabolites in the rhizosphere: interactions with plants. FEMS Microbiol Ecol 92:fiw036

De Lorenzo G, Brutus A, Savatin DV, Sicilia F, Cervone F (2011) Engineering plant resistance by constructing chimeric receptors that recognize damage-associated molecular patterns (DAMPs). FEBS Lett 585:1521–1528

De Meyer G, Bigirimana J, Elad Y, Hofte M (1998) Induced systemic resistance in *Trichoderma harzianum* T39 biocontrol of *Botrytis cinerea*. Eur J Plant Pathol 104:279–286

Djonovic S, Pozo MJ, Dangott LJ, Howell CR, Kenerley CM (2006) Sm1, a proteinaceous elicitor secreted by the biocontrol fungus *Trichoderma virens* induces plant defence responses and systemic resistance. Mol Plant Microbe Interact 19:838–853

Djonovic S, Vargas WA, Kolomiets MV, Horndeski M, Wiest A, Kenerley CM (2007) A proteinaceous elicitor Sm1from the beneficial fungus *Trichoderma virens* is required for induced systemic resistance in maize. Plant Physiol 145:875–899

Druzhinina IS, Seidl-Seiboth V, Herrera-Estrella A (2011) *Trichoderma*: the genomics of opportunistic success. Nat Rev Microbiol 9:749–759

Freitas RS, Steindorff AS, Ramada MH, de Siqueira SJ, Noronha EF, Ulhoa CJ (2014) Cloning and characterization of a protein elicitor Sm1 gene from *Trichoderma harzianum*. Biotechnol Lett 36:783–788

Godio RP, Fouces R, Martin JF (2007) A squalene epoxidase is involved in biosynthesis of both the antitumor compound Clavaric acid and Sterols in the Basidiomycete *H. sublateritium*. Chem Biol 14:1334–1346

Gulijimila M, Fan HJ, Liu ZH, Wang N, Dou K, Huang Y, Wang ZY (2012) Cloning and sequence analysis of small molecular hydrophobin protein hyb2 gene from Trichoderma asperellum T4. Chin Agric Sci Bull 28:85–91

Hanson LE, Howell CR (2004) Elicitors of plant defence responses from biocontrol strains of *Trichoderma virens*. Phytopathology 94:171–176

Harman GE, Howell CR, Viterbo A, Chet I, Lorito M (2004) *Trichoderma* species-opportunistic, avirulent plant symbionts. Nat Rev Microbiol 2:43–56

Hermosa R, Botella L, Keck E, Jiménez JA, Montero-Barrientos M, Arbona V, Gómez-Cadenas A, Monte E, Nicolás C (2011) The overexpression in *Arabidopsis thaliana* of a *Trichoderma harzianum* gene that modulates glucosidase activity, and enhances tolerance to salt and osmotic stresses. J Plant Physiol 168:1295–1302

Hermosa R, Viterbo A, Chet I, Monte E (2012) Plant-beneficial effects of *Trichoderma* and of its genes. Microbiology 158:17–25

Hermosa R, Belén Rubio M, Cardoza RE, Nicolás C, Monte E, Gutiérrez S (2013) The contribution of *Trichoderma* to balancing the costs of plant growth and defence. Int Microbiol 16:69–80

Hidangmayum A, Dwivedi P (2018) Plant responses to *Trichoderma* spp. and their tolerance to abiotic stresses: a review. J Pharma Phytochem 7:758–766

Howell CR, Hanson LE, Stipanovic RD, Puckhaber LS (2000) Induction of terpenoid synthesis in cotton roots and control of *Rhizoctonia solani* by seed treatment with Trichoderma virens. Phytopathology 90:248–252

Jones JD, Dangl JL (2006) The plant immune system. Nature 444:323–329

Kashyap PL, Kumar S, Srivastava AK (2017) Nano diagnostics for plant pathogens. Environ Chem Lett 15:7–13

Kubicek CP, Herrera-Estrella A, Seidl-Seiboth V et al (2011) Comparative genome sequence analysis underscores mycoparasitism as the ancestral life style of *Trichoderma*. Genome Biol 12:R40

Levy NO, Meller HY, Haile ZM, Elad Y, David E, Jurkevitch E, Katan J (2015) Induced resistance to foliar diseases by soil solarization and *Trichoderma harzianum*. Plant Pathol 64:365–374

Li MF, Li GH, Ke-Qin Zhang KQ (2019) Non-volatile metabolites from *Trichoderma* spp. Metabolites 9:58

Lopez Mondejar R, Ros M, Pascual JA (2011) Mycoparasitism-related genes expression of *Trichoderma harzianum* isolates to evaluate their efficacy as biological control agent. Biol Control 56:59–66

López-Bucio J, Pelagio-Flores R, Herrera-Estrella A (2015) *Trichoderma* as biostimulant: exploiting the multilevel properties of a plant beneficial fungus. Sci Hortic 196:109–123

Luo Y, Zhang DD, Dong XW, Zhao PB, Chen LL, Song XY, Wang XJ, Chen XL, Shi M, Zhang YZ (2010) Antimicrobial peptaibols induce defence responses and systemic resistance in tobacco against tobacco mosaic virus. FEMS Microbiol Lett 313:120–126

Malmierca MG, Cardoza RE, Alexander NJ, McCormick SP, Hermosa R, Monte E, Gutiérrez S (2012) Involvement of *Trichoderma* trichothecenes in the biocontrol activity and induction of plant defence-related genes. Appl Environ Microbiol 78:4856–4868

Malmierca MG, Cardoza RE, Alexander NJ, McCormick SP, Collado IG, Hermosa R, Monte E, Gutiérrez S (2013) Relevance of trichothecenes in fungal physiology: disruption of *tri5* in *Trichoderma arundinaceum*. Fungal Genet Biol 53:22–33

Marra R, Ambosino P, Carbone V, Vinale F, Woo SL (2006) Study of the three-way interaction between *Trichoderma atroviride*, plant and fungal pathogens using a proteome approach. Curr Genet 50:307–321

Martinez C, Blanc F, Le Claire E, Besnard O, Nicole M, Baccou JC (2001) Salicylic acid and ethylene pathways are differentially activated in melon cotyledons by active or heat-denatured cellulase from *Trichoderma longibrachiatum*. Plant Physiol 127:334–344

Masunaka A, Hyakumachi M, Takenaka S (2011) Plant growth promoting fungus, *Trichoderma koningii* suppresses isoflavonoid phytoalexin vestitol production for colonization on/in the roots of *Lotus japonicus*. Microb Environ 26:128–134

Matarasso N, Schuster S, Avni A (2005) A novel plant cysteine protease has a dual function as a regulator of 1-aminocyclopropane-1-carboxylic acid synthase gene expression. Plant Cell 17:1205–1216

Mathys J, De Cremer K, Timmermans P, Van Kerckhove S, Lievens B, Vanhaecke M, Cammue BP, De Coninck B (2012) Genome-wide characterization of ISR induced in *Arabidopsis thaliana* by *Trichoderma hamatum* T382 against *Botrytis cinerea* infection. Front Plant Sci 3:108

Meller HY, Haile MZ, David D, Borenstein M, Shulchani R, Elad Y (2013) Induced systemic resistance against grey mould in tomato (*Solanum lycopersicum*) by benzothiadiazole and *Trichoderma harzianum* T39. Phytopathology 104:150–157

Mendoza-Mendoza A, Pozo MJ, Grzegorski D, Martinez P, Garcia JM, Olmedo-Monfil V, Cortes C, Kenerly C, Herrera-Estrella A (2003) Enhanced biocontrol activity of *Trichoderma* through inactivation of a mitogen- activated protein kinase. Proc Natl Acad Sci U S A 100:15965–15970

Mendoza-Mendoza A, Rosales-Saavedra T, Cortes C (2007) The MAP kinase TVK1 regulates conidiation, hydrophobicity and the expression of genes encoding cell wall proteins in the fungus *Trichoderma virens*. Microbiology 153:2137–2147

Migheli Q, Gonzalez-Candelas L, Dealessi L, Camponogara A, Ramon-Vidal D (1998) Transformants of *Trichoderma longibrachiatum* overexpressing the beta-1,4-endoglucanase gene egl1 show enhanced biocontrol of *Pythium ultimum* on cucumber. Phytopathology 8:673–677

Morán-Diez E, Hermosa R, Ambrosino P, Cardoza RE, Gutiérrez S, Lorito M, Monte E (2009) The ThPG1 endopolygalacturonase is required for the *Trichoderma harzianum*-plant beneficial interaction. Mol Plant Microbe Interact 22:1021–1031

Mukherjee PK, Latha J, Hadar R, Horwitz BA (2003) TmkA, a mitogen-activated protein kinase of *Trichoderma virens*, is involved in biocontrol properties and repression of conidiation in the dark. Eukaryot Cell 2:446–455

Mukherjee PK, Latha J, Hadar R, Horwitz BA (2004) Role of two G-protein alpha subunits, TgaA and TgaB, in the antagonism of plant pathogens by *Trichoderma virens*. Appl Environ Microbiol 70:542–549

Mukherjee M, Mukherjee PK, Kale SP (2007) cAMP signalling is involved in growth, germination, mycoparasitism and secondary metabolism in *Trichoderma virens*. Microbiology 153:1734–1742

Mukherjee PK, Horwitz BA, Kenerley CM (2012) Secondary metabolism in *Trichoderma*: a genomic perspective. Microbiology 158:35–45

Nemcovic M, Farkas V (1998) Stimulation of conidiation by derivatives of cAMP in *Trichoderma viride*. Folia Microbiol 43:399–402

Nielsen KF, Gräfenhan T, Zafari D, Thrane U (2005) Trichothecene production by *Trichoderma brevicompactum*. J Agric Food Chem 53:8190–8196

Nogueira KMV, Costa MDN, de Paula RG, Mendonça-Natividade FC, Ricci-Azevedo R, Nascimento SRN (2015) Evidence of cAMP involvement in cellobiohydrolase expression and secretion by *Trichoderma reesei* in presence of the inducer sophorose. BMC Microbiol 15:195

Perazzolli M, Roatti B, Bozza E, Pertot I (2011) Trichoderma harzianum T39 induces resistance against downy mildew by priming for defence without costs for grapevine. Biol Control 58:74–82

Perazzolli M, Moretto M, Fontana P, Ferrarini A, Velasco R, Moser C, Delledonne M, Pertot I (2012) Downy mildew resistance induced by *Trichoderma harzianum* T39 in susceptible grapevines partially mimics transcriptional changes of resistant genotypes. BMC Genomics 13:660

Petutschnig EK, Jones AME, Serazetdinova L, Lipka U, Lipka V (2010) The lysin motif receptor-like kinase (LysM-RLK) CERK1 is a major chitin-binding protein in *Arabidopsis thaliana* and subject to chitin-induced phosphorylation. J Biol Chem 285:28902–28911

Piel J, Atzorn R, Gäbler R, Kühnemann F, Boland W (1997) Cellulysin from the plant parasitic fungus *Trichoderma viride* elicits volatile biosynthesis in higher plants via the octadecanoid signalling cascade. FEBS Lett 416:143–148

Pineda A, Zheng SJ, Van Loon JJA, Pieterse CMJ, Dicke M (2010) Helping plants to deal with insects: the beneficial soil-borne microbes. Trends Plant Sci 15:507–514

Pozo MJ, Baek JM, García JM, Kenerley CM (2004) Functional analysis of tvsp1, a serine protease-encoding gene in the biocontrol agent *Trichoderma virens*. Fungal Genet Biol 41:336–348

Reithner B, Brunner K, Schuhmacher R, Peissl I, Seidl V, Krska R, Zeilinger S (2005) The G protein alpha subunit Tga1 of *Trichoderma atroviride* is involved in chitinase formation and differential production of antifungal metabolites. Fungal Genet Biol 42:749–760

Reithner B, Schuhmacher R, Stoppacher N, Pucher M, Brunner K, Zeilinger S (2007) Signaling via the *Trichoderma atroviride* mitogen-activated protein kinase Tmk1 differentially affects mycoparasitism and plant protection. Fungal Genet Biol 44:1123–1133

Rocha-Ramirez V, Omero C, Chet I, Horwitz BA, Herrera-Estrella A (2002) *Trichoderma atroviride* G-protein alpha-subunit gene *tga1* is involved in mycoparasitic coiling and conidiation. Eukaryot Cell 1:594–605

Ron M, Avni A (2004) The receptor for the fungal elicitor ethylene- inducing xylanase is a member of a resistance-like gene family in tomato. Plant Cell 16:1604–1615

Rosado IV, Rey M, Codón AC, Govantes J, Moreno-Mateos MA, Benítez T (2007) QID74 cell wall protein of *Trichoderma harzianum* is involved in cell protection and adherence to hydrophobic surfaces. Fungal Genet Biol 44:950–964

Rotblat B, Enshell-Seijffers D, Gershoni JM, Schuster S, Avni A (2002) Identification of an essential component of the elicitation active site of the EIX protein elicitor. Plant J 32:1049–1055

Rubio MB, Hermosa R, Reino JL, Collado IG, Monte E (2009) *Thctf1* transcription factor of *Trichoderma harzianum* is involved in 6-pentyl80 2H-pyran-2-one production and antifungal activity. Fungal Genet Biol 46:17–27

Ruiz N, Wielgosz-Collin G, Poirier L, Grovel O, Petit KE, Mohamed-Benkada M, Pont TR, Bissett J, Vérité P, Barnathan G, Pouchus YF (2007) New Trichobrachins,11-residue peptaibols from a marine strain of *Trichoderma longibrachiatum*. Peptides 28:1351–1358

Salas-Marina MA, Silva-Flores MA, Uresti-Rivera EE, Castro-Longoria E, Harrera-Estrella A, Casas-Flores S (2011) Colonization of *Arabidopsis* roots by *Trichoderma atroviride* promotes growth and enhances systemic disease resistance through jasmonate and salicylate pathways. Eur J Plant Pathol 131:15–26

Salas-Marina MA, Isordia-Jasso M, Islas-Osuna MA, Delgado-Sánchez P, Jiménez-Bremont JF, Rodríguez-Kessler M, Rosales-Saavedra MT, Herrera-Estrella A, Casas-Flores S (2015) The Epl1 and Sm1 proteins from *Trichoderma atroviride* and *Trichoderma virens* differentially modulate systemic disease resistance against different life style pathogens in *Solanum lycopersicum*. Front Plant Sci 23:77

Samolski I, Rincón A, Pinzón LM (2012) The *qid74* gene from *Trichoderma harzianum* has a role in root architecture and plant biofertilization. Microbiology 158:129–138

Seaman A (2003) Efficacy of OMRI-approved products for tomato foliar disease control. NY State Integr Pest Manag Progr Publ 129:164–167

Segarra G, Casanova E, Bellido D (2007) Proteome, salicylic acid, and jasmonic acid changes in cucumber plants inoculated with *Trichoderma asperellum* strain T34. Proteomics 7:3943–3952

Seidl V, Schmoll M, Scherm B, Balmas V, Seiboth B, Migheli Q, Kubicek CP (2006) Antagonism of *Pythium* blight of zucchini by *Hypocrea jecorina* does not require cellulase gene expression but is improved by carbon catabolite derepression. FEMS Microbiol Lett 257:145–151

Sesták S, Farkas V (1993) Metabolic regulation of endoglucanase synthesis in *Trichoderma reesei*: participation of cyclic AMP and glucose-6-phosphate. Can J Microbiol 39:342–347

Sharma P, Kumar PV, Ramesh R, Saravanan K, Deep S, Shrama M, Mahesh S, Singh D (2011) Biocontrol genes from Trichoderma species: a review. Afr J Biotech 10(86):19898–19907

Shoresh M, Harman GE (2008a) The relationship between increased growth and resistance induced in plants by root colonizing microbes. Plant Signal Behav 3:737–739

Shoresh M, Harman GE (2008b) The molecular basis of shoot responses of maize seedlings to *Trichoderma harzianum* T22 inoculation of the root: a proteomic approach. Plant Physiol 147:2147–2163

Shoresh M, Yedidia I, Chet I (2005) Involvement of jasmonic acid/ethylene signaling pathway in the systemic resistance induced in cucumber by *Trichoderma asperellum* T203. Phytopathology 95:76–84

Shoresh M, Harman GE, Mastouri F (2010) Induced systemic resistance and plant responses to fungal biocontrol agents. Annu Rev Phytopathol 48:21–43

Simarmata T, Hersanti, Turmuktini T, Fitriatin BN, Setiawati MR, Purwanto (2015) Application of bioameliorant and biofertilizers to increase the soil health and rice productivity. Hayati J Biosci 23:181–184

Singh HB (2006) *Trichoderma*: a boon for biopesticide industry. J Mycol Plant Pathol 36:373–384

Singh HB, Singh BN, Singh SP, Sarma BK (2012) Exploring different avenues of *Trichoderma* as a potent biofungicidal and plant growth promoting candidate – an overview. In: Bagyanarayana G, Gour HN, Manoharachary C, Kunwar IK (eds) Review of plant pathology, vol 5. Scientific Publishers, Jodhpur, pp 315–426

Steyaert JM, Ridgway HJ, Elad Y, Stewart A (2003) Genetic basis of mycoparasitism: a mechanism of biological control by species of *Trichoderma*. J Crop Hortic Sci 31:281–291

Tijerino A, Cardoza RE, Moraga J, Malmierca MG, Vicente F, Aleu J, Collado IG, Gutiérrez S, Monte E, Hermosa R (2011) Overexpression of the trichodiene synthase gene *tri5* increases trichodermin production and antimicrobial activity in *Trichoderma brevicompactum*. Fungal Genet Biol 48:285–296

Tisch D, Kubicek CP, Schmoll M (2011) New insights into the mechanism of light modulated signaling by heterotrimeric G-proteins: ENVOY acts on gna1 and gna3 and adjusts cAMP levels in *Trichoderma reesei* (*Hypocrea jecorina*). Fungal Genet Biol 48:631–640

Tucci M, Ruocco M, De Masi L, De Palma M, Lorito M (2011) The beneficial effect of *Trichoderma* spp. on tomato is modulated by the plant genotype. Mol Plant Pathol 12:341–354
Van Wees SCM, van der Ent S, Pieterse CMJ (2008) Plant immune responses triggered by beneficial microbes. Curr Opin Plant Biol 11:443–448
Vargas WA, Mandawe JC, Kenerley CM (2009) Plant-derived sucrose is a key element in the symbiotic association between *Trichoderma virens* and maize plants. Plant Physiol 151:792–808
Vinale F, Sivasithamparam K, Ghisalberti EL, Marra R, Barbetti MJ, Li H, Woo SL, Lorito M (2008a) A novel role for *Trichoderma* secondary metabolites in the interactions with plants. Physiol Mol Plant Pathol 72:80–86
Vinale F, Sivasithamparam K, Ghisalberti EL, Marra R, Woo SL, Lorito M (2008b) *Trichoderma*-plant-pathogen interactions. Soil Biol Biochem 40:1–10
Viterbo A, Chet I (2006) TasHyd1, a new hydrophobin gene from the biocontrol agent *Trichoderma asperellum*, is involved in plant root colonization. Mol Plant Pathol 7:249–258
Viterbo A, Harel M, Horwitz BA, Chet I, Mukherjee PK (2005) *Trichoderma* mitogen activated protein kinase signalling is involved in induction of plant systemic resistance. Appl Environ Microbiol 71:6241–6246
Viterbo A, Wiest A, Brotman Y, Chet I, Kenerley C (2007) The 18mer peptaibols from *Trichoderma virens* elicit plant defence responses. Mol Plant Pathol 8:737–746
Vizcaíno JA, Cardoza RE, Hauser M, Hermosa R, Rey M, Llobell A, Becker JM, Gutiérrez S, Monte E (2006) ThPTR2, a di/tri-peptide transporter gene from *Trichoderma harzianum*. Fungal Genet Biol 43:234–246
Walters DR, Ratsep J, Havis ND (2013) Controlling crop diseases using induced resistance: challenges for the future. J Exp Bot 64:1263–1280
Wu Q, Sun R, Ni M, Yu J, Li Y, Yu C, Dou K, Ren J, Chen J (2017) Identification of a novel fungus, *Trichoderma asperellum* GDFS1009, and comprehensive evaluation of its biocontrol efficacy. PLoS One 12(6):e0179957
Yasmeen R, Siddiqui ZS (2017) Physiological responses of crop plants against *Trichoderma harzianum* in saline environment. Acta Bot Croat 76:154–162
Yedidia I, Shoresh M, Kerem Z, Benhamou N, Kapulnik Y, Chet I (2003) Concomitant induction of systemic resistance to *Pseudomonas syringae* pv. *lachrymans* in cucumber by *Trichoderma asperellum* (T-203) and accumulation of phytoalexins. Appl Environ Microbiol 69:7343–7353
Yoshioka Y, Ichikawa H, Naznin HA, Kogure A, Hyakumachi M (2012) Systemic resistance induced in *Arabidopsis thaliana* by *Trichoderma asperellum* SKT-1, a microbial pesticide of seedborne diseases of rice. Pest Manag Sci 68:60–66
Zamioudis C, Pieterse CMJ (2012) Modulation of host immunity by beneficial microbes. Mol Plant Microbe Interact 25:139–150
Zeilinger S, Omann M (2007) *Trichoderma* biocontrol: signal transduction pathways involved in host sensing and mycoparasitism. Gene Regul Syst Bio 8:227–234
Zeilinger S, Gruber S, Bansal R, Mukherjee PK (2016) Secondary metabolism in *Trichoderma*-chemistry meets genomics. Fungal Biol Rev 30:74–90
Zhang J, Zhou JM (2010) Plant immunity triggered by microbial molecular signatures. Mol Plant 3:783–793

Chapter 8
Host–Pathogen–*Trichoderma* Interaction

Harekrushna Swain and Arup K. Mukherjee

Abstract On our planet, the customary agricultural operations are influenced by different issues (i.e., disease, pests, drought, diminished soil fertility) because of the utilization of hazardous chemical pesticides, pollution, and climate change. Among these, the fungal genus *Trichoderma* carries out noteworthy activity in controlling plant maladies. *Trichoderma* spp. are the best biocontrol agents, as more than 60% of the enlisted biofungicides utilized in present agribusinesses are based on *Trichoderma*. Plant–*Trichoderma*–pathogen is a complex community having multiple mechanisms. Proteome and genome examinations have greatly upgraded the capacity for comprehensive and genome-based useful investigations as they have identified and determined the role of a variety of novel genes and gene products, including ABC transporters, enzymes, and various proteins that produce novel elicitors of induced resistance, proteins responsible for a gene-for-gene avirulent interaction between *Trichoderma* spp. and plants, antagonism-related genes, and plant proteins specifically induced by *Trichoderma*, but there remains much more to be learned than has already been discovered. Recent methodologies based on molecular science can easily identify, clone, sequence, and express the activity of the genes and can observe their capacities and function in the biocontrol system. Herein we present the comprehensive genetics of interactions of *Trichoderma* with plants and phytopathogens and their role through different modes of action. The aforementioned variables are required to enhance their efficiency and applications and also demonstrate the possibilities for development of strains of *Trichoderma*.

Keywords Antibiosis · Biocontrol genes · Induced resistance · *Trichoderma* · Mycoparasitism · PGPF · MAMP

H. Swain · A. K. Mukherjee (✉)
Molecular Plant Pathology Laboratory, Division of Crop Protection, ICAR-National Rice Research Institute, Cuttack, Odisha, India

A. K. Sharma, P. Sharma (eds.), *Trichoderma*, Rhizosphere Biology,
https://doi.org/10.1007/978-981-15-3321-1_8

8.1 Introduction

The great increase in food grain production has helped provide food security, but we are still confronted with yield losses in the crops. To manage or control these biotic constraints (i.e., fungi, bacteria, virus, nematodes, weeds, insects), pesticide use has increased year by year. In addition to negatively affecting both the environment and nontarget organisms, the continued tremendous use of chemical pesticides creates high selection pressure on pathogens. These agricultural practices force mutation within the pathogens and the development of pesticide-resistant races such as dodine and metalaxyl resistance in *Venturia inaequalis* and *Phytophthora infestans*, respectively (Meszka et al. 2008). The injudicious use of chemical pesticides for disease management urges the introduction of new alternatives in biological control. Besides the classic mycorrhizal fungi and bacteria, plant growth-promoting fungi such as *Trichoderma* spp. (division Ascomycota, subdivision Pezizomycotina, class Sordariomycetes, order Hypocreales, family Hypocreaceae) can stimulate plant growth by suppressing plant diseases (Van Wees et al. 2008). As omnipresent and regularly prevalent parts of the mycoflora in local and horticultural soils all through every climatic zone, *Trichoderma* spp. assume a significant function in the well-being of biological systems (Kubicek et al. 2011). These species colonize above-ground and subterranean plant organs and develop within living cells (endophytes), and they show up in plant litter, natural soil issue (saprophytes), and mammalian tissues (human pathogens). However, the capacity of these fungi to detect, invade, and decimate other fungi has been the major driving force behind their commercial success as biopesticides (more than 60% of all enlisted biopesticides are *Trichoderma* based) (Verma et al. 2007). These fungi not only protect plants by killing other fungi and certain nematodes but induce resistance against plant pathogens, impart abiotic stress tolerance, amend plant growth and vigour, solubilize plant nutrients, and bioremediate metals and environmental pollutants (Hermosa et al. 2012; Lorito et al. 2010; Mastouri et al. 2010; Shoresh et al. 2006). The capacity to stop other growth, above all that of plant pathogens, has commanded enthusiasm for *Trichoderma* for a long time (Benitez et al. 2004; Harman 2000, 2006; Herrera-Estrella and Chet 2004; Viterbo and Horwitz 2010). Investigation of the genomes of the mycoparasites *Trichoderma atroviride* and *Trichoderma virens* contrasted with the cellulose-corrupting, basically saprophytic *Trichoderma reesei* showed that qualities explicit to the mycoparasites emerged in a typical progenitor yet were lost from the *Trichoderma reesei* lineage (Kubicek et al. 2011). Although various *Trichoderma* spp. have been explored for their mycotrophy, research with *Trichoderma reesei* has mainly fixated on the engenderment of plant cell wall-degrading enzymes (CWDEs) (Kubicek et al. 2011; Seiboth et al. 2012), for which this species has become a model system. On the off chance that mycotrophy was the first way of life of *Trichoderma* spp., they similarly coevolved with land plants to form symbioses with roots. Recent years have seen a rush of enthusiasm for plant ailment obstruction [induced systemic resistance (ISR); somewhat systemic acquired resistance (SAR)] initiated by the advantageous *Trichoderma*–root interaction

(Harman et al. 2004; Hermosa et al. 2012). These plant-centered mechanisms have rivaled mycoparasitism as an explanation for how *Trichoderma* controls plant diseases (Lorito et al. 2010; Mastouri et al. 2010; Shoresh et al. 2006). In India alone, about 250 *Trichoderma*-based products are available for field applications but the percentage share of biofungicides is only a small part of the fungicide market, which is dominated by synthetic chemicals (Singh et al. 2009). Here, genetic manipulation of design could possibly prove beneficial in developing beneficial strains. It could be done via gaining information on the molecular mechanisms of interactions of those organisms with different biotic and abiotic elements.

8.2 Connection Between *Trichoderma* and Plants

8.2.1 Effect on Growth Promotion

Trichoderma spp. produces auxins that are chargeable for plant bloom and root improvement in each symbiotic and pathogenic interaction with plants (Gravel et al. 2007; Losane and Kumar 1992; Shayakhmetov 2001). Indole-3-acetic acid (IAA) engendered by *Trichoderma virens* and *Trichoderma atroviride* was found to stimulate the growth of *Arabidopsis* plants in vitro (Contreras-Cornejo et al. 2009). The role of *Trichoderma* spp. as plant symbionts for upgraded nutrient uptake, expanded root and shoot development, improved plant life, and biotic/abiotic stress resistance has been extensively studied (Inbar et al. 1994; Yedidia et al. 2001; Harman et al. 2008; Harman 2011). Swain et al. (2018) demonstrated that *Trichoderma erinaceum-treated seeds* induced comparatively better effects as compared to other isolates in two direct seeded rice varieties under field conditions. An excellent impact on plant improvement has been proven for several *Trichoderma* secondary metabolites. Koninginins, 6-pentyl-α-pyrone, trichocaranes A–D, harzianopyridone, cyclonerodiol, harzianolide, and harzianic acid are examples of compounds that have an effect on plant growth in an attention-dependent way (Vinale et al. 2014).

8.2.2 Effect on Germination and Early Vigour

Seed biopriming and seed treatment with *Trichoderma* spp. trigger the production of phytohormones that are involved in seed germination and also enhance the speed of germination and seedling vigour. Enhanced germination percentages have been found in okra, maize, beans, mustard, chilli, soyabean, chickpea, tomato, etc. (Mukhtar 2008; Okoth et al. 2011; Rehman et al. 2012; Kumar et al. 2014; Babychan and Simon 2017). Many seed-invading pathogens such as *Pythium* are unable to attack these hosts because of their faster seed germination and seedling vigour (Mastouri et al. 2010). Bezuidenhout et al. (2012) reported that gliotoxin, a metabolite produced by *Trichoderma harzianum*, may copy the plant development hormone gibberellic acid, which is included in the seed germination process. Swain et al. (2018) reported significant differences in mean germination time and vigour indices among different *Trichoderma*-treated rice seeds. Doni et al. (2014) reported that seven isolates of *Trichoderma* spp. upgraded germination rate and seedling vigour in rice. Various phenolics or auxiliary metabolites with phytohormones known to be specific auxins discharged by the fungi might be liable for better seedling vigour. Early vigour with better root development in *Trichoderma* medications helped the plant take up more supplements and remain green for a longer period, in this manner boosting yield. Seed or soil treatment with *Trichoderma* accelerates seed germination percentage directly by activating enzymes and phytohormones and indirectly by altering soil microflora and nutrient availability. Thus, farmers attain a better field stand.

8.2.3 Effect on Plant Morphology

Trichoderma spp. are able to enhance rice growth parameters including root dry weight, shoot dry weight, root fresh weight, shoot fresh weight, leaf number, plant height, root length, shoot length, and tiller number (Swain et al. 2018). In maize (*Zea mays*) plants, *Trichoderma* inoculation affects root system architecture, including stronger root biomass and multiplied root hair development (Bjorkman et al. 1998; Harman et al. 2004). The root system is responsible for plant fitness because it provides anchorage, contributes to water use efficiency, and enables the acquisition of mineral nutrients from the soil. Two *Trichoderma* spp. (i.e., *Trichoderma harzianum* and *Trichoderma virens*) have been found to have growth promotional activity, which was correlated with the prolific formation of lateral roots (Contreras-Cornejo et al. 2009). The potential of *Trichoderma* spp. has been well reported to provide phytohormones, the important component to increase rice plant tops (Chowdappa et al. 2013). *Trichoderma virens* can deliver auxins such as indole-3-acetic acid (IAA), indole-3-acetaldehyde (IAAld), and indole-3-ethanol (IEt), which have roles in plant improvement and development (Contreras-Cornejo et al. 2009). Cai et al. (2013) reported that harzianolide engendered by *Trichoderma* spp. improved the early stage of plant development through the enhancement of root length. Evidence for enhancement of the number of leaves, tillers, branches, flowers, fruits, and plant height by *Trichoderma* spp. is available (Neumann and Laing 2006; Sajeesh 2015). These morphological modifications are conceivable in view of the capacity of *Trichoderma* spp. to act through several components, for example, natural buffering (against pH, drought, waterlogging, cold, and heat), phosphorus solubilization, organic matter decomposition, chelation, and siderophore production.

8.2.4 Impact on Plant Physiology

Trichoderma spp. have the capacity to alter several physiological operations including net photosynthetic rate, transpiration, stomatal conductance, internal CO_2 concentration, water use efficiency, and nutrient uptake. Higher nutrient uptake will benefit the physiological pathways within plants treated with *Trichoderma* spp., leading to excellent increases in growth (Doni et al. 2014; Saba et al. 2012). A good-sized difference was found in various parameters related to chlorophyll content among different treatments in two rice varieties. Further, seed treatment with *Trichoderma atroviride* induced comparatively larger effects on chlorophyll content as compared to other isolates (Swain et al. 2018). Micro-nutrients such as nitrogen and magnesium are components of chlorophyll and enzymes of mRNA synthesis and gene regulation engaged in photosynthesis. *Trichoderma harzianum* drastically extended the potential of rice flora to tolerate drought stress and increased the water-conserving capacity for rice (Shukla et al. 2012). An increase in net photosynthetic rate and stomatal conductance of approximately threefold and a twofold increase in water use efficiency in *Trichoderma*-treated rice plants as compared to plants treated with nitrogen (N), phosphorus (P), and potassium (K) (NPK) has been observed (Doni et al. 2014). High photosynthetic rates coupled with low transpiration rates in *Trichoderma*-treated plants indicate high water use efficiency (Thakur et al. 2010; Doni et al. 2014). Low internal CO_2 concentration and high photosynthetic rate in *Trichoderma*-treated plants are correlated with the activity of carboxylation by CO_2 fixation for glucose production during carbohydrate metabolism being very active (Thakur et al. 2010). Interest in *Trichoderma* spp. that contribute to the enhancement of root growth and distribution became further considered as a key component in extended photosynthetic activity and delayed senescence in rice vegetation (Mishra and Salokhe 2011). *Trichoderma* releases cellulases, which degrade cellulose and enhance the organic matter and nutrients in the rhizosphere (Jiang et al. 2011). Solubilization and chelation of minerals can enhance nutrient availability that is engaged in plant metabolism, leading to the enhancement of plant physiological activity (Harman et al. 2004).

8.2.5 Impact on the Yield and Nature of Products

The effects of *Trichoderma* on seed germination, plant morphology, and physiology lead to better field stands and also accelerate the vegetative and reproductive growth of plants. The numbers of branches, spikes, flowers, and fruits per plant are enhanced. In many cases, the average weight of individual fruits is also comparatively higher. Higher yield has been found following the application of *Trichoderma* spp. in mustard, wheat, corn, tuberose, sugarcane, tomato, okra, etc. (Haque et al. 2012; Naznin et al. 2015; Srivastava et al. 2006; Tucci et al. 2011; Idowu et al. 2016).

8.2.6 Promoted Opposition Against Biotic and Abiotic Stress

Plants have a resistant framework that can distinguish motifs or domains with preserved basic qualities typical of whole classes of microorganisms called microbe-associated molecular patterns (MAMPs) (Hermosa et al. 2012). The ability of *Trichoderma* spp. hyphae to release MAMPs for molecular recognition may contribute to the signal cascade by signalling molecules within the plant such as salicylic acid (SA), jasmonic acid (JA), and ethylene (ET). Earlier it was believed that JA or ET is the signalling molecule for *Trichoderma*-induced resistance, but recent findings show the involvement of SA also. Colonization of *Arabidopsis* roots by *Trichoderma asperellum* produces an unmistakable induced systematic resistance (ISR) through a salicylic acid flagging course, and both the salicylic acid and jasmonic acid/ethylene flagging pathways join in the ISR activated without cell culture filtrates of *Trichoderma* (Yoshioka et al. 2012). *Trichoderma* can trigger a durable upregulation of SA quality markers in plants unchallenged by pathogens, in spite of the fact that when plants are tainted by the pathogen *Botrytis cinerea*, the pretreatment with *Trichoderma* may tweak the salicylic acid-subordinate gene expression and, not long after contamination, the expression of defence genes is prompted through the JA signal transduction pathway, causing an ISR increment after some time (Tucci et al. 2011). Elicitors discharged by *Trichoderma* spp. are additionally associated with activating expression of defence proteins inside the plant (Thakur and Sohal 2013). Along these lines, plant susceptibility to pathogens is reduced, thusly improving plant development. A common feature of induced resistance responses of helpful microorganisms is preparing for improved protection. In primed plants, defence responses are not initiated legitimately, yet are quickened on assault by pathogens or different stresses, bringing about quicker and more grounded protection from the attacker encountered (Van Wees et al. 2008). *Trichoderma* act locally and systemically, involving the signalling cascade and activation and accumulation of defence-related antimicrobial compounds, including enzymes such as phenyl ammonia lyase (PAL), peroxidase, polyphenol oxidase, lipoxygenase, proteins such as PR (pathogenesis-related protein), and terpenoid, and phytoallexin such as rishitin, lubimin, phytotuberol, coumarin, solevetivone, resveratrol, and antioxidants such as ascorbic acid and glutathione (Howell et al. 2000). Swain et al. (2018) reported that higher expression of stress-related enzymes [catalase (CAT), peroxidase (PER), and superoxide dismutase (SOD)] was observed in *Trichoderma*-treated plants as compared to untreated plants in two rice varieties. Enzymes such as SOD, CAT, and PER are important in protection against oxidative damage.

8.2.7 Effect on Uptake of Nutrients (Nitrogen, Phosphorus, Potassium)

The advantage of *Trichoderma* spp. in improving rice plant development could be clarified through upgraded solubilization and uptake of inorganic soil supplements such as N, P, K, and Zn, and the improvement of root hair growth (Lorito et al. 2010). Doni et al. (2014) observed taller plants in *Trichoderma* spp.-treated rice plants as contrasted with NPK treatment alone. Swain et al. (2018) additionally found higher N, P, and K uptake in both roots and shoots in *Trichoderma*-treated plants when contrasted with controls, which findings were similar to the observations made by Saba et al. (2012). Better uptake of nitrogen, phosphorus, and potassium subsequently improves physiological pathways, prompting good crop development. *Trichoderma* spp. are known to improve development and supplement uptake by the generation of phytohormones, natural buffering (against pH, dry spells, water logging), P solubilization, and siderophore creation (Newman 2006). Soil utilization of *Trichoderma* spp. was earlier found to expand P and Zn uptake by rice in the fields of farmers in the Philippines by discharging soil-fixed P and emitting advantageous hormones (Cuevas 2006).

8.2.8 Linkage of Plants and Trichoderma

Plant resistance and growth are interconnected in a system of hormone signalling pathways (Pieterse et al. 2009). There is a cross-correspondence between salicylic acid, jasmonic acid, and ethylene, the focal players in guarding and the reaction pathways of different hormones such as abscisic acid, gibberellic acid, and auxin that are generally connected with plant improvement and abiotic stress, IAA, related to plant development and horizontal root advancement, and gibberellins, which control plant development by managing the degradation of development-curbing DELLA proteins (Hermosa et al. 2012). In *Trichoderma*, ACC deaminase activity reduces the accessibility of ACC important for ET biosynthesis. Decrease in ethylene promotes plant development by means of gibberellin signalling by expanding the degradation of DELLA proteins. Further, gibberellins may control the beginning of the jasmonic acid and salicylic acid subordinate resistance reactions of the plant through DELLA protein debasement (Brotman et al. 2013). Ethylene and indole acetic acid in the roots can equally direct each other's biosynthesis (Stepanova et al. 2007). In *Trichoderma*, IAA adds to exogenous auxin-invigorated ET biosynthesis through ACC synthase (Hansen and Grossmann 2000), which thus triggers an increase in abscisic acid biosynthesis. Contingent upon planning and the results of *Trichoderma* upgrades, phytohormone homeostasis can regulate plant advancement and immune responses.

8.3 *Trichoderma*–Pathogen Interactions

Trichoderma show antagonistic interaction with pathogens by the following modes of action.

8.3.1 *Mycoparasitism*

Mycoparasitism is one of the main mechanisms involved in the antagonisms of *Trichoderma*. The events leading to mycoparasitism are complex, and take place as follows. First is the chemotropic growth of *Trichoderma*, recognition of the host, coiling and appressoria formation, secretion of different hydrolytic enzymes, penetration of hyphae, and lysis of the host. This complex process includes sequential events, involving a cycle of recognition by the binding of carbohydrates in the *Trichoderma* cell wall to lectins on the target fungus, hyphal coiling and appressoria formation that contains a higher amount of osmotic solutes such as glycerol and induces penetration, an attack on cellular machinery via producing several fungitoxic cell wall-degrading enzymes glucanases, chitinases, and proteases (Harman et al. 2004). The cumulative action of these compounds result in parasitism of the target fungus and dissolution of the cell walls. There are in any event 20–30 genes, proteins, and different metabolites that are straightforwardly engaged with this interaction. The elements of various glucanases and chitinases during the time used in mycoparasitism are very much concentrated from *Trichoderma* spp. utilizing gene-for-gene experiments, and future investigations will assist us with understanding this complex process (Daguerre et al. 2014).

8.3.2 *Antibiosis*

Antibiosis is an antagonistic interaction involving low molecular weight diffusible secondary metabolites or antibiotics produced by a microorganism that are detrimental for the growth of pathogens. *Trichoderma* spp. produces a variety of antibiotics, such as gliovirin, gliotoxin, viridin, viridol, koninginins, pyrones, and peptaibols, against fungal phytopathogens (Howell 2003; Harman et al. 2004). Trichokonin VI, a type of peptaibol from *Trichoderma pseudokoningii* SMF2, exhibited antibiotic activities by inducing extensive apoptotic programmed cell death in fungal pathogens (Shi et al. 2012). Gliotoxin and gliovirin are the two most important *Trichoderma* secondary metabolites belonging to P and Q group strains, respectively. The P group strains of *Trichoderma* (*Gliocladium*) *virens* are active against *Pythium ultimum*, but not against *Rhizoctonia solani*, and the Q group is very active against *Rhizoctonia solani*, but less so against *Pythium ultimum* (Howell et al. 2000). The *T. virens* veA ortholog *vel1* (VELVET protein Vel1) is associated with regulation of gliotoxin biosynthesis, biocontrol action, and numerous other secondary metabolism-related genes (Mukherjee et al. 2012). Koninginin D also hindered the development of soil-borne plant pathogens (Dunlop et al. 1989). Viridins isolated from diverse *Trichoderma* spp. (*Trichoderma koningii*, *Trichoderma viride*, *Trichoderma virens*) prevent spore germination of *Colletotrichum lini*, *Botrytis allii*, *Fusarium coeruleum*, *Aspergillus niger*, *Penicillium expansum*, and *Stachybotrys atra* (Singh et al. 2005). Harzianic acid isolated from a *Trichoderma harzianum* strain showed in vitro antibiotic activity against *Pythium irregulare*, *Sclerotinia sclerotiorum*, and *Rhizoctonia solani* (Vinale et al. 2009). The segregation and over-articulation of the tri5 (trichodiene synthase) gene in *Trichoderma brevicompactum* Tb41tri5 transformant expanded the trichodermin creation and fungal activity against *Aspergillus fumigatus* and *Fusarium* spp. (Tijerino et al. 2011a, b). A *Trichoderma asperellum* strain produces two asperelines (A and E) and five trichotoxins (T5D2, T5E, T5F, T5G, 1717A) that can be associated with antibiosis (Mukherjee et al. 2013).

8.3.3 Competition

The best known explanation behind the cell death of numerous microorganisms developing in the region of *Trichoderma* strains is starvation because of the shortage of restricting supplements and rhizospheric colonization. Root exudates and the rhizosphere are rich sources of nutrients such as sugar, amino acids, iron, vitamins, and organic acids. Competition for carbon is a successful mode in *Trichoderma* as well as some other organisms, for example, strains of *Rhizoctonia solani* and *Fusarium oxysporum* (Sarrocco et al. 2009; Alabouvette et al. 2009). The capable preparation of immobile supplements and their use makes it more effective and aggressive than numerous other soil microorganisms. This procedure could also be connected to the generation of organic acids, for example, gluconic, citrus, and fumaric acids, which diminish soil pH and permit the solubilization of phosphates, micronutrients, and mineral cations such as iron, manganese, and magnesium (Vinale et al. 2008). In the aerobic environment iron exists mainly as Fe^{3+} and tends to form insoluble ferric oxide, making it unavailable for root absorption and microbial growth (Miethke 2013). Iron acts as a cofactor of numerous enzymes and is an essential nutrient for the growth of plants and other organisms. *Trichoderma* secrete siderophore, an iron-chelating compound that binds with insoluble iron (Fe(III)) and is converted to soluble form (Fe(II)) for plant absorption and inhibits the growth of plant pathogens by depriving them of iron sources (Leong 1986). Siderophore-bound Fe(III) decreases to Fe(II), which is catalyzed by free extracellular or membrane-standing ferric chelate reductases. If not already discharged extracytoplasmatically, the iron must be expelled from the Fe(III)–siderophore complex in the cytosol: this is interceded either by intracellular ferric-siderophore reductases or, in a couple of cases, by ferric-siderophore hydrolases. Siderophores produced by some *Trichoderma* isolates are highly efficient chelators for iron. Consequently, *Trichoderma* spp. outcompetes with *Pythium* for accessible iron in the soil and successfully controls its development. There are many more models about the successful utilization of rivalry for the biocontrol of pathogens such as *Botrytis cinerea* (Latorre et al. 2001). These reports recommend that the molecular and proteomic gathering of *Trichoderma* is progressively productive to prepare and take up soil supplements when contrasted with numerous different pathogens and different organisms.

8.3.4 Trichoderma*–Pathogen Association*

The seven *trans*-membrane G-protein-coupled receptor (*Gpr1*) is involved in sensing nearby fungal prey (Omann et al. 2012). *Vel1* of *Trichoderma virens* is involved in hydrophobin expression for adherence to the host and mycoparasitism (Zeilinger et al. 2005). Ligand binding with such receptors leads to downstream signalling events via activation of G-protein cascades. *Trichoderma* spp. have three MAPK cascades: MAPKKK, MAPKK, and MAPK; the MAPK signalling pathways may also act in mycoparasitism and biocontrol (Reithner et al. 2007; Kumar et al. 2010). Production and release of cell wall-degrading enzymes (CWDEs) and antibiotics are the most important members of the chemical arsenal deployed by *Trichoderma* to kill pathogens. Glucan and chitin synthases are produced by *Trichoderma* species to repair their own cell wall damage by pathogens during *Trichoderma*–pathogen interaction. Simultaneously, genes encoding hydrolytic enzymes such as chitinases and glucanases, and those for secondary metabolism such as NRPSs, are expressed to kill the pathogens (Kubicek et al. 2011). Earlier evidence suggested the involvement of *chit42*, *chit3*, *bgn13.1*, *Bgn2*, *Bgn3*, and *prb1* in chitinases, glucanases, and proteases in biocontrol (Mondejar et al. 2011). Certain species such as *Trichoderma atroviride* (*T. virens*) produce the unstable metabolite 6-pentyl-2*H*-pyran-2-one (6-PP), which assumes a significant task in *Trichoderma*–fungal associations (Hasan et al. 2008; Vinale et al. 2009). Recently, hereditary proof has been accumulated about the gathering of 11- and 14-module peptaibols by a solitary NRPS Tex2 of *T. virens* (Mukherjee et al. 2011). The species of *Trichoderma* possess many qualities and attributes and have great potential for use in agriculture such as plant growth development by improving germination, alteration of morphological features and physiological functions by increasing the uptake of nutrients in plants, enhancing nitrogen use efficiency in different crops, and improving photosynthetic efficiency. A few procedures have been applied to recognize the primary genes and compounds associated with this complex, three-way crosstalk between the fungal antagonist, the plant, and microbial pathogens. The fungi also stimulate defence in response to biotic and abiotic stress and antagonism including mycoparasitism, antibiosis, and competition against pathogens. The genome of *Trichoderma* spp. has been broadly explored and has been demonstrated to contain numerous useful genes with the capacity to create an incredible assortment of expression patterns, which enables these organisms to adjust to a wide range of conditions, including soil, water, dead tissues, and inside the plants. The metabolomics of *Trichoderma* spp. are unfathomably complex, particularly as antiinfection agents and generating auxiliary metabolites, yet with the assistance of cutting-edge genomic and proteomic approaches it is possible to investigate new pathways and novel functions of compounds created by this family. The presence of *Trichoderma* genes in plants has advantageous outcomes, essentially in the control of plant infections and protection from adverse ecological conditions. The trial results show that *Trichoderma*–plant connections share, practically speaking, affiliations with other advantageous organisms but also show their own qualities because of the specific way of life of *Trichoderma*. All things considered, more studies are required to acquire knowledge about the signalling transduction pathways related to defence and improvement that result from *Trichoderma*–plant connections within the aspects of pathogens as well as various abiotic stresses (Table 8.1).

8.4 Future Prospects

The genetics of three-way crosstalk among plant–*Trichoderma*–pathogen interactions should be investigated, including the induction and regulation of enzyme expression. Moreover, the natural prompts in the transmission of the respective signals to the indicators of the output pathways should also be considered. Such understanding of the relevant physiology will further improve enzyme production in microbial cell manufacturing plants and upgrade the mycoparasitic capability of biocontrol strains. The distinguishing proof of the plant receptors for *Trichoderma* elicitor proteins and effectors that may reconstruct a host's hereditary hardware can function as the establishment for seeing how these symbionts trigger the host protection response. *Trichoderma* compatibility with chemical fungicides as a part of integrated disease management should be evaluated. There is a need to enhance the popularity of *Trichoderma*-based formulations among farmers for ecofriendly management of diseases. The ecological impact of large-scale applications of a single fungal species and their secondary metabolites for biocontrol should be evaluated to establish a knowledge base for the safe and sustainable use of *Trichoderma*.

8.5 Conclusion

With the expanding risks to nature and to our food security, selection of *Trichoderma* spp. as a biological control agent (BCA) has been assuming significance in providing security for plant protection and growth. To achieve this, unexplored parts of the world should be investigated to locate appropriate and efficient *Trichoderma* strains to function extraordinarily as biocontrol agents. Different species of *Trichoderma* are engaged in biocontrol action, and their

Table 8.1 Biocontrol genes of *Trichoderma* spp.

Trichoderma spp.	Genes	Product/function	Pathogen/ abiotic stress	References
T. virens	*Tvsp1*	Serine protease in cotton seedlings	*Rhizoctonia solani*	Pozo et al. (2004)
T. virens	*TgaA, TgaB*	Antagonist activity	*Rhizoctonia solani* and *Sclerotium rolfsii*	Mukherjee et al. (2004)
T. virens	*tac1*	Mycoparasitic activity	*Rhizoctonia solani* and *Phythium ultimum*	Abbas et al. (2017)
T. harzianum	*tri5*	Trichothecene inhibitor of protein and DNA synthesis	Soil-borne pathogens	Cardoza et al. (2011)
T. harzianum	*erg1*	Squalene epoxidase for ergosterol synthesis and *erg1* silencing provides resistance to terbinafine, an antifungal compound	–	Godio et al. (2007)
T. harzianum	*exc1, exc2, chit42, chit3, bgn13.1, prb1*	NAGases, chitinases, glucanases, proteases synthesis	*Fusarium oxysporum*	Mondejar et al. (2011)
T. harzianum	*ThPG1*	Endopolygalacturonase for cell wall degradation	*Rhizoctonia solani* and *Phythium ultimum*	Moran-Diez et al. (2009)
T. harzianum	*Th-Chit42*	Mycoparasitic activity	*Sclerotinia sclerotiorum*	Steyaert et al. (2004)
T. harzianum	*Sm1*	Elicitor for triggering of plant defence	Biotic/abiotic stress	Freitas et al. (2014)
T. harzianum	*Thkel1*	Regulate glucosidase in *Arabidopsis thaliana*	Salt, osmotic stresses	Hermosa et al. (2011)
T. harzianum	*qid74*	Surface adherence in mycoparasitism	*R. solani*	Rosado et al. (2007)
T. brevicompactum	*tri5*	Trichodermin for antifungal activity	*Candida albicans* and *Aspergillus fumigatus*	Tijerino et al. (2011a, b)
T. longibrachiatum	*egl1*	Biocontrol activity	*Phythium ultimum*	Migheli et al. (1998)
T. reesei	*ac1, TrCCD1*	Carotenoid metabolism for conidial and hyphal growth	–	Zhong et al. (2009)

mechanisms of activity are outstanding in their qualities, as has been seen and reported. The genes isolated from these biocontrol specialists have been found to assume a basic function in biocontrol actions, with the assistance of genetic engineering procedures to progressively enumerate more valuable genes that should be found to enhance farming for ourselves and the future. Epic discoveries will then show the advances and reception of *Trichoderma* as a BCA to help in the improvement of our harvests and in overcoming infestations.

Acknowledgments The authors acknowledge the Director, ICAR-National Rice Research Institute, Cuttack, for providing technical support, and the Department of Science and Technology, Govt. of India, New Delhi, for providing an INSPIRE Fellowship (IF140749) to Harekrushna Swain. The authors are also immensely thankful to Dr. Prasun K. Mukherjee, Scientist G and Head of the Environmental Biotechnology Section and group leader of Agricultural Microbiology, Nuclear Agricultural Biotechnology Division, Bhabha Atomic Research Centre, Trombay, Mumbai, for his tremendous support and guidance.

References

Abbas A, Jiang D, Fu Y (2017) *Trichoderma* spp. as antagonist of *Rhizoctonia solani*. J Plant Pathol Microb 8:1000402. https://doi.org/10.4172/2157-7471.1000402

Alabouvette C, Olivain C, Migheli Q, Steinberg C (2009) Microbiological control of soil-borne phytopathogenic fungi with special emphasis on wilt-inducing *Fusarium oxysporum*. New Phytol Trust 184:529–544

Babychan M, Simon S (2017) Efficacy of *Trichoderma* spp. against *Fusarium oxysporum* f. sp. *lycopersici* (FOL) infecting pre-and post-seedling of tomato. J Pharmacogn Phytochem 6:616–619

Benitez T, Rincon AM, Limon MC, Codon AC (2004) Biocontrol mechanisms of *Trichoderma* strains. Int Microbiol 7:249–260

Bezuidenhout CN, Van Antwerpen R, Berry SD (2012) An application of principal component analyses and correlation graphs to assess multivariate soil health properties. Soil Sci 177:498–505

Bjorkman T, Blanchard LM, Harman GE (1998) Growth enhancement of shrunken-2 sweet corn when colonized with *Trichoderma harzianum* 1295-22: effect of environmental stress. J Am Soc Hort Sci 123:35–40

Brotman Y, Landau U, Cuadros-Inostoza A, Takayuki T, Fernie AR et al (2013) *Trichoderma*–plant root colonization: escaping early plant defense responses and activation of the antioxidant machinery for saline stress tolerance. PLoS Pathog 9:e1003221. https://doi.org/10.1371/journal.ppat.1003221

Cai F, Yu G, Wang P, Wei Z, Fu L, Shen Q, Chen W (2013) Harzianolide, a novel plant growth regulator and systemic resistance elicitor from *Trichoderma harzianum*. Plant Physiol Biochem 73:106–113

Cardoza RE, Malmierca MG, Hermosa MR, Alexander NJ, McCormick SP, Proctor RH, Tijerino AM, Rumbero A, Monte E, Gutierrez S (2011) Identification of loci and functional characterization of trichothecene biosynthesis genes in filamentous fungi of the genus *Trichoderma*. Appl Environ Microbiol 77:4867–4877

Chowdappa P, Kumar SPM, Lakshmi MJ, Upreti KK (2013) Growth stimulation and induction of systemic resistance in tomato against early and late blight by *Bacillus subtilis* OTPB1 or *Trichoderma harzianum* OTPB3. Biol Control 65:109–117

Contreras-Cornejo HA, Macías-Rodríguez L, Cortés-Penagos C (2009) *Trichoderma virens*, a plant beneficial fungus enhances biomass production and promotes lateral root growth through an auxin dependent mechanism in *Arabidopsis*. Plant Physiol 149:1579–1592

Cuevas VC (2006) Soil inoculation with *Trichoderma pseudokoningii* Rifai enhances yield of rice. Philipp J Sci 135(1):31–37

Daguerre Y, Siegel K, Edel-Hermann V, Steinberg C (2014) Fungal proteins and genes associated with biocontrol mechanisms of soil-borne pathogens: a review. Fungal Biol Rev 28:97–125

Doni F, Isahak A, Radziah C, Zain CM, Ariffin SM, Nurashiqin W, Mohamad W, Mohtar W, Yusoff W (2014) Formulation of *Trichoderma* sp. SL2 inoculants using different carriers for soil treatment in rice seedling growth. SpringerPlus 3:1–5

Dunlop RW, Simon A, Siwasithamparam D, Ghisalberti EL (1989) An antibiotic from *Trichoderma koningii* active against soilborne plant pathogens. J Nat Prod 52:67–74

Freitas RS, Steindorff AS, Ramada MHS, Siqueira SJL, Noronha EF, Ulhoa CJ (2014) Cloning and characterization of a protein elicitor Sm1 gene from *Trichoderma harzianum*. Biotechnol Lett 36(4):783–788

Godio RP, Fouces R, Martín FJ (2007) A squalene epoxidase is involved in biosynthesis of both the antitumor compound clavaric acid and sterols in the basidiomycete *H. sublateritium*. Cell Press 14:1334–1346

Gravel V, Antoun H, Tweddell RJ (2007) Growth stimulation and fruit yield improvement of greenhouse tomato plants by inoculation with *Pseudomonas putida* or *Trichoderma atroviride*: possible role of indole acetic acid (IAA). Soil Biol Biochem 39:1968–1977

Hansen H, Grossmann K (2000) Auxin induced ethylene triggers abscisic acid biosynthesis and growth inhibition. Plant Physiol 124:1437–1448

Haque MM, Ilias GNM, Molla AH (2012) Impact of *Trichoderma*-enriched biofertilizer on the growth and yield of mustard (*Brassica rapa* L.) and tomato (*Solanum lycopersicum* Mill.). Agriculturists 10:109–119

Harman GE (2000) Myths and dogmas of biocontrol changes in perceptions derived from research on *Trichoderma harzianum* T-22. Plant Dis 84:377–393

Harman GE (2006) Overview of mechanisms and uses of *Trichoderma* spp. Phytopathology 96:190–194

Harman GE (2011) Multifunctional fungal plant symbionts: new tools to enhance plant growth and productivity. New Phytol 189:647–649

Harman GE, Howell CR, Viterbo A, Chet I, Lorito M (2004) *Trichoderma* species opportunistic, avirulent plant symbionts. Nat Rev Microbiol 2:43–56

Harman GE, Björkman T, Ondik K, Shoresh M (2008) Changing paradigms on the mode of action and uses of *Trichoderma* spp. for biocontrol. Outlooks Pest Manag 19:24–29

Hasan EA, Walker F, Buchenauer H (2008) *Trichoderma harzianum* and its metabolite 6-pentyl-alpha-pyrone suppress fusaric acid produced by *Fusarium moniliforme*. J Phytopathol 156:79–87

Hermosa R, Botella L, Keck M, Jimenez JA, Montero Barrientos M, Arbona V, Gomez Cadenas A, Monte E, Nicolas C (2011) The over expression in *Arabidopsis thaliana* of a *Trichoderma harzianum* gene that modulates glucosidase activity and enhances tolerance to salt and osmotic stresses. J Plant Physiol 168:1295–1302

Hermosa R, Viterbo A, Chet I, Monte E (2012) Plant-beneficial effects of *Trichoderma* and of its genes. Microbiology 158:17–25

Herrera-Estrella A, Chet I (2004) The biological control agent *Trichoderma* from fundamentals to applications. In: Arora D (ed) Handbook of fungal biotechnology. Dekker, New York, NY, pp 147–156

Howell CR (2003) Mechanisms employed by *Trichoderma* species in the biological control of plant diseases: the history and evolution of current concepts. Plant Dis 87:4–10

Howell CR, Hanson LE, Stipanovic RD, Puckhaber LS (2000) Induction of terpenoid synthesis in cotton roots and control of *Rhizoctonia solani* by seed treatment with *Trichoderma virens*. Phytopathology 90:248–252

Idowu OO, Olawole OI, Idumu OO, Salami AO (2016) Bio-control effect of *Trichoderma asperellum* (Samuels) Lieckf. and *Glomus intraradices* Schenk on okra seedlings infected with *Pythium aphanidermatum* (Edson) Fitzp. and *Erwinia carotovora* (Jones). Am J Exp Agric 10:1–12

Inbar J, Abramsky M, Cohe D, Chet I (1994) Plant growth enhancement and disease control by *Trichoderma harzianum* in vegetable seedlings grown under commercial conditions. Eur J Plant Pathol 100:337–346

Jiang X, Geng A, He N, Li Q (2011) New isolate of *Trichoderma viride* strain for enhanced cellulolytic enzyme complex production. J Biosci Bioeng 111:121–127

Kubicek CP, Herrera-Estrella A, Seidl-Seiboth V, Martinez DA et al (2011) Comparative genome sequence analysis underscores mycoparasitism as the ancestral life style of *Trichoderma*. Genome Biol 12:1–15

Kumar A, Scher K, Mukherjee M, Pardovitz-Kedmi E, Sible GV, Singh US, Kale SP, Mukherjee PK, Horwitz BA (2010) Overlapping and distinct functions of two *Trichoderma virens* MAP kinases in cell wall integrity, antagonistic properties and repression of conidiation. Biochem Biophys Res Commun 398:765–777

Kumar V, Shahid M, Singh A, Srivastava M, Mishra A, Srivastava YK, Pandey S, Sharma A (2014) Effect of biopriming with biocontrol agents *Trichoderma harzianum* (Th. Azad) and *Trichoderma viride* on chickpea genotype (Radhey). J Plant Pathol Microb 5:1–4

Latorre BA, Lillo C, Rioja ME (2001) Eficacia de los tratamientos fungicidas para el control de *Botrytis cinerea* de la vid en función de la época de aplicación. Cienc Invest Agraria 28:61–66

Leong J (1986) Siderophores: their biochemistry and possible role in biocontrol of plant pathogen. Annu Rev Phytopathol 24:187–209

Lorito M, Woo SL, Harman GE, Monte E (2010) Translational research on *Trichoderma*: from 'omics to the field. Annu Rev Phytopathol 48:395–417

Losane BK, Kumar PKR (1992) Fungal plant growth regulators. In: Arora DK, Elander KG, Mujerji KG (eds) Handbook of applied mycology: fungal biotechnology, vol 4. Dekker, New York, NY, pp 565–602

Mastouri F, Björkman T, Harman GE (2010) Seed treatment with *Trichoderma harzianum* alleviates biotic, abiotic and physiological stresses in germinating seeds and seedlings. Phytopathology 100:1213–1221

Meszka B, Broniarek-Niemiec A, Bielenin A (2008) The status of dodine resistance of *Venturia inaequalis* populations in Poland. Phytopathol Pol 47:57–61

Miethke M (2013) Molecular strategies of microbial iron assimilation: from high affinity complexes to cofactor assembly systems. Metallomics 5:15–28

Migheli Q, González-Candelas L, Dealessi L, Camponogara A, Ramon-Vidal D (1998) Transformants of *Trichoderma longibrachiatum* over-expressing the beta-1,4-endoglucanase gene egl1 show enhanced biocontrol of *Pythium ultimum* on cucumber. Phytopathology 88:673–677

Mishra A, Salokhe VM (2011) Rice growth and physiological responses to SRI water management and implications for crop productivity. Paddy Water Environ 9:41–52

Mondejar LR, Ros M, Pascual JA (2011) Mycoparasitism related genes expression of *Trichoderma harzianum* isolates to evaluate their efficacy as biological control agent. Biol Control 56:59–66

Moran-Diez E, Hermosa R, Ambrosino P, Cardoza RE, Gutierrez S, Lorito M, Monte E (2009) The ThPG1 endopolygalacturonase is required for the *Trichoderma harzianum*–plant beneficial interaction. Mol Plant Microbe Interact 22:1021–1031

Mukherjee PK, Latha J, Hadar R, Horwitz BA (2004) Role of two G-protein alpha subunits, TgaA and TgaB, in the antagonism of plant pathogens by *Trichoderma virens*. Appl Environ Microbiol 70:542–549

Mukherjee PK, Wiest A, Ruiz N, Keightley A, Moran-Diez ME, McCluskey K, Pouchus YF, Kenerley CM (2011) Two classes of new peptaibols are synthesized by a single non-ribosomal peptide synthetase of *Trichoderma virens*. J Biol Chem 286:4544–4554

Mukherjee PK, Horwitz BA, Kenerley CM (2012) Secondary metabolism in *Trichoderma*: a genomic perspective. Microbiology 158:35–45

Mukherjee PK, Horwitz BA, Singh US, Mukherjee M, Schmoll M (2013) *Trichoderma*: biology and applications. CAB International, Wallingford. (www.cabi.org)

Mukhtar I (2008) Influence of *Trichoderma* species on seed germination in okra. Mycopathology 6:47–50

Naznin A, Hossain MM, Anju-ManAra K, Hoque A, Islam M, Hasan T (2015) Influence of organic amendments and bio- control agent on yield and quality of tuberose. J Hort 2:4. https://doi.org/10.4172/2376-0354.1000156

Neumann B, Laing M (2006) *Trichoderma*: an ally in the quest for soil system sustainability. In: Uphoff N, Fernandes E, Herren H, Husson O, Laing M, Palm C, Pretty J, Sanchez P, Sanginga N, Thies J (eds) Biological approaches to sustainable soil system. Taylor & Francis, Boca Raton, FL, pp 491–500

Newman ME (2006) Modularity and community structure in networks. Proc Natl Acad Sci U S A 103(23):8577–8582

Okoth SA, Otadoh JA, Ochanda OJ (2011) Improved seedling emergence and growth of maize and beans by *Trichoderma harziunum*. Trop Subtrop Agroecosyst 13:65–71

Omann MR, Lehner S, Escobar-Rodríguez C, Brunner K, Zeilinger S (2012) The seven-transmembrane receptor Gpr1 governs process relevant for the antagonistic interaction of Trichoderma atroviride with its host. Microbiol 158:107–118. https://doi.org/10.1099/mic.0.052035-0

Pieterse CM, Leon-Reyes A, Vander Ent S, Van Wees SC (2009) Networking by small molecule hormones in plant immunity. Nat Chem Biol 5:308–316

Pozo MJ, Baek JM, García JM, Kenerley CM (2004) Functional analysis of tvsp1, a serine protease-encoding gene in the biocontrol agent *Trichoderma virens*. Fungal Genet Biol 41:336–348

Rehman SU, Lawrence R, Kumar EJ, Badri ZA (2012) Comparative efficacy of *Trichoderma viride*, *T. harzianum* and carbendazim against damping-off disease of cauliflower caused by *Rhizoctonia solani* Kuehn. J Biopest 5:23–27

Reithner B, Schuhmacher R, Stoppacher N, Pucher M, Brunner K, Zeilinger S (2007) Signaling via the *Trichoderma atroviride* mitogen-activated protein kinase Tmk1 differentially affects mycoparasitism and plant protection. Fungal Genet Biol 44:1123–1133

Rosado IV, Rey M, Codón AC, Govantes J, Moreno-Mateos MA (2007) QID74 cell wall protein of *Trichoderma harzianum* is involved in cell protection and adherence to hydrophobic surfaces. Fungal Genet Biol 44:950–964

Saba H, Vibhash D, Manisha M, Prashant KS, Farhan H, Tauseef A (2012) *Trichoderma*: a promising plant growth stimulator and biocontrol agent. Mycosphere 3:524–531

Sajeesh PK (2015) Cu-Chi-Tri: a triple combination for the management of late blight disease of potato (*Solanum tuberosum* L.). Dissertation, GBPUA&T, Pantnagar, India

Sarrocco S, Guidi L, Fambrini S, Des I'Innocenti E, Vannacci G (2009) Competition for cellulose exploitation between *Rhizoctonia solani* and two *Trichoderma* isolated in the decomposition of wheat straw. J Plant Pathol 91:331–338

Seiboth B, Herold S, Kubicek CP (2012) Metabolic engineering of inducer formation for cellulase and hemicellulase gene expression in *Trichoderma reesei*. Subcell Biochem 64:367–390

Shayakhmetov IF (2001) Biological activity of metabolites from culture filtrates of *Cochliobolus sativus* and *Fusarium oxysporum* in connection with in vitro cellular selection of crop plants for resistance to phytopathogens. Mycol Phytopathol 35:66–71

Shi M, Chen L, Wang XW, Zhang T, Zhao PB, Song XY, Sun CY, Chen X-L, Zhou BC, Zhang YZ (2012) Antimicrobial peptaibols from *Trichoderma pseudokoningii* induce programmed cell death in plant fungal pathogens. Microbiology 158:166–175

Shoresh M, Gal-On A, Leibman D, Chet I (2006) Characterization of a mitogen-activated protein kinase gene from cucumber required for *Trichoderma*-conferred plant resistance. Plant Physiol 142:1169–1179

Shukla N, Awasthi RP, Rawat L, Kumar J (2012) Biochemical and physiological responses of rice (*Oryza sativa* L.) as influenced by *Trichoderma harzianum* under drought stress. Plant Physiol Biochem 54:78–88

Singh S, Dureja P, Tanwar RS, Singh A (2005) Production and antifungal activity of secondary metabolites of *Trichoderma virens*. Pest Res J 17:26–29

Singh HB, Singh BN, Singh SP, Singh SR, Sarma BK (2009) Biological control of plant diseases: current status and future prospects. In: Johri JK (ed) Recent advances in biopesticides: biotechnological applications. New India, New Delhi

Srivastava SN, Singh V, Awasthi SK (2006) *Trichoderma* induced improvement in growth, yield and quality of sugarcane. Sugar Technol 8:166. https://doi.org/10.1007/BF02943654

Stepanova AN, Yun J, Likhacheva AV, Alonso JM (2007) Multilevel interactions between ethylene and auxin in *Arabidopsis* roots. Plant Cell 19:2169–2185

Steyaert JM, Stewart A, Jaspers MV, Carpenter M, Ridgway HJ (2004) Co-expression of two genes, a chitinase (chit42) and proteinase (prb1), implicated in mycoparasitism by *Trichoderma hamatum*. Mycologia 96:1245–1252

Swain H, Adak T, Mukherjee AK, Mukherjee PK, Bhattacharya P, Behera S, Bagchi T, Patro R, Shasmita KAK, Bag MK, Danger TK, Lenka S, Jena M (2018) Novel *Trichoderma* strains isolated from tree barks as potential biocontrol agents and biofertilizers for direct seeded rice. Microbiol Res 214:83–90

Thakur M, Sohal BS (2013) Role of elicitors in inducing resistance in plants against pathogen infection: a review. ISRN Biochem 2013:762412. https://doi.org/10.1155/2013/762412. pp 1–10

Thakur AK, Uphoff N, Antony E (2010) An assessment of physiological effects of system of rice intensification (SRI) practices compared with recommended rice cultivation practices in India. Exp Agric 46:77–98

Tijerino A, Cardoza RE, Moraga J, Malmierca MG, Vicente F, Aleu J, Collado IG, Gutierrez S, Monte E, Hermosa R (2011a) Overexpression of the trichodiene synthase gene tri5 increases trichodermin production and antimicrobial activity in *Trichoderma brevicompactum*. Fungal Genet Biol 48:285–296

Tijerino A, Hermosa R, Cardoza RE, Moraga J, Malmierca MG, Aleu J, Collado IG, Monte E, Gutierrez S (2011b) Overexpression of the *Trichoderma brevicompactum* tri5 gene: effect on the expression of the trichodermin biosynthetic genes and on tomato seedlings. Toxins 3:1220–1232

Tucci M, Ruocco M, De Masi L, De Palma M, Lorito M (2011) The beneficial effect of *Trichoderma* spp. on tomato is modulated by the plant genotype. Mol Plant Pathol 12 (4):341–354. https://doi.org/10.1111/j.1364-3703.2010.00674

Van Wees SCM, Van der Ent S, Pieterse CMJ (2008) Plant immune responses triggered by beneficial microbes. Curr Opin Plant Biol 11:443–448

Verma M, Brar SK, Tyagi RD, Surampalli RY, Val'ero JR (2007) Antagonistic fungi, *Trichoderma* spp.: panoply of biological control. Biochem Eng J 37:1–20

Vinale F, Sivasithamparam K, Ghisalberti EL, Marra R, Woo SL, Lorio M (2008) *Trichoderma*–plant–pathogen interactions. Soil Biol Biochem 40:1–10

Vinale F, Ghisalberti EL, Sivasithamparam K, Marra R, Ritieni A, Ferracane R, Woo S, Lorito M (2009) Factors affecting the production of *Trichoderma harzianum* secondary metabolites during the interaction with different plant pathogens. Lett Appl Microbiol 48:705–711

Vinale F, Marra R, Ruocco M (2014) *Trichoderma* secondary metabolites active on plants and fungal pathogens. Open Mycol J 8:127–139

Viterbo A, Horwitz BA (2010) Mycoparasitism. In: Borkovich K, Ebbole DJ (eds) Cellular and molecular biology of filamentous fungi. ASM Press, Washington, DC, pp 676–693

Yedidia I, Srivastva AK, Kapulnik Y, Chet I (2001) Effect of *Trichoderma harzianum* on microelement concentrations and increased growth of cucumber plants. Plant and Soil 235:235–242

Yoshioka Y, Ichikawa H, Naznin HA, Kogure A, Hyakumachi M (2012) Systemic resistance induced in *Arabidopsis thaliana* by *Trichoderma asperellum* SKT-1, a microbial pesticide of seed borne diseases of rice. Pest Manag Sci 68:60. https://doi.org/10.1002/ps.2220

Zeilinger S, Reithner B, Scala V, Peiss I, Lorito M, Mach RL (2005) Signal transduction by Tga3, a novel G-protein-alpha subunit of *Trichoderma atroviride*. Appl Environ Microbiol 71:1591–1597

Zhong YH, Wang TH, Wang XL, Zhang GT, Yu HN (2009) Identification and characterization of a novel gene, *TrCCD1*, and its possible function in hyphal growth and conidiospore development of *Trichoderma reesei*. Fungal Genet Biol 46:255–263

Chapter 9
Trichoderma–Fusarium Interactions: A Biocontrol Strategy to Manage Wilt

Ishwar Prakash Sharma and Anil K. Sharma

Abstract Soil-borne microorganisms are either beneficial or harmful to proper plant growth and crop yield. All these microorganisms share the same niche in the rhizosphere and interact with each other. *Fusarium* is one of the most important fungal pathogen in the soil, which is found nearly in every agricultural crop and causes yield loss due to suppression in plant growth and development. Various chemical methods are being used for its management, which are expensive and hazardous to the environment for sustainable development due to their toxicity. Biological control is one of the most effective and popular ways to manage it because it has very low level of toxicity in the environmental point of view and is cost-effective. Among the biocontrol agents, *Trichoderma* is a popular symbiotic fungus that inhibits or controls the *Fusarium* wilt through the mechanism of mycoparasitism, completion, production of various lytic enzymes, and other antimicrobial activities along with enhancement in the growth of the host plant by the production of plant growth hormones. In the soil, *Trichoderma* releases various antimycoparasitic proteins, enzymes, volatile and nonvolatile compounds, and other secondary metabolites that have the potency to solubilize nutrients and defend against pathogens. Due to these qualities, *Trichoderma* enhances growth in plants and protect them from soil pathogenic microorganisms. A mechanism from hyphal attraction to cell lysis via attachment and coiling performed in a series in the interaction of *Trichoderma* and *Fusarium*. Thus *Trichoderma* induces systemic resistance in the host plant due to pathogen inhibition and would be a potent biocontrol agent to attain sustainable agriculture, for better growth, production, and yield.

Keywords Biocontrol · *Fusarium* · Metabolites · Sustainable · *Trichoderma*

I. P. Sharma (✉) · A. K. Sharma
Biological Sciences, CBSH, G.B. Pant University of Agriculture & Technology, Pantnagar, Uttarakhand, India

A. K. Sharma, P. Sharma (eds.), *Trichoderma*, Rhizosphere Biology,
https://doi.org/10.1007/978-981-15-3321-1_9

9.1 Introduction

Soil-borne microorganisms are either important or harmful for human beings directly and indirectly, among them the pathogenic microorganisms create the most daunting problems in crop protection. Fungi are more difficult to observe and control, and avoiding their occurrence is nearly impossible as they can be hardly identified due to their low concentrations (Cornejo et al. 2015). Among the pathogenic fungi, the *Fusarium* spp. can damage more than a 100 different crop species by producing a series of toxic secondary metabolites, which directly involve in the threat of agriculture biosafety, food security, and plant and animal health (Berges et al. 2013). This fungus is common almost in all farm soils where it survives through chlamydospores that germinate after contacting with susceptible plant roots, and it can be considered as a limiting factor in the cultivation of all the agricultural crops throughout the world. It colonizes in the plant through the root system, basically root tip and lateral roots, via xylem causing wilt and prevents the plant from optimal development (Gawehns et al. 2013; Giurgiu et al. 2018). In onion *F. oxysporum* f. sp. *cepae* and *F. proliferatum* (Stankovic et al. 2007), pigeon pea *F. udum* (Dehariya et al. 2015), tomato *F. oxysporum* f. sp. *Lycoperssici* (Christopher et al. 2010; Kumar et al. 2011), and in cucumber *F. oxysporum* and *F. solani* (Al-Tuwaijri 2009) are found to be the most destructive patho-agents. These pathogens cause serious damage in both farm and store conditions. Various controlling methods have been proposed by the researchers to manage them, but biological control is one of the most effective methods, which is eco-friendly as well as cost-effective. Biologically it can be controlled through various bioagents such as various strains of *Trichoderma*, fluorescent Pseudomonas, arbuscular mycorrhizal fungi, and *Bacillus*, *Streptomyces griseoviridis*. (Kumar et al. 2011; Dehariya et al. 2015). The biological control by these biocontrol agents is considered important in studying the genetic diversity within and between species or strains. Among the bioagents, the *Trichoderma* spp. has immense potential to control the disease caused by *Fusarium* spp., whose application in agriculture is still limited. Therefore the characteristics and interactions of *Trichoderma* spp. with *Fusarium* spp. and other pathogens such as *Phytophthora*, *Rhizoctonia*, *Sclerotium*, and *Pythium* need to be studied, which can help farmers use it for the proper biocontrol of pathogen-related diseases (Ghanbarzadeh et al. 2014; Redda et al. 2018). Biocontrol is gaining popularity among farmers and researchers due to its efficacy in disease management and its lesser negative impact on the environment. *Trichoderma* sp. is one of the most popular bioagents for biocontrolling pathogens that can be found in all the soils, and it symbiotically associates with the plants' roots for nutrients flux and enhances plants' development by competing with soil-borne pathogens such as *Fusarium*, *Pythium*, and *Sclerotium* for nutrients and space (Althaf and Srinivas 2013; Lace et al. 2014). In a symbiotic relationship *Trichoderma* spp. derive sucrose or other nutrients from plant roots, while in return they boost the plant's immunity and improve the photosynthetic abilities, which ultimately coordinate transcriptomic, proteomic, and metabolomic response in the plant (Vargas et al. 2011; Brotman et al.

2012; Moran-Diez et al. 2012). In this regard, *Trichoderma* is considered as the most common rhizospheric and endophytic fungus and is reported as a plant growth promoter and biocontrol agent in several economically important cropping systems (Harman et al. 2004). In plant roots, *Trichoderma* spp. can colonize both externally and internally, where they biologically interact and attract toward plant roots which results chemical signal by both the partners. The characteristics of *Trichoderma* such as mycoparasitizing, rhizosphere competition, antibiotic and enzyme production, and induced resistance are the major driving forces behind their commercial success as biofungicides (Dehariya et al. 2015). The application of *Trichoderma* spp. in the field increase the secondary metabolism in the root systems, which acts as a chemical factory against the pathogens and in the conventional soil cultivation enhances the plant biomass, controls the diseases or pathogens, and improves the plant rooting systems (Ming et al. 2013; Lopez-Bucio et al. 2015). These secondary metabolites not only facilitate the survival of the plants but also play a defensive role against pathogens, which results in higher concentrations of bioactive substances (Rejeb et al. 2014; Abdelrahman et al. 2016; Giurgiu et al. 2018). Mycoparasitic activity can be performed by *Trichoderma* spp. through antagonistic interactions before meeting with pathogen, chemo-attraction, or overcoming and parasitism (Vinale et al. 2008). Systematically, both *Trichoderma* and *Fusarium* are filamentous fungi, belong to the class Ascomycetes and family Hypocreaceae (Pitt et al. 1994). Earlier it was only a hypothesis that *Trichoderma* is involved in plant growth promotion, which directly resulted in the control of minor pathogens; this hypothesis suggests that the plants grown in nonfertile soil which faces the degree of disease unable to reach their maximum growth since, *Trichoderma* used to disease suppression so that plant extent their maximum growth. On this basis various studies continue to control soil-borne disease in crop plants through this fungus species. Since, plant diseases have major effects on agricultural production and food demand so that we need to sustainable strategies for disease management in order to reduce the uses of chemical pesticides in agricultural land, thus, *Trichoderma* spp. is an effective and useful biocontrol agent in the place of chemical pesticides for disease management in sustainable agriculture.

9.2 Effect of *Trichoderma* and *Fusarium* on Plant Growth

Many of the plant growth-promoting fungi are effective in vitro, but in field they do not perform significantly. The survival rate and population density of these fungi are prerequisites for fungus effectiveness in the rhizosphere (Raaijmakers et al. 2009; Zhang et al. 2013a). Among all the plant growth-promoting fungi, *Trichoderma* spp. are the most renown, which enhance plant growth via nutrient uptake by solubilization of soil nutrients, significantly increase the root length and the number of root hairs for exploring a larger area in the soil, facilitate growth hormones (indole acetic acids, cytokinins, gibberellins, and zeatins) production, and protect against biotic (pathogen infections) and abiotic stress (Contreras-Cornejo et al. 2009; Zhang et al.

2013b, 2018). Till date, a total of 1100 strains of *Trichoderma* have been identified from 75 molecularly characterized species, among which many are reported for their antimicrobial activities (Contreras-Cornejo et al. 2016). The application of *Trichoderma* spp. in soil improves fertility, which significantly altered the fungal diversity and community in the rhizosphere of the crop plants, ultimately improving the crop yields (Zhang et al. 2018). The nutrient solubilization especially P and Fe availability enhance to plants by *T. asperellum* with significantly increasing morpho-physiological parameters such as dry weight, shoot length, and leaf area (Yedidia et al. 2001). Soil-borne pathogens are some of the most daunting problems in crop protection. *Fusarium* spp. are one of the most important soil-borne pathogens that can damage more than 100 plant species by producing various toxic secondary metabolites and reducing the agricultural yields, which creates problems of bio-safety, food security, and health in plants and animals (Berges et al. 2013). *Fusarium* produces varieties of mycotoxins, mainly trichothecenes A (diacetoxyscirpenol—DAS, HT-2 toxin, T-2 toxin, and deoxynivalenol—DON) and trichothecenes B (nivalenol—NIV) along with fumonisins, enniatins, zearalenone, beauvericins, moniliformins, fusarins, fusaric acids, and fusaproliferin, that are responsible for plant diseases and yield losses in agricultural lands (Dinolfo et al. 2017). In maize, *F. verticillioides* is the most destructive pathogen responsible for stalk rot and cob rot, which reduces the yield and grain quality by producing fumonisins (a mycotoxin), which create harmful effects on consumers and cause leukoencephalomalacia in horses (Silva et al. 2006). In the crops of economic importance, *Fusarium* causes vascular wilts. The parasitic stage of this fungus involves penetration, colonization in root cortex/endodermis, movement toward xylem, colonization in xylem where it expresses the symptoms, and finally death of the host plant (Okungbowa and Shittu 2012). The general symptoms of wilt caused by *Fusarium* are stunted growth, wilting of leaves, reddish discoloration in xylem vessels, and white, pink, and orange fungal growth on the outside of affected stems (Mueller and Beckman 1988; Miller et al. 2011). Based on these above studies, *Trichoderma* spp. have potential toward the plant health and growth while *Fusarium* spp. oppose it.

9.3 *Trichoderma*–Plant Interactions

A systematic interaction of *Trichoderma* with a plant and *Fusarium* is represented in Fig. 9.1. *Trichoderma* spp. grow in the rhizosphere, and they have the capability to penetrate plant roots and colonize themto absorb sucrose or other nutrients, while in return they boost the plant's immunity against pathogens and improve the plant's photosynthetic abilities (Vargas et al. 2009, 2011). In the rhizosphere, *Trichoderma* directly involves in the coordination of transcriptomic, proteomic, and metabolomic responses in the plant, which is often important for the plant's growth, yield, and resistance against pathogens. *Trichoderma* also increases the sugar content, phenolic compounds, photosynthetic rate, transpiration, and water content in plants

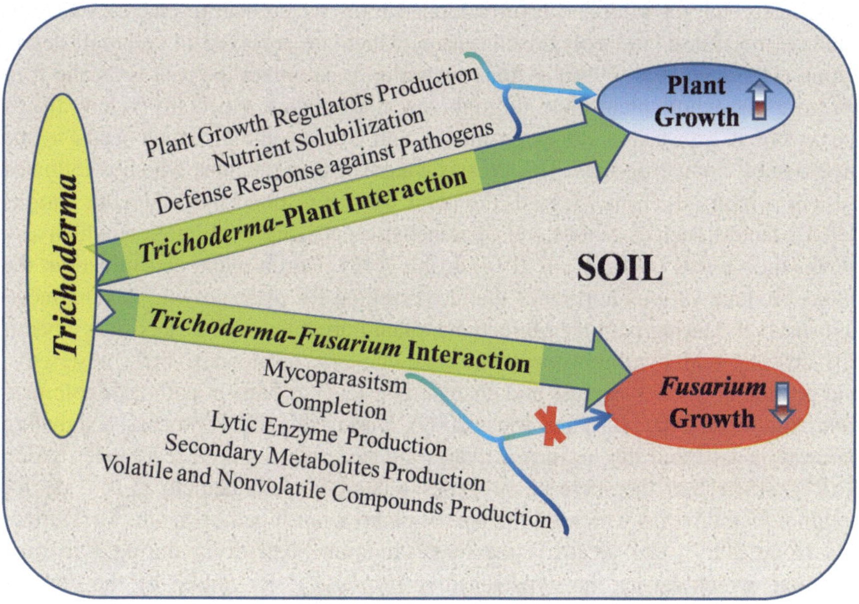

Fig. 9.1 Representation of *Trichoderma* interaction with plant and *Fusarium*

(Mukherjee et al. 2012a). To establish interaction with plants, *Trichoderma* first involves the roots for colonization both externally and internally. Both the partners, *Trichoderma* and plants interconnected by the chemical signals; the fungal partner produces the hormonal signals in order to colonization in the roots primarily it produces auxins that promote root growth for the fungal colonization due to increasing surface area (Contreras-Cornejo et al. 2009). After colonization and fungal growth inside the tissues, plants immediately respond with ion fluxes followed by callose deposition and polyphenols synthesis, and subsequently the salicylic acid and jasmonic acid signaling pathways are activated resulting in the entire plant acquiring varying degrees of tolerance against pathogen invasion (Shoresh et al. 2010). Inside the root system, this fungus lives as a true endophyte, where it induces the transcriptomic changes and protects the plants from various biotic and abiotic stresses (Bae et al. 2009; Druzhinina et al. 2011). Some species of this fungus colonize on the surface of glandular trichomes and form appressoria-like structures. This is an example of nonroot mode of entry of *Trichoderma* in plants (Bailey et al. 2009; Mukherjee et al. 2012a).

9.4 *Fusarium*–Plant Interactions

Fusarium is a parasitic fungus that interacts with plants and causes wilt. This fungus entirely depends on plant tissues for its growth and development. During parasitism this fungus develops hyphae that attach to the plant roots and penetrate the roots via epidermal cells via wounded or unwounded roots by producing enzymes such as cutinase, pectinase, and polygalacturonase, which are involved in cell wall degradation (Mendgen et al. 1996). After penetration it colonizes the root cells and then spread to the whole plant body through vascular bundles, especially xylem. Inside the xylem it grows and develops micro- and macroconidia, which leads to the blockage of vascular tissues. This creates water stress in the host, which is followed by wilt symptoms. In this regard, the plant tries to respond by producing various defense factors such as tyloses, gels, parenchyma cells, and proliferation, which also blocks the vessels (Ortiz et al. 2014; Joshi 2018). During pathogen infection, the plants produce various hormones that are essential for plant growth, development, and survival. During pathogen infections, plant hormones such as auxins, gibberellins, abscisic acid, cytokinins, salicylic acid, ethylene, jasmonates, brassinosteroids, and peptide hormones express and change their level to prevent pathogen colonization (Adie et al. 2007; Bari and Jones 2009). During this whole process, a signaling pathway is activated that in turn activates the systemic acquired resistance (SAR). SAR characterizes the increase of genes related to pathogenicity that code for antimicrobial proteins which immunize the plants against pathogen attacks (Durrant and Dong 2004). Cell death is the most common plant event during *Fusarium* infection, which carries the hypersensitive response (HR) caused by the reactive oxygen species (ROS) that are produced during tissue necrosis and *Fusarium* development (Dinolfo et al. 2017).

9.5 *Trichoderma–Fusarium* Interactions

Mycoparasitism is apparently an ancestral trait of *Trichoderma* (Kubicek et al. 2011; Druzhinina et al. 2011). The ability of *Trichoderma* to kill other pathogenic fungi is a major driving force toward commercial success. A typical interaction of *Tricoderma* with *Fusarium* involves attraction, attachment, coiling, and lysis by hydrolytic enzymes or secondary metabolites (Mukherjee et al. 2012a). For the sensing of pathofungus, G protein-coupled receptor Gpr1 involves in sensing the prey which silences by *gpr1* gene that is reported in *T. atroviride* (Omann et al. 2012). Attachment on *Fusarium* by *Trichoderma* is accompanied by the formation of papillae-like structures known as appressoria, which coil around the host hyphae. Attachment of genetic evidences from both partners (*Trichoderma* and *Fusarium*) are not well understood; probably, hydrophobins might be involved, which is experimentally still not clear, but indirectly the expression of hydrophobins reported from *T. virens* mutants decreases in the transcriptional regulator of secondary metabolism and morphogenesis Vel1 which are defective in mycoparasitism and hydrophobicity (Mukherjee and Kenerley 2010; Druzhinina et al. 2011). Hydrolytic enzymes, such as chitinases and glucanases, and antibiotics are the most important chemical factors that are produced by *Trichoderma* to kill the pathogenic fungi. These are cell wall-lytic enzymes that play a major role in mycoparasitism or biocontrol. Other such chemicals are β-glucanases (Bgn2 and Bgn3) reported in *T. virens* against *R. solani*

(Djonovic et al. 2007). Apart from chitinases and glucanases, proteases (Prb1/Sp1) are induced during mycoparasitism and are useful in biocontrol (Viterbo and Horwotz 2010). Another class of enzymes called laccases are reported in the sclerotia of *T. virens* that are also involved in mycoparasitism (Catalano et al. 2011). Certain species of *Trichoderma* as *T. pseudokoningii* involve in programmed cell death in *Fusarium* (Shi et al. 2012). A volatile compound metabolite called 6-pentyl-2*H*-pyran-2-one (6-PP) isolated from *T. atroviride* plays an important role in *Trichoderma–Fusarium* interactions (El-Hasan et al. 2008; Vinale et al. 2009). In a recent study, the interaction of *Fusarium* with *Trichoderma* involved in direct parasitism of *Fusarium* via production of exoenzymes and volatile organic compounds of *T. harzianum* TE5; these antifungal traits are induced and upregulated in the presence of *F. oxysporum* f. sp. *Cucumerinum* (Zhang et al. 2014). On the basis of these interactions, various studies are done by rhizosphere biologists in various crops that are generally very susceptible to *Fusarium* wilt; the *Trichoderma* spp. that control them are mentioned in Table 9.1.

9.6 Genetic and Biochemical Biocontrol of *Fusarium* Through *Trichoderma*

The main biochemicals from *Trichoderma* are hydrolytic enzymes and antibiotics, which are responsible for killing the pathogens. The mycoparasitic *Trichoderma* spp. are rich in genes that mainly encode the biocontrolling enzymes such as chitinases, glucanases, and many other secondary metabolites such as nonribosomal peptide synthestases (NRPSs) (Reino et al. 2008; Kubicek et al. 2011). Many of the earlier studies suggested that the chitinases involve in biocontrol through the deletion of *chit42/ech42*, which is largely a reservoir of compensatory genes (Druzhinina et al. 2011). Another group of lytic enzymes are glucanases play vital role in mycoparasitism in which the *tvbgn3* (β-1,6-glucanase-encoding) coding region is deleted that directly involve to the reduction of mycoparasitism through the *T. virens* (Djonovic et al. 2006). Two β-glucanases named Bgn2 and Bgn3 co-express and involve in the improvement in mycoparasitism of *T. virens* against *R. solani* and other parasitic fungi (Djonovic et al. 2007). Along with these enzymes, proteases such as Prb1/Sp1are also induced during mycoparasitism and play an important role in biocontrol (Viterbo and Horwotz 2010). In the hyphal parasitism study, laccase enzyme reported to parasitism whose expression observes in *T. virens* for sclerotial colonization (Catalano et al. 2011). Various mycoparasitism-related secondary metabolite genes are expressed during parasitism, which suppress the pathogenicity-related metabolites such as gliotoxin and gliovirin of *R. solani* and other pathogenic fungi (Mukherjee et al. 2012a; Howell 2006). Recently, the nonribosomal peptide synthetase (NRPS) in *T. virens*, Tex1 [assembly of 18 peptaibol (trichovirin II) residues among them Δ*tex1* mutants the trichovirin II type peptaibols] and Tex2 [assembly of 11- and 14-modules peptaibols] observed to

Table 9.1 Mycoparasitism of various *Trichoderma* strains against *Fusarium* in different crops

Crops	*Trichoderma* strains	*Fusarium* strains	References
Cucumis sativus L.	*T. viride* TM15	*F. oxysporum* f. sp. *radiciscucumerinum*, *F. solani* f. sp. *cucurbitae*	Al-Tuwaijiri (2008, 2009)
Solanum lycopersicum L.	*T. virens, T. harzianum, T. viride, T. asperellum*	*F. oxysporum* f. sp. *lycopersici*, *F. oxysporum* f. sp. *Radicis-lycopersici*	Christopher et al. (2010), Rini and Sulochana (2007), Verma et al. (2017, 2018), Jogaiah et al. (2018), Altinok and Erdogan (2015), Wani et al. (2009)
Musa × *paradisiaca* L.	*T. asperellum*, *T. harzianum.swn-1*, *T. harzianum.swn-2*, *T. harzianum.psr-1*, *T. asperellum*, *T. gamsii*, *T. koningiopsis*, *T. harzianum*	*F. oxysporum* f. sp. *cubense*	Chaves et al. (2016), Taribuka et al. (2017), Castillo et al. (2019), Khan et al. (2017)
Solanum melongina L.	*T. asperellum* CP, *T. viride*, *T. harzianum*, *T. virens*, *T. atroviride*, *T. koningii*, *T. pseudokoningii*, *T. reesei*	*F. oxysporum* f. sp. *melongenae*	Adhikary et al. (2017), Cherkupally et al. (2017), Altinok and Erdogan (2015)
Cajanuscajan (L.) Huth	*T. harzianum*, *T. virens*, *T. viride*	*F. udum*	Dehariya et al. (2015)
Cucumis melo L.	*T. harzianum*	*F. oxysporum* f. sp. *melonis*	Martınez-Medina et al. (2009), Altinok and Erdogan (2015)
Cicer arietinum L.	*T. viride*, *T. harzianum*, *T. hamatum*, *T. viride*, *T. polysporum*, *T. koningii*	*F. oxysporum* f. sp. *ciceri*	Patole et al. (2017), Khan et al. (2014)
Lens culinaris Medikus	*T. harzianum*, *T. viride*, *T. koningii*	*F. oxysporum* f. sp. *lentis*	Choudhary and Reena (2012)
Pinus radiate D. Don	*Trichoderma* strains UDC-23, UDC-408	*F. circinatum*	Moraga-Suazo et al. (2011)
Cucumis melo L., *Citrullus lanatus* (Thunb.) Matsum. & Nakai	*T. erinaceum*, *T. viride*, *T. helicum*, *T. harzianum*	*F. oxysporum* f. sp. *niveum*, *F. oxysporum* f. sp. *melonis*, *F. solani* f. sp. *cucurbitae*	Boughalleb-M'Hamdi et al. (2018)
Lens culinaris Medic.	*T. harzianum*, *T. asperellum*, *T. virens*	*F. oxysporum* f. sp. *lentis*	Akrami et al. (2011)

(continued)

Table 9.1 (continued)

Crops	*Trichoderma* strains	*Fusarium* strains	References
Capsicum annuum L.	*T. viride* PDBCHTV 24, *T. viride* PDBCHTV 23, *T. harzianum* PDBCTH 8, *T. harzianum* PDBCTH 10, *T. viride* PDBCTV 10	*F. solani*	Naik et al. (2009), Sinha et al. (2018)
Dioscoreacayenensis subsp. *rotundata (Poir.) J. Miège*	*T. harzianum*	*F. oxysporum*	Nwankiti and Gwa (2018)
Solanum tuberosum L.	*T. harzianum*	*F. oxysporum* f. sp. *tuberosi*, *F. solani*, *F. graminearum*, *F. sambucinum*	Daami-Remadi et al. (2006)
Zea mays L.	*T. asperellum* CCTCC-SBW0102, *T. aureoviride* CCTCC-SBW0005, *T. harzianum* CCTCC-SBW0101, *T. asperellum* CCTCC-SBW0109, *T. atroviride* CCTCC-SBW0016, *T. asperellum* CCTCC-SBW0052, *T. asperellum* CCTCC-SBW0091, *T. tawa*, *T. harzianum* CCTCC-RW0024, *T. harzianum* CCTCC-SBW0181	*F. graminearum*	Saravanakumar et al. (2017)
Arachis hypogaea L.	*T. viride*, *T. harzianum*	*F. oxysporum*	Rajeswari (2019)
Phaseolus vulgaris L.	*T. harzianum* (CEN287, CEN288, CEN289, CEN290, and CEN316)	*F. oxysporum* f. sp. *phaseoli*	Carvalho et al. (2014)
Pisum sativum L.	*T. harzianum*, *T. asperellum*, *T. longibrachiatum*	*F. oxysporum* f. sp. *pisi*	Coral et al. (2017)
Allium cepa L.	*T. harzianum*	*F. oxysporum*, *F. solani*	Dabire et al. (2016)
Cucumber	*T. asperellum* 525, *T. harzianum* 610, *T. pseudokoningii* 886	*F. oxysporum* f. sp. *cucumerinum*	Mei et al. (2019)
Tuberose bulbs	*T. harzianum* T22	*F. oxysporum* f. sp. *tuberosi*	Nosir (2016)

induce resistance in plants (Viterbo et al. 2007; Mukherjee et al. 2011). Another peptaibol trichokonin VI reported in *T. pseudokoningii* was observed to induce programmed cell death in *F. oxysporum* (Shi et al. 2012). These peptaibiotics are antimicrobial in nature because they have the capacity to form voltage-gated membrane channels that facilitate interaction between one fungus hyphae with another and also involve in fungus-to-fungus interactions (Mukherjee et al. 2012a). A total of five species of *Trichoderma*, namely *T. reesei*, *T. atroviride*, *T. virens*, *T. harzianum*, and *T. asperellum*, have been sequenced right now. Among them, *T. reesei* is found on dead and decaying wood due to the large amount of secretion of cellulases and hemicellulases compared with the other species, and all the five sequenced *Trichoderma* spp. have smaller genomes (34.1 Mb, 9129 gene models), which might be due to the loss of mycoparasitism-specific genes, while for mycoparasitism the genome size must range from 36.1 Mb (*T. atroviride*, 11,863 gene models), 37.4 Mb (*T. asperellum*, 12,586 gene models), 38.8 Mb (*T. virens*, 12,427 gene models) to 40.98 Mb (*T. harzianum*, 14,095 gene models) (Kubicek et al. 2011; Druzhinina et al. 2011; Mukherjee et al. 2012a). According to phylogenetic studies, *T. atroviride* and *T. asperellum* are from the same clade; hence, both have powerful antagonistic activity against pathogenic fungi and stimulate defense responses in plants (Kubicek et al. 2011). Various genes are involved in specific mycoparasitism such as synthesis of chitinases, glucanases and other secondary metabolites. Recent study on the basis of secretome analysis confirms that *Trichoderma* is one of the major fungi that have a largest set of proteases such as dipeptidyl and tripeptidyl peptidases, which are involved in the mycoparasitism of pathogenic fungi (Druzhinina et al. 2012). *Trichoderma* spp. produce systemic resistance in plants through mycoparasitism via various signaling pathways that are activated by the contact of receptors with the host-derived ligands; in this mechanism, mainly heterotrimeric G proteins and mitogen-activated protein (MAP) kinases such as cyclic adenosine monophosphate (cAMP) play an important role.

9.6.1 Role of G Protein in Mycoparasitism

During mycoparasitism, normally an interaction develops between biomolecules of one fungus that acts as a messenger and received by the other participating fungus through its receptors. During this interaction, a signal generated at the reception site than a downstream action originated via signal transduction mechanism through various receptors present in the cell (Sarma et al. 2014). The heterotrimeric G proteins act as signal transducers that bind with the receptors that are present in the cell surface and then move to the effector protein in the cytoplasm. In this signaling mechanism, basically three components, i.e., G protein-coupled receptor (GPCR), heterotrimeric G protein (α, β, ϒ-subunits), and an effector, are involved (Neer 1995; Pandey et al. 2016). At the time of mycoparasitism, the ligands bind with GPCR, and a conformational change occurs that involves in the liberation of G-α-subunit from the G protein due to this GDP exchange with GTP; subsequently,

G-β-, and ϒ-subunits are disassociated that regulate the downstream activities of the effectors (Sarma et al. 2014). Previously, from the biochemical study of G protein from *T. troviride* on the parasitism of pathogenic fungus *R. solani* it was observed that the overexpression of this protein promotes vegetative proliferation, which in turn leads to enhanced mycoparasitism (Rocha-Ramirez et al. 2002). G protein also involves in the production of antifungal metabolites such as extracellular chitinases (Reithner et al. 2005). In the molecular study it was reported that the mutants of two G protein α-subunit genes, i.e., *tgaA* and *tgaB*, in *T. virens* suppress the mycelial parasitism of *R. solani* (Mukherjee et al. 2004).

9.6.2 Role of Mitogen-Activated Protein (MAP) Kinases in Mycoparasitism

MAP kinases are phosphorylated proteins that defend organisms from environmental stress, both biotic and abiotic. In *Trichoderma* spp. these enzymes are present in an adequate amount that affect the processes involve in mycoparasitism and produces systemic resistance in plants. During mycoparasitism, MAP kinase kinase kinase (MAPKKK) receives the signal and is phosphorylated into MAP kinase kinase (MAPKK) then phosphorylates into MAP kinase (MAPK) suddenly this MAPK phosphorylates and a signal transferred to downstream effectors (transcription factor or other molecules) that activates and repress the corresponding target genes which is essential for adaptive response and provide systemic resistance to plants during *Trichoderma*–plant interaction (Cargnello and Roux 2011; Pandey et al. 2016). In *Trichoderma* spp., three MAPKs, Tmk1 (homologous to yeast Fus3/Kss1), Tmk2 (homologous to yeast Slt2), and Tmk3 (homologous to yeast Hog1), have been reported. Among them, Tmk1 involves in the regulation of mycoparasitic host attack, aerial hyphae formation, and conidiation; Tmk2 involves in cell wall integrity, sporulation, and repression of cellulase formation; and Tmk3 participates in high osmolarity resistance along with the promotion of cellulase production, hence they play a crucial role in mycoparasitism (Schmoll 2008; Wang et al. 2014; Gruber and Zeilinger 2014). In *Trichoderma*, cyclic adenosine monophosphate (cAMP) is such a MAP kinase that involves in the growth, germination, secondary metabolites production, and mycoparasitism which produced from ATP by adenylate cyclase.

9.7 Secondary Metabolites from *Trichoderma* Against *Fusarium*

Trichoderma spp. are prolific producers of secondary metabolites with more than 100 reported compounds, which include low-molecular-mass nonpolar compounds such as terpenoids, steroids, polyketides, ethylene, hydrogen cyanide, aldehydes, pyrones, sesquiterpenes, and ketones (Mukherjee et al. 2012b; Chen et al. 2015). These metabolites are volatile in nature and are used to control pathogens including pathogenic fungi (Qualhato et al. 2013; Zhang et al. 2014; Zeilinger et al. 2016; Meena et al. 2017). Some nonvolatile secondary metabolites have been studied by Kavitha and Nelson (2013) and Petrisor et al. (2017) which were isolated from the various isolates of *Trichoderma* and evaluated against *F. javanicum*, *F. solani*, and *F. oxysporum*. In the study by Li et al. (2018), a total of 15 volatile compounds (2-Methyl-1-propanol, Isopentyl acetate, 3-Octanone, 1-Pentanol, 3-Methyl-1-butanol, Acetoin, 1-Octen-3-one, Methyl (2E)-4,4dimethyl-2-pentenoate, 1-Hexanol, 3-Methylcyclopentanol, Acetic acid, 1-Octen-3-ol, 1-Heptanol, 1-Octanol, and 2-Phenylethyl alcohol) secreted from various strains of *T. virens, T. harzianum*, and *T. viride* were identified, which significantly inhibited the different strains of *F. oxysporum*. Peptaibols are an important class of metabolites produced largely by the same fungus that are ecologically and commercially important for their antimicrobial and anticancer activities, with the ability to induce systemic resistance in plants against pathogens attack (Mukherjee et al. 2012b). Alamethicin, a peptaibol produced by *T. viride*, is a mixture of 12 compounds each with 20 amino acid residues is responsible for the biosynthesis of homoterpenes and methyl salicylate volatile compounds, and it sends local and long-distance electrical signals in plants during pathogen attacks (Engelberth et al. 2001; Rippa et al. 2010; Maischak et al. 2010). Iron acquisition is an important component in microbial competition. As a proof, all *Trichoderma* spp. that have been completely sequenced were found to have a single gene for ferricrocin synthesis in a secondary metabolism gene cluster (Kubicek et al. 2011). These extracellular siderophores have biocontrol properties and hence can be considered as virulence factors.

9.8 Conclusion

Various plant pathogens such as *Fusarium* affect the normal life of plants and cause severe diseases. Though several methods that are available for crop protection are being used to control such diseases, they are very expensive and considered hazardous to the environment and for sustainable development. Hence, biocontrol is a safer and the cheapest method to control these severe diseases and to facilitate better plant growth, yield, and production. The pathogenic fungi *Fusarium* spp. generally cause wilt nearly in all the crop plants from various groups and cause a severe threat to food demand. Among the various biocontrol agents, *Trichoderma* spp. are considered to be the most beneficial, and they assure increased level of safety with minimum environmental impacts. This symbiotic fungus shows significant biocontrol activities against a wide range of plant pathogenic fungi through different mechanisms such as mycoparasitism, completion, production of various lytic enzymes, and other antimicrobial activities. *Trichoderma* spp. are well-known fungi that enhance plant growth by nutrient solubilization in the soil, growth hormones production in plants, and defense against pathogens. Due to these characteristics, *Trichoderma* protect the plants against various biotic as well as abiotic stresses. To defend against pathogenic fungi, various mycotoxins have been reported to be produced in this fungus. In the interaction of *Trichoderma* with *Fusarium* a mechanism involves in a series from attraction to cell lysis via attachment and coiling of hyphae through different hydrolytic enzymes and secondary metabolites. These fungal biochemicals directly involve in mycoparasitism and cause programmed cell death in *Fusarium*. Various volatile and nonvolatile organic compounds are also reported in the different strains of *Trichoderma* that directly or indirectly involve in the mycoparasitism on *Fusarium*. Hence, by these mechanisms *Trichoderma* spp. induce systemic resistance in the host plant and help the plant for better growth, yield, and productivity.

References

Abdelrahman M, Abdel-Motaal F, El-Saved M, Jogaiah S, Shigyo M, Ito S, Tran LS (2016) Dissection of *Trichoderma longibrachiatum*-induced defense in onion (*Allium cepa* L.) against *Fusarium oxysporum* f. Sp. *cepa* by target metabolite profiling. Plant Sci 245:128–138

Adhikary MC, Begum HA, Meah MB (2017) Possibility of recovering *Fusarium* wilt affected eggplants by *Trichoderma*. Int J Agric Res Innov Technol 7(1):38–42

Adie BAT, Perez-Perez J, Perez-Perez MM et al (2007) ABA is an essential signal for plant resistance to pathogens affecting JA biosynthesis and the activation of defenses in Arabidopsis. Plant Cell 19:1665–1681

Akrami M, Golzary H, Ahmadzadeh M (2011) Evaluation of different combinations of *Trichoderma* species for controlling *Fusarum* rot of lentil. Afr J Biotechnol 10:2653–2658

Althaf HSK, Srinivas P (2013) Association of arbuscular mycorrhizal fungi and other rhizosphere microbes with different medicinal plants. Res J Biotechnol 8(6):24–28

Altinok HH, Erdogan O (2015) Determination of the *in vitro* effect of *Trichoderma harzianum* on phytopathogenic strains of *Fusarium oxysporum*. Notulae Botanicae Horti Agrobotanici Cluj-Napoca 43(2):494–500

Al-Tuwaijiri MMY (2008) Biological control of *Fusarium* root-rot of cucumber (*Cucumis sativus* L.) by rhizospheric isolates of *Bacillus subtilis* and *Trichoderma viride*. Egypt J Exp Biol (Bot) 4:79–86

Al-Tuwaijri MMY (2009) Role of the bio-control agents, *Trichoderma viride* and *Bacillus subtilis*, in elimination of the deteriorative effects of the root-rot pathogens, *Fusarium oxysporum* and *F. Solani*, on some metabolic and enzyme activities of cucumber plants. Egypt J Exp Biol (Bot) 5:29–35

Bae H, Sicher RC, Kim MS, Kim SH, Strem MD, Melnick RL, Bailey BA (2009) The beneficial endophyte *Trichoderma hamatum* isolate DIS 219b promotes growth and delays the onset of the drought response in *Theobroma cacao*. J Exp Bot 60:3279–3295

Bailey BA, Stream MD, Wood D (2009) *Trichoderma* species form endophytic associations within *Theobroma cacao* trichomes. Mycol Res 113:1365–1376

Bari R, Jones JDG (2009) Role of plant hormones in plant defense responses. Plant Mol Biol 69:473–488

Berges MSL, Hera C, Sulyok M, Schafer K, Capilla J, Guarro J, Di Pietro A (2013) The velvet complex governs mycotoxin production and virulence of *Fusarium oxysporum* on plant and mammalian hosts. Mol Microbiol 86(1):49–65

Boughalleb-M'Hamdi N, Salem IB, M'Hamdi M (2018) Evaluation of the efficiency of *Trichoderma*, *Penicillium*, and *Aspergillus* species as biological control agents against four soil-borne fungi of melon and watermelon. Egypt J Biol Pest Cont 28:25. https://doi.org/10.1186/s41938-017-0010-3

Brotman Y, Lisec J, Meret M, Chet I, Willmitzer L, Viterbo A (2012) Transcript and metabolite analysis of the *Trichoderma* induced systemic resistance response to *Pseudomonas syringe* in *Arabidopsis thaliana*. Microbiology 158:139–146

Cargnello M, Roux PP (2011) Activation and function of the MAPKs and their substrates, the MAPK-activated protein kinases. Microbiol Mol Biol Rev 75(1):50–83

Carvalho DDC, Junior ML, Martins I, Inglis PW, Mello SCM (2014) Biological control of *Fusarium oxysporum* f. sp. *phaseoli* by *Trichoderma harzianum* and its use of common bean seed treatment. Trop Plant Pathol 39(5):384–391

Castillo AG, Puig CG, Cumagun CJR (2019) Non-synergistic effect of *Trichoderma harzianum* and *Glomus* spp. in reducing infection of *Fusarium* wilt in banana. Pathogens 8. https://doi.org/10.3390/pathogens8020043

Catalano V, Vergara M, Hauzenberger JR et al (2011) Use of a nonhomologous end-joining-deficient strain (delta-ku70) of the biocontrol fungus *Trichoderma virens* to investigate the function of the laccase gene lcc1 in sclerotia degradation. Curr Genet 57:13–23

Chaves NP, Staver C, Dita A (2016) Potential of *Trichoderma asperellum* for biocontrol of *Fusarium* wilt in banana. Acta Hort 1114:261–266

Chen LJ, Sun SZ, Miao CP, Wu K, Chen YW, Xu LH, Guan HL, Zhao LX (2015) Endophytic *Trichoderma gamsii* YIM PH30019: a promising biocontrol agent with hyperosmolar, mycoparasitism and antagonistic activities of induced volatile organic compounds on root-rot pathogenic fungi of *Panax notoginseng*. J Ginseng Res 38:1–10

Cherkupally C, Amballa H, Reddy BN (2017) *In vitro* antagonistic activity of *Trichoderma* species against *Fusarium oxysporum* f. sp. *melongenae*. Int J Appl Agric Res 12(1):87–95

Choudhary S, Reena M (2012) *In-vitro* antagonism of indigenous *Trichoderma* isolates against phytopathogen causing wilt of lentil. Int J Life Sci Pharma Res 2(3):195–202

Christopher DJ, Raj TS, Rani SU, Udhayakumar R (2010) Role of defense enzymes activity in tomato as induced by *Trichoderma virens* against *Fusarium* wilt caused by *Fusarium oxysporum* f sp. *lycopersici*. J Biopest 3:158–162

Contreras-Cornejo H, Macias-Rodriguez L, Cortes-Penagos C, Lopez-Bucio J (2009) *Trichoderma virens*, a plant beneficial fungus, enhances biomass production and promotes lateral root growth through an auxin dependent mechanism in *Arabidopsis*. Plant Physiol 149:1579–1592

Contreras-Cornejo HA, Macias-Rodriguez L, del-Val E, Larsen J (2016) Ecological functions of *Trichoderma* spp. and their secondary metabolites in the rhizosphere: interactions with plants. FEMS Microbiol Ecol 92(4):1–17

Coral OEC, Criollo MT, Vallejo JD (2017) Antagonism of *Trichoderma* spp. strains against pea (*Pisum sativum* L.) *Fusarium* wilt caused by *Fusarium oxysporum* f. sp. *pisi*. Acta Agron 66 (3):442–448

Cornejo HAC, Rodriguez LM, Vergara AG, Bucio JL (2015) *Trichoderma* modulates stomatal aperture and leaf transpiration through an abscisic acid-dependent mechanism in Arabidopsis. J Plant Growth Regul 34(2):425–434

Daami-Remadi M, Hibar K, Jabnoun-Khiareddine H, Ayed F, El Mahjoub M (2006) Effect of two *Trichoderma* species on severity of potato tuber dry rot caused by Tunisian *Fusarium* complex. Int J Agric Res 1:432–441

Dabire TG, Bonzi S, Somda I, Legreve A (2016) Evaluation of the potential of *Trichoderma harzianum* as a plant growth promoter and biocontrol agent against *Fusarium* damping-off in onion in Burkina Faso. Asian J Plant Pathol 10:49–60

Dehariya K, Shukla A, Sheikh IA, Vyas D (2015) *Trichoderma* and arbuscular mycorrhizal fungi based biocontrol of *Fusarium udum* butler and their growth promotion effects on pigeon pea. J Agric Sci Technol 17:505–517

Dinolfo MI, Castanares E, Stenglei SA (2017) *Fusarium*–plant interaction: state of the art - a review. Plant Protect Sci 53(2):61–70

Djonovic S, Pozo MJ, Kenerley CM (2006) Tvbgn3, a beta-1,6-glucanase from the biocontrol fungus *Trichoderma virens*, is involved in mycoparasitism and control of *Pythium ultimum*. Appl Environ Microbiol 72:7661–7670

Djonovic S, Vargas WA, Kolomiets MV, Horndeski M, Wiest A, Kenerley CM (2007) A proteinaceous elicitor Sm1 from the beneficial fungus *Trichoderma virens* is required for induced systemic resistance in maize. Plant Physiol 145:875–889

Druzhinina IS, Seidl-Seiboth V, Herrera-Estrella A et al (2011) *Trichoderma*: the genomics of opportunistic success. Nat Rev Microbiol 9:749–759

Druzhinina IS, Shelest E, Kubicek CP (2012) Novel traits of *Trichoderma* predicted through the analysis of its secretome. FEMS Microbiol Lett 337:1. https://doi.org/10.1111/j.1574-6968.2012.02665.x

Durrant WE, Dong X (2004) Systemic acquired resistance. Annu Rev Phytopathol 42:185–209

El-Hasan A, Walker F, Buchenauer H (2008) *Trichoderma harzianum* and its metabolite 6-pentyl-alpha-pyrone suppress fusaric acid produced by *Fusarium moniliforme*. J Phytopathol 156:79–87

Engelberth J, Koch T, Schuler G, Bachmann N, Rechtenbach J, Boland W (2001) Ion channel-forming alamethicin is a potent elicitor of volatile biosynthesis and tendril coiling. Cross talk between jasmonate and salicylate signaling in lima bean. Plant Physiol 125:369–377

Gawehns F, Houterman PM, Ichou FA et al (2013) The *Fusarium oxysporum* effector six6 contributes to virulence and suppresses i-2-mediated cell death. Mol Plant-Microbe Interact 27(4):336–348

Ghanbarzadeh B, Safaie N, Goltapeh EM (2014) Antagonistic activity and hyphal interactions of *Trichoderma* spp. against *Fusarium proliferatum* and *F. oxysporum in vitro*. Arch Phytopathol Plant Protect 47(16):197–198

Giurgiu RM, Dumitras A, Morar G, Scheewe P, Schroder FG (2018) A study on the biological control of *Fusarium oxysporum* using *Trichoderma* spp., on soil and rockwool substrates in controlled environment. Not Bot Horti Agrobo 46(1):260–269

Gruber S, Zeilinger S (2014) The transcription factor ste12 mediates the regulatory role of the tmk1 map kinase in mycoparasitism and vegetative hyphal fusion in the filamentous fungus *Trichoderma atroviride*. PLoS One 9(10):e111636. https://doi.org/10.1371/journal.pone.0111636

Harman GE, Howell CR, Viterbo A, Chet I, Lorito M (2004) *Trichoderma* species - opportunistic, avirulent plant symbionts. Nat Rev Microbiol 2:43–56

Howell CR (2006) Understanding the mechanisms employed by *Trichoderma virens* to effect biological control of cotton diseases. Phytopathology 96:178–180

Jogaiah S, Abdelrahman M, Tran LP, Ito SI (2018) Different mechanisms of *Trichoderma virens*-mediated resistance in tomato against *Fusarium* wilt involve the jasmonic and salicylic acid pathways. Mol Plant Pathol 19(4):870–882

Joshi R (2018) A review of *Fusarium oxysporum* on its plant interaction and industrial use. J Med Plants Stud 6(3):112–115

Kavitha T, Nelson R (2013) Exploiting the biocontrol activity of *Trichoderma* spp against root rot causing phytopathogens. ARPN J Agric Biol Sci 8(7):571–574

Khan M, Ashraf S, Rasool F, Salati K (2014) Field performance of *Trichoderma* species against wilt disease complex of chickpea caused by *Fusarium oxysporum* f. sp. *ciceri* and *Rhizoctonia solani*. Turkish J Agric For 38(4):447–454

Khan B, Akash Z, Asad S et al (2017) Antagonistic potential of *Trichoderma harzianum* against *Fusarium oxysporum* f. sp. *Cubense* associated with panama wilt of banana. Pak J Phytopathol 29(1):111–116

Kubicek CP, Herrera-Estrella A, Seidl-Seiboth V et al (2011) Comparative genome sequence analysis underscores mycoparasitism as the ancestral life style of *Trichoderma*. Genome Biol 12:R40

Kumar DP, Thenmozhi R, Anupama A et al (2011) Selection of potential antagonistic *Bacillus* and *Trichoderma* isolates from tomato rhizosphere soil against *Fusarium oxysporum* f. sp. *lycoperscisi*. Res J Biol Sci 6(10):523–331

Lace B, Genre A, Woo S, Faccio A, Lorito M, Bonfante P (2014) Gate crashing arbuscular mycorrhizas: *in vivo* imaging shows the extensive colonization of both symbionts by *Trichoderma atroviride*. Environ Microbiol Rep 7(1):64–77

Li N, Alfiky A, Wang W, Islam M et al (2018) Volatile compound-mediated recognition and inhibition between *Trichoderma* biocontrol agents and *Fusarium oxysporum*. Front Microbiol 9:2614. https://doi.org/10.3389/fmicb.2018.02614

Lopez-Bucio J, Pelagio-Flores R, Herrera-Estrella A (2015) *Trichoderma* as biostimulant: exploiting the multilevel properties of a plant beneficial fungus. Sci Hortic 196:109–123

Maischak H, Zimmermann MR, Felle HH, Boland W, Mithofer A (2010) Alamethicin-induced electrical long distance signaling in plants. Plant Signal Behav 5:988–990

Martınez-Medina A, Pascual JA, Lloret E, Roldan A (2009) Interactions between arbuscular mycorrhizal fungi and *Trichoderma harzianum* and their effects on *Fusarium* wilt in melon plants grown in seedling nurseries. J Sci Food Agric 89:1843–1850

Meena M, Swapnil P, Zehra A, Dubey MK, Upadhyay RS (2017) Antagonistic assessment of *Trichoderma* spp. by producing volatile and nonvolatile compounds against different fungal pathogens. Arch Phytopathol Plant Protect 50:629–648

Mei LI, Guang-shu MA, Hua L, Xiao-lin SU, Ying T, Wen-kun H, Jie M, Xiliang J (2019) The effects of *Trichoderma* on preventing cucumber *Fusarium* wilt and regulating cucumber physiology. J Integr Agric 18(3):607–617

Mendgen K, Hahn M, Deising H (1996) Morphogenesis and mechanisms of penetration by plant pathogenic fungi. Annu Rev Phytopathol 34(1):367–386

Miller SA, Rowe RC, Riedel RM (2011) *Fusarium* and *Verticillium* wilts of tomato, potato, pepper, and eggplant. Fact Sheet. Ohio State University, Columbus

Ming Q, Su C, Zheng C, Jia M, Zhang Q, Zhang H, Rahman K, Han T, Qin L (2013) Elicitors from the endophytic fungus *Trichoderma atroviride* promote *Salvia* miltiorrhiza hairy root growth and tanshinone biosynthesis. J Exp Bot 64(18):5687–5694

Moraga-Suazo P, Opazo A, Zaldua S, Gonzalez G, Sanfuentes E (2011) Evaluation of *Trichoderma* spp. and *Clonostachys* spp. strains to control *Fusarium circinatum* in *Pinus radiata* seedlings. Chil J Agric Res 71:412–417

Moran-Diez E, Rubio B, Dominguez S, Hermosa R, Monte E, Nicolas C (2012) Transcriptomic response of *Arabidopsis thaliana* after 24h incubation with the biocontrol fungus *Trichoderma harzianum*. J Plant Physiol 169:614–620

Mueller WC, Beckman CH (1988) Correlated light and electron microscope studies of callose deposits in vascular parenchyma cells of tomato plants inoculated with *Fusarium oxysporum* f. sp. *lycopersici*. Physiol Mol Plant Pathol 33:201–208

Mukherjee PK, Kenerley CM (2010) Regulation of morphogenesis and biocontrol properties in *Trichoderma virens* by a VELVET protein, Vel1. Appl Environ Microbiol 76:2345–2352

Mukherjee PK, Latha J, Hadar R, Horwitz BA (2004) Role of two G-protein alpha subunits, TgaA and TgaB, in the antagonism of plant pathogens by *Trichoderma virens*. Appl Environ Microbiol 70:542–549

Mukherjee PK, Wiest A, Ruiz N et al (2011) Two classes of new peptaibols are synthesized by a single non-ribosomal peptide synthetase of *Trichoderma virens*. J Biol Chem 286:4544–4554

Mukherjee M, Mukherjee PK, Horwitz BA et al (2012a) *Trichoderma*–plant–pathogen interactions: advances in genetics of biological control. Indian J Microbiol 52(4):522–529

Mukherjee PK, Horwitz BA, Kenerley CM (2012b) Secondary metabolism in *Trichoderma* – a genomic perspective. Microbiology 158:35–45

Naik MK, Madhukar HM, Rani GSD (2009) Evaluation of biocontrol efficacy of *Trichoderma* isolates and methods of its application against wilt of chilli (*Capsicum annuum* L.) caused by *Fusarium solani* (Mart) Sacc. J Biol Cont 23(1):31–36

Neer EJ (1995) Heterotrimeric G proteins: organizers of transmembrane signals. Cell 80 (2):249–257

Nosir WS (2016) *Trichoderma harzianum* as a growth promoter and bio-control agent against *Fusarium oxysporum* f. sp. *tuberosi*. Adv Crop Sci Tech 4:217. https://doi.org/10.4172/2329-8863.1000217

Nwankiti A, Gwa V (2018) Evaluation of antagonistic effect of *Trichoderma harzianum* against *Fusarium oxysporum* causal agent of white yam (*Dioscorea rotundata* poir) tuber rot. Trends Tech Sci Res 1(1):555554

Okungbowa FI, Shittu HO (2012) *Fusarium* wilts: an overview. Environ Res J 6(2):83–102

Omann MR, Lehner S, Rodriguez CE, Brunner K, Zeilinger S (2012) The seven-transmembrane receptor Gpr1 governs processes relevant for the antagonistic interaction of *Trichoderma atroviride* with its host. Microbiology 158:107–118

Ortiz E, Cruz M, Melgarejo LM, Marquinez X, Hoyos-Carvajal L (2014) Histopathological features of infections caused by *Fusarium oxysporum* and *F. solani* in purple passion fruit plants (*Passiflora edulis* Sims). Summa Phytopathol 40(2):134–140

Pandey V, Shukla A, Kumar J (2016) Physiological and molecular signalling involved in disease management through *Trichoderma*: an effective biocontrol paradigm. In: Kumar P, Gupta V, Tiwari A, Kamle M (eds) Current trends in plant disease diagnostics and management practices. Fungal biology. Springer, Cham, pp 317–346

Patole SP, Dhore SB, Pradhan RS, Shankara K (2017) *In vitro* evaluation of *Trichoderma viride* and *Trichoderma harzianum* against *Fusarium* wilt of chickpea. Int J Pure App Biosci 5(5):460–464

Petrisor C, Paica A, Constantinescu F (2017) Effect of secondary metabolites produced by different *Trichoderma* spp. isolates against *Fusarium oxysporum* f. sp. *radicis-lycopersici* and *Fusarium solani*. Sci Pap B Hortic 61:407–411

Pitt JI, Hocking AD, Bhudhasamai K et al (1994) The normal mycoflora of commodities from Thailand. 2. Beans, rice, small grains and other commodities. Int J Food Microbiol 23:35–43

Qualhato FT, Cardoso-Lopes FA, Steindorff AS et al (2013) Mycoparasitism studies of *Trichoderma* species against three phytopathogenic fungi: evaluation of antagonism and hydrolytic enzyme production. Biotechnol Lett 35:1461–1468

Raaijmakers JM, Paulitz TC, Steinberg C et al (2009) The rhizosphere: a playground and battlefield for soil borne pathogens and beneficial microorganisms. Plant Soil 321:341–361

Rajeswari P (2019) Combination of *Trichoderma viride* and *Pseudomonas fluorescens* for the enhanced control of *Fusarium* wilt disease caused by *Fusarium oxysporum* infecting *Arachis hypogaea* L. J Appl Nat Sci 11(1):138–143

Redda ET, Ma J, Mei J, Li M, Wu B, Jiang X (2018) Antagonistic potential of different isolates of *Trichoderma* against *Fusarium oxysporum*, *Rhizoctonia solani*, and *Botrytis cinerea*. Eur J Exp Biol 8:1–8

Reino JL, Guerrero RF, Hernandez-Galan R, Collado IG (2008) Secondary metabolites from species of the biocontrol agent *Trichoderma*. Phytochem Rev 7:89–123

Reithner B, Brunner K, Schuhmacher R, Peissl I, Seidl V, Krska R, Zeilinger S (2005) The G protein alpha subunit Tga1 of *Trichoderma atroviride* is involved in chitinase formation and differential production of antifungal metabolites. Fungal Genet Biol 42:749–760

Rejeb IB, Pastor V, Mauch-Mani B (2014) Plant responses to simultaneous biotic and abiotic stress: molecular mechanisms. Plan Theory 3(4):458–475

Rini CR, Sulochana KK (2007) Usefulness of *Trichoderma* and *Pseudomonas* against *Rhizoctonia solani* and *Fusarium oxysporum* infecting tomato. J Trop Agric 45:21–28

Rippa S, Eid M, Formaggio F, Toniolo C, Beven L (2010) Hypersensitive-like response to the pore-former peptaibol alamethicin in *Arabidopsis thaliana*. ChemBioChem 11:2042–2049

Rocha-Ramirez V, Omero C, Chet I, Horwitz BA, Herrera-Estrella A (2002) *Trichoderma atroviride* G-protein alpha-subunit gene tga1 is involved in mycoparasitic coiling and conidiation. Eukaryot Cell 1:594–605

Saravanakumar K, Li Y, Yu C, Wang Q, Wang M, Sun J, Gao J, Chen J (2017) Effect of *Trichoderma harzianum* on maize rhizosphere microbiome and biocontrol of *Fusarium* stalk rot. Sci Rep 7:1771. https://doi.org/10.1038/s41598-017-01680-w

Sarma BK, Yadav SK, Patel JS, Singh HB (2014) Molecular mechanisms of interactions of *Trichoderma* with other fungal species. Open Mycol J 8:140–147

Schmoll M (2008) The information highways of a biotechnological workhorse-signal transduction in *Hypocrea jecorina*. BMC Genomics 9:430

Shi M, Chen L, Wang XW, Zhan T et al (2012) Antimicrobial peptaibols from *Trichoderma pseudokoningii* induce programmed cell death in plant fungal pathogens. Microbiology 158:166–175

Shoresh M, Harman GE, Mastouri F (2010) Induced systemic resistance and plant responses to fungal biocontrol agents. Annu Rev Phytopathol 48:21–43

Silva VN, Fernandes FMC, Cortez A et al (2006) Characterization and genetic variability of *Fusarium verticillioides* strains isolated from corn and sorghum in Brazil based on fumonisins production, microsatellites, mating type locus, and mating crosses. Can J Microbiol 52:798–804

Sinha A, Harshita Singh R, Verma A (2018) Bioefficacy of *Trichoderma harzianum* and *Trichoderma viride* against *Fusarium oxysporum* f. sp. *capsici* causing wilt disease in chilli. J Pharmacog Phytochem 7(5):965–966

Stankovic S, Levic J, Petrovic T, Logrieco A, Moretti A (2007) Pathogenicity and mycotoxin production by *Fusarium* proliferatum isolated from onion and garlic in Serbia. Eur J Plant Pathol 118:165–172

Taribuka J, Wibowo A, Widyastuti SM, Sumardiyono C (2017) Potency of six isolates of biocontrol agents endophytic *Trichoderma* against *Fusarium* wilt on banana. J Degrade Min Land Manage 4:723–731

Vargas WA, Mandawe JC, Kenerley CM (2009) Plant-derived sucrose is a key element in the symbiotic association between *Trichoderma virens* and maize plants. Plant Physiol 151:792–808

Vargas WA, Crutcher FK, Kenerley CM (2011) Functional characterization of a plant-like sucrose transporter from the beneficial fungus *Trichoderma virens*. Regulation of the symbiotic association with plants by sucrose metabolism inside the fungal cells. New Phytol 189:777–789

Verma NP, Kaur I, Masih H, Singh AK, Singla A (2017) Efficacy of *Trichoderma* in controlling *Fusarium* wilt in tomato (*Solanum lycopersicum* L.). Res Environ Life Sci 10(7):636–639

Verma NP, Kuldeep YK, Sinha BK (2018) Efficacy of indigenous *Trichoderma* strain bio-control against of *Fusarium* sp. tomato plant causal agent of (*Solanum lycopersicon* L.) *in vitro* condition. Int J Curr Microbiol App Sci 7(3):1578–1584

Vinale F, Sivasithamparam K, Ghisalberti EL, Marra R, Woo SL, Lorito M (2008) *Trichoderma*-plant-pathogen interactions. Soil Biol Biochem 40:1–10

Vinale F, Ghisalberti EL, Sivasithamparam K et al (2009) Factors affecting the production of *Trichoderma harzianum* secondary metabolites during the interaction with different plant pathogens. Lett Appl Microbiol 48:705–711

Viterbo A, Horwotz BA (2010) Mycoparasitism. In: Borkovich KA, Ebbole DJ (eds) Cellular and molecular biology of filamentous fungi. ASM Press, Herndon, pp 676–694

Viterbo A, Wiest A, Brotman Y, Chet I, Kenerley C (2007) The 18mer peptaibols from *Trichoderma virens* elicit plant defence responses. Mol Plant Pathol 8:737–746

Wang M, Dong Y, Zhao Q, Wang F, Liu K, Jiang B, Fang X (2014) Identification of the role of a MAP kinase Tmk2 in *Hypocreajecorina* (*Trichoderma reesei*). Sci Rep 4:6732. https://doi.org/10.1038/srep06732

Wani AH, Taskeen-Un-Nisa, Mir RA (2009) Biological control of *Fusarium oxysporum* f. sp. *lycopercici* on tomato with fungal antagonists. J Biol Cont 23(2):169–173

Yedidia I, Srivastva AK, Kapulnik Y, Chet I (2001) Effect of *Trichoderma harzianum* on micro-element concentrations and increased growth of cucumber plants. Plant Soil 235:235–242

Zeilinger S, Gruber S, Bansal R, Mukherjee PK (2016) Secondary metabolism in *Trichoderma*-chemistry meets genomics. Fungal Biol Rev 30:74–90

Zhang F, Zhu Z, Yang XM, Ran W, Shen Q (2013a) *Trichoderma harzianum* T-E5 significantly affects cucumber root exudates and fungal community in the cucumber rhizosphere. Appl Soil Ecol 72:41–48

Zhang F, Yuan J, Yang X, Cui Y, Chen L, Ran W, Shen Q (2013b) Putative *Trichoderma harzianum* mutant promotes cucumber growth by enhanced production of indole acetic acid and plant colonization. Plant Soil 368:433–444

Zhang F, Yang X, Ran W, Shen Q (2014) *Fusarium oxysporum* induces the production of proteins and volatile organic compounds by *Trichoderma harzianum* TE-5. FEMS Microbiol Lett 359:116–123

Zhang F, Huo Y, Cobb AB, Luo G, Zhou J, Yang G, Wilson GWT, Zhang Y (2018) *Trichoderma* biofertilizer links to altered soil chemistry, altered microbial communities, and improved grassland biomass. Front Microbiol 9:848. https://doi.org/10.3389/fmicb.2018.00848

Chapter 10
Use of *Trichoderma* in the Management of Diseases in North American Row Crops

Shrishail S. Navi and X. B. Yang

Abstract North America is a continent having the largest row crop productions in the world. Disease management is critical to ensure the success and profitability of the producers each year. Traditionally, disease management in row crops has been through synthetic fungicides use. The number of biofungicides in market for row crop disease management is relatively few compared with other parts of the world, due to higher regulation standards for environmental protection. In this chapter, we review the current status of biological controls for row crops in which *Trichoderma* spp. and other biological agents are used. We discuss the major diseases of row crops and yield losses in the United States. A list of biopesticides with *Trichoderma* spp. as microbial active ingredient is complied with the latest registration by EPA as of July 25, 2018. We also discuss management options (seed treatments and foliar and soil applications) for some of the major diseases of a few row crops using biocontrol agents with a distinct reference to *Trichoderma* spp.

Keywords Biopesticides · *Trichoderma* application · Row crops

10.1 Introduction

An alternative to synthetic fungicides in crop production is the use of biocontrol agents-based products. Conventional crop protection practices using chemical pesticides to control plant diseases can have a negative impact on pollinators, beneficial predators, and microorganisms (Woo et al. 2014). The EPA of the United States (www.epa.gov) has listed a vast number of biopesticides. The majority of microbial active ingredient-based biopesticides are from *Bacillus* spp. (47), followed by *Pseudomonas* spp. (9), but only 8 *Trichoderma* spp.-based biopesticides have been registered with the EPA as of July 25, 2018. Woo et al. (2014) have provided a

S. S. Navi (✉) · X. B. Yang
Department of Plant Pathology and Microbiology, Iowa State University, Ames, IA, USA
e-mail: ssnavi@iastate.edu

A. K. Sharma, P. Sharma (eds.), *Trichoderma*, Rhizosphere Biology,
https://doi.org/10.1007/978-981-15-3321-1_10

comprehensive list of *Trichoderma*-based products in the world, of which, at least 29 commercial products are listed in North America. Out of the 29, 16 products carry a single species of *Trichoderma* and 13 carry a mixture of more than one *Trichoderma* spp. as their active ingredient. *Trichoderma*-based products that are labeled for major row crops in the US are not abundant in the literature. However, we have tried to compile the biofungicides marketed in the US targeting different pathogens and have labeled the details in this chapter.

10.2 *Trichoderma* spp.

To control plant diseases caused by major soil-borne pathogens, *Trichoderma* spp. have been used for decades (Weindling 1932; Papavizas 1985; Lifshitz et al. 1986; Harman and Taylor 1988; Nelson et al. 1988; Harman 1991; Lewis and Larkin 1997; Harman et al. 1980, 2004; Menendez and Godeas 1998; Gardener and Fravel 2002; Matroudi and Zamani 2009; Mastouri et al. 2010; Samuels and Hebbar 2015). The *Trichoderma* spp. based biocontrol agents (BCA) has progressively balancing some of the chemicals (Templeton and Heiny 1989; Chet 1993; Whipps and Lumsden 2001).

According to Whipps and Lumsden (2001), nearly 60% of fungal BCA market is shared by *Trichoderma* spp. and are important contestants to explore.

The activity or mode of action of *Trichoderma* spp. is as follows:

1. Production of trichodermin, trichothecenes, trichorzianins, or gliotoxins (Howell 2002; Chaube et al. 2003; Brozóvá 2004).
2. Competing for nutrition and space (Celar 2003; Lewis and Papavizas 1991a; Sivan and Chet 1986).
3. Antibiosis (Schirmböck et al. 1994).
4. Mycoparasitic abilities—a relationship in which one living fungus acts as a nutrient source for another (Chet et al. 1981; Jeffries 1995; Punja and Utkhede 2003).

Trichoderma spp. are highly reproductive and also produce chlamydospores (Fig. 10.1) to survive under unfavorable conditions (Dal et al. 1997; García-Garza et al. 1997; Harman et al. 2004; García et al. 2005). Recently, Jahagirdar et al. (2019) have described *Trichoderma* spp. as plant growth-promoting fungi and have documented diversity of *Trichoderma* spp. based on geography, soil edaphic factors, phylloplane survivability, and rhizosphere competence.

The BCA potential of *Trichoderma* spp. provides growth enhancement of plants and agricultural crops (Verma et al. 2007), can help control diseases, and increase yield (Mastouri et al. 2010). An extensive list of commercially available BCA products used against plant diseases are documented in the literature (Gardener and Fravel 2002; Navi and Bandyopadhyay 2002; Verma et al. 2007; Samuels and Hebbar 2015). However, in the most recent list of biopesticides, only the

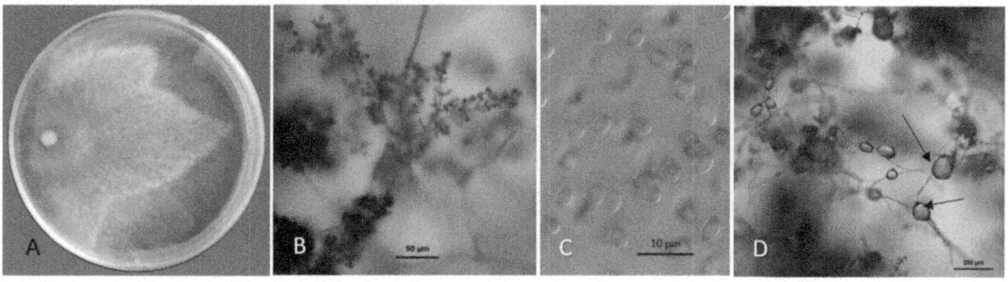

Fig. 10.1 *Trichoderma* spp. (**a**) colony morphology, (**b**) branching, (**c**) conidia and (**d**) chlamydospores

Table 10.1 List of biopesticides with *Trichoderma* spp. as microbial active ingredient registered by the EPA as of July 25, 2018

Active ingredient name	PC code	Year first registered	Trade name
Trichoderma asperellum (ICC 012)	119208	2010	Bio-Tam®
T. gamsii (ICC 080)	119207	2010	Bioten WP; Tenet WP; Remedier WP
T. hamatum isolate 382	119205	2010	Floraguard
T. harzianum Rifai (variety); KRL-AG2	119202	1993	Rootshield® WP
T. harzianum Rifai strain T-22	119202	2000	Rootshield® Plus, T-22ᴹ HC, RootShield® AG
T. polysporum (ATCC 20475)	128902	1989	Binabᴹ WP
T. virens strain G-41 and *T. harzianum* Rifai strain T-22	1776604	2012	RootShield® Plus WP
T. viride (ATCC 20476) 128903 1989	128903	1989	Binabᴹ WP

Trichoderma spp. has been registered by the EPA as microbial active ingredient as of July 25, 2018, listed in Table 10.1.

10.3 Major Row Crops in the United States

The major row crops grown vary with the geographical location (country or continent), and the categories of row crops also vary, such as food crops (wheat, corn, rice, millets, and pulses), cash crops (sugarcane, tobacco, cotton, jute, and oilseeds), plantation crops (coffee, coconut, tea, and rubber), and horticultural crops (fruits and vegetables). The basic definition of a row crop is that a crop grown in rows with suitable spacing to allow regular crop production operations during the cropping season, such as planting, spraying to protect crop from weeds, pest, and diseases, and harvesting. Irrespective of their classification, most of these crops do suffer from

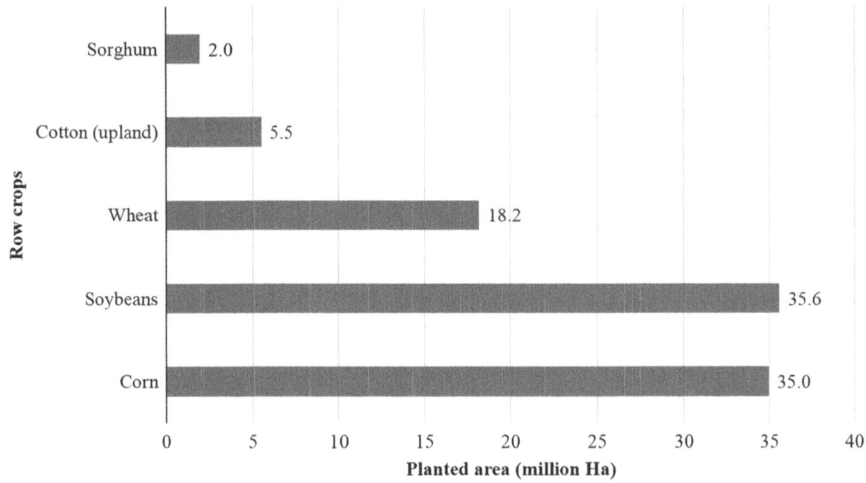

Fig. 10.2 Crop acreage data 2018 (as of January 28, 2019), the United States. (Data source: www.nass.usda.gov)

diseases (root, stem, and foliar) caused by plant pathogens (seed-, soil-, and air borne). The United States has the ability to grow a wide variety of crops with diverse climatic conditions. The 2018 crop year acreage data reported (as of Jan 28, 2019) for row crops like corn, soybeans, wheat, cotton, and sorghum in million hectare (Mha) is given in Fig. 10.2. These row crops account for 96.3 Mha of the 247.5 Mha planted to crops (nass.usda.gov).

10.4 Major Diseases and Yield Losses in Row Crops

10.4.1 Corn

The major diseases of corn are gray leaf spot (GLS), southern rust (SR), anthracnose stalk rot and top dieback (ASR/TD), seedling blights (SB), goss's wilt (GW), *Fusarium* stalk rot (FSR), diplodia ear rot (DER), nematodes (N), *Fusarium* ear rot (FER), northern leaf blight (NLB), and charcoal rot (CR) (Mueller et al. 2018). Of these diseases, GLS, SR, ASR/TD, SB, GW, FSR, and DER are major yield-reducing diseases in 12 northernmost states, and GLS, SR, FSR, nematodes, FER, NLB, and CR in 13 southernmost states. Out of these, only GLS, SR, and FSR are reported both in northern and southern states (Fig. 10.2). Yield loss due to these diseases in the northern states varies from 1.2 million ton to 4.2 million ton and in southern states from 0 to 1.8 million ton (Fig. 10.3).

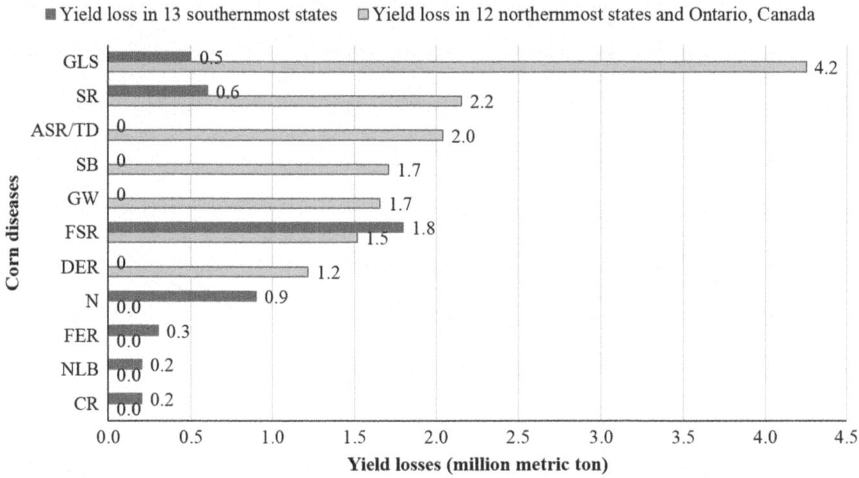

Fig. 10.3 Estimated corn yield losses due to diseases (*GLS* gray leaf spot, *SR* southern rust, *ASR/TD* anthracnose stalk rot and top dieback, *SB* seedling blights, *GW* Goss's wilt, *FSR Fusarium* stalk rot, *DER* diplodia ear rot, *N* nematodes, *FER Fusarium* ear rot, *NLB* northern leaf blight, *CR* charcoal rot) in 12 northernmost U.S. states (Illinois, Indiana, Iowa, Michigan, Minnesota, Nebraska, New York, North Dakaota, Ohio, Pennsylvania, South Dakota and Wisconsin) and Ontario, Canada, and 13 southernmost U.S. Sates (Arkansas, Colorado, Delaware, Kansas, Kentucky, Louisiana, Maryland, Mississippi, Missouri, North Carolina, Tennessee, Texas, and Virginia) in 2017. (Data Source: Mueller et al. 2017)

10.4.2 Soybean

According to Hartman (2015), more than 200 pathogens are affecting soybean worldwide, of which 35 have been reported economically vital. A comprehensive report of soybean diseases in major soybean growing countries has been compiled by Hartman et al. (2015). Rhizoctonia root rot, Pythium damping off and Phytophthora root rot, sudden death syndrome, and white mold cause yield losses up to 35% (Wrather and Koenning 2009). Out of these diseases, SDS is one the major diseases reported (Fig. 10.4 revised after Navi and Yang 2016), and the yield loss due to the SDS was US$3.06 billion (Navi and Yang 2016).

Recently, Navi et al. (2019) have compiled losses due to seedling diseases. The average annual yield loss due to seedling diseases was 0.90 mmt (range 0.34–1.51 mmt) and economic losses range from $67.60 million in 1999 to $779.92 million in 2008 (Navi et al. 2019). The economic impact of charcoal rot in the U.S. ranks fourth (Smith and Carvil 1997) with an average estimated yield loss of 0.7344 mmt/year (Wrather and Koenning 2010). Yield losses due to white mold vary from 10% to 50%; sometimes total loss depends on severity, conditions, and control practices. The yield losses due to major soybean diseases are given in Fig. 10.5.

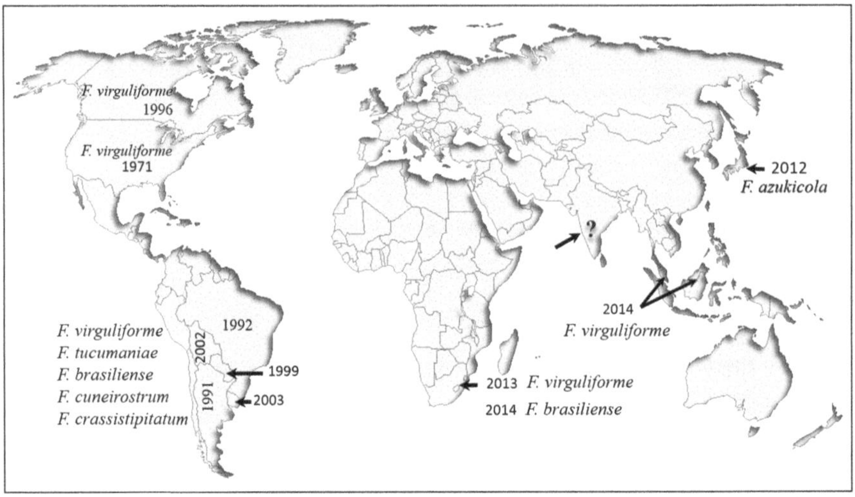

Fig. 10.4 Occurrence of soybean sudden death syndrome (*Fusarium* spp.) in different continents. (Revised version of Navi and Yang 2016)

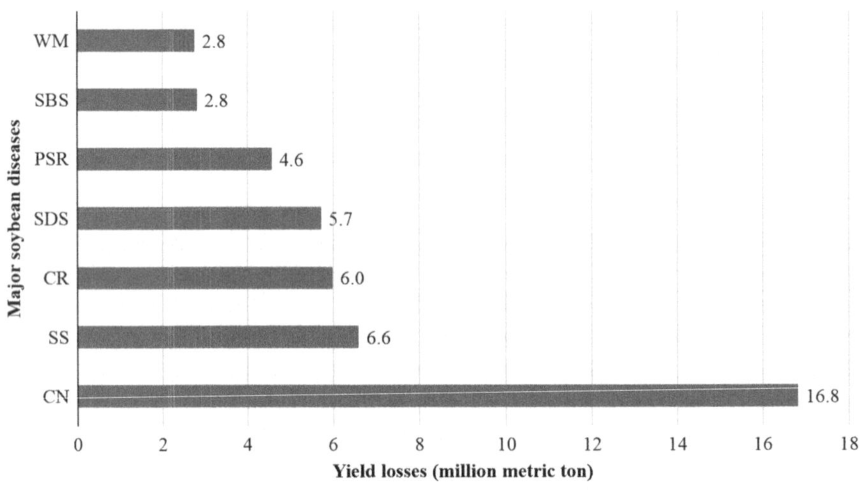

Fig. 10.5 Total estimated yield losses due to some major diseases (*WM* white mold or Sclerotima stem rot, *SBS* Septoria brown spot, *PSR* Phytophthora stem and root rot, *SDS* sudden death syndrome, *CR* charcoal rot, *SS* seedling diseases (caused by *Rhizoctonia, Pythium, Fusarium* and *Phomopsis)* and *CN* cyst nematode) in 28 US states (Alabama, Arkansas, Delaware, Florida, Georgia, Illinois, Indiana, Iowa, Kansas, Kentucky, Louisiana, Maryland, Michigan, Minnesota, Mississippi, Missouri, Nebraska, North Carolina, Ohio, Oklahoma, Pennsylvania, South Carolina, South Dakota, Tennessee, Texas, Virginia, and Wisconsin) and Ontario Canada, from 2010 to 2014. (Data Source: Allen et al. 2017)

10.4.3 Wheat

Acreage planted for wheat in the U.S. has reduced to 47.8 million acres in 2018 from about 65 million acres in 1998, and harvested acreage reduced to 39.6 million acres in 2018 from 64 million acres in 1998 (nass.usda.gov). The estimated yield loss due to fungal diseases was reported as 15–20% per annum (Figueroa et al. 2018).

10.4.4 Cotton

The National Cotton Council Disease Loss Committee submitted estimates of the losses due to each disease during the 2017 growing season. However, we have used yield loss estimates data of some of the major diseases across 14 US states (Fig. 10.6). Cotton specialists in each state have estimated disease incidence by discussing disease incidence observed across each state in 2017 (Lawrence et al. 2018).

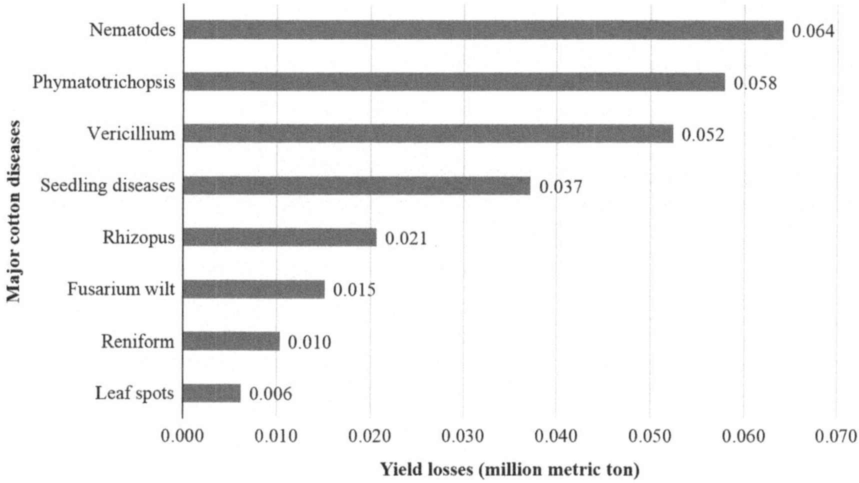

Fig. 10.6 Estimated yield losses (million metric ton) due to some of the major diseases in 14 US states (Alabama, Arizona, Arkansas, California, South Carolina, Florida, Georgia, Louisiana, Mississippi, Missouri, New Mexico, Oklahoma, Tennessee, and Texas) in 2017 cropping season. (Data Source: Lawrence et al. 2018)

10.5 Assessments of *Trichoderma* spp.

Several species of *Trichoderma* have been evaluated (in vitro or in vivo) against species of (1) *Colletotrichum* (Padder and Sharma 2011; Jagtap et al. 2012; Sawant et al. 2012; Navi et al. 2018), (2) *Fusarium* (Milanesi et al. 2013a, b; Moreno et al. 2009; Santos et al. 2011; Abd-El-Khair et al. 2010; Hajieghrari et al. 2010; Navi et al. 2018), (3) *Pythium* (Bailey and Melnick 2013; Kandula et al. 2015; Chet et al. 1981; Scheuerell et al. 2005; Navi et al. 2018), (4) *Septoria* (Broekaert et al. 1990), *Macrophomia* (Bailey and Melnick 2013; Druzhinina et al. 2011; Dwivedi 2014; Larralde-Corona et al. 2008; Mendoza et al. 2015; Etebarian 2010; Hajieghrari et al. 2010; Navi et al. 2018), (5) *R. solani* (Abd-El-Khair et al. 2010; Chet et al. 1981; Hajieghrari et al. 2010; Kandula et al. 2015; Lewis and Larkin 1997; Lewis and Papavizas, 1991b; Scheuerell et al. 2005; Zinati 2005; Bailey and Melnick 2013; Druzhinina et al. 2011; Navi et al. 2018), and (6) *S. sclerotiorum* (Kandula et al. 2015; Lopes et al. 2012; Ojaghian 2011; Elad et al. 1982; García-Garza et al. 1997; Matroudi and Zamani 2009; Whipps and Budge 1990; Navi et al. 2018).

10.6 Commercially Available Biopesticides in the United States

10.6.1 Bio-Tam® 2.0

According to the webpage http://www.isagro-usa.com/bio-tam.html or the product link at http://www.isagro-usa.com/assets/204032-55-base-booklet-a-7march2016.pdf BIO-TAM® 2.0 contains 2% each of *T. asperellum* ICC 012 and *T. gamsii* ICC 080 with a minimum of 5×10^6 cfu per gram dry weight of each strain. Application methods, dilution information, tank mixing, compatibility, storage and disposal details, and target pathogens of this product can be obtained from the link.

Labeled crops and application rates: This product is labeled for several vegetable and fruit crops, but for row crops like corn and cotton in-furrow at 42.5–85.04 g per 1000 row feet, ground banded at 2.80–3.36 kg/ha, broadcast at 2.80–5.60 kg/ha at planting and Chemigation at 2.80–5.60 kg/ha.

10.6.2 Tenet® WP

Other alternate names are Bioten™ WP and Remedier WP. According to the Isagro USA (www.isagro-usa.com), or the product link at http://www.blacksmithbiosciences.com/tenet-wp.html, Tenet® WP contains *T. asperellum* ICC 012 at 2% (minimum of 5×10^6 cfu per gram dry weight) and *T. gamsii* ICC 080 at 2% (minimum of 5×10^6 cfu per gram dry weight). This product is used in the management of soil-borne diseases caused by *Fusarium, Phytophthora, Pythium, Rhizoctonia, Sclerotinia, Verticillium, Armillaria, Rosellinia, Sclerotium rolfsii, and Thielaviopsis basicola.*

Labeled crops and application rates: Although this product is labeled for several vegetable and fruit crops, but for row crops like corn and cotton in-furrow at 42.5–85.04 g per 1000 row feet, ground banded at 2.80–3.36 kg/ha, broadcast at 2.80–5.60 kg/ha at planting and Chemigation at 2.80–5.60 kg/ha.

10.6.3 RootShield® WP

According to Bioworks Inc. (https://www.bioworksinc.com), or the product link at https://www.bioworksinc.com/products/rootshield-wp/rootshield-wp-label.pdf, RootShield WP contains *T. harzianum* Rifai strain KRL-AG2 at 1.15% (1×10^7 cfu per gram dry weight). This product is recommended for several fruits, vegetables, bulb, and flowering crops but not for row crops that are discussed in this chapter. However, methods of application and rates can be found on their web.

10.6.4 RootShield® Plus⁺ WP

According to Bioworks Inc. (https://www.bioworksinc.com), or the product link at https://www.bioworksinc.com/products/rootshield-plus-wp/rootshield-plus-wp-label.pdf, RootShield® Plus$^+$ WP contains *T. harzianum* Rifai strain T-22 at 1.15% (1×10^7 cfu per gram dry weight) and *T. virens* strain G-41 at 0.61% (5.3×10^6 cfu per gram dry weight). This product can be applied to seeds, to soil and planting mixes, transplants, or other propagative materials.

Labeled crops and application rates: This product is recommended for several fruits, vegetables, and flowering crops. Also, recommended for corn, wheat and sorghum as field chemigation at 85.04–226.8 g mixed in 378.54 L water. This product can be sprayed in-furrow or as transplant starter solution at 1.12–2.24 kg/ha.

10.6.5 RootShield® Plus$^+$ Granules

According to Bioworks Inc. (https://www.bioworksinc.com), or the product link at https://www.bioworksinc.com/products/rootshield-plus-granules/rootshield-plus-granules-label.pdf, the active substances of RootShield® Plus$^+$ WP are *T. harzianum* Rifai strain T-22 at 1.15% (1×10^7 cfu per gram dry weight) and *T. virens* strain G-41 at 0.61% (5.3×10^6 cfu per gram dry weight). This product can be applied as an inoculant to soil or to planting mixes. Target pathogens of this product can be obtained from the product link.

Labeled crops and application rates: This product is recommended for several fruits, vegetables, and flowering crops. Also, recommended for corn, cotton in-furrow spray at 5.60–11.2 kg/ha.

10.6.6 RootShield® AG

According to Bioworks Inc. (https://www.bioworksinc.com), or the product link at https://www.bioworksinc.com/products/rootshield-ag/RootShieldAG-Label.pdf, the active substance of RootShield® AG is *T. harzianum* Rifai strain T-22 at 1.15% (at least 1×10^7 cfu per gram dry weight). *This product can be used either alone or tank mixed with other fungicides. For application methods, target pathogens, and compatibility details with synthetic fungicides visit the product link.*

Labeled crops and application rates: This product is recommended for several fruits, vegetables, and flowering crops. It is also recommended for corn, wheat, and sorghum as planter box treatment at 42.5–85.04 g/50.80 kg seed, commercial seed treatment at 0.181–240 g/50.80 kg seed, and field chemigation at 85.04–141.74 g/ 378.54 L water. For cotton, it is recommended as planter box treatment at 42.5–85.04 g/50.80 kg seed, commercial seed treatment at 7.98–240 g/50.80 kg seed, and field chemigation at 85.04–141.74 g/378.54 L water. Agricultural crops seed treatments: field corn at 0.0016–0.32 g/453.6 g seed; soybeans and sorghum at 0.08–2.4 g/453.6 g seed.

10.6.7 Other Products

Over the last 5 years, we have field-tested the efficacies of a few *Trichoderma* spp.-based BCA from various private institutes, including those in markets and preregistration products. We tested biopesticides (registered or preregistered) against WM and SDS. Our data showed increased soybean yields in fields severely infested with SDS and WM (Figs. 10.7, 10.8, and 10.9).

Biopesticides can be an effective tool to manage both WM and SDS in North Central Region of the US. Commercial field trials in growers' fields (Figs. 10.8 and 10.9) showed excellent control of SDS or white mold. A preregistered product named PHD-1 (Plant Health Defense Inc.) showed 0.67 t/ha yield advantage compared with unsprayed control. Similarly, a biological control agent named PHD-2 showed 0.47 t/ha yield increase. These preregistered products could deliver a yield benefit compared with unsprayed control. They can be applied with various forms such as sprays with tank mix and seed treatments (Figs. 10.8 and 10.9). Seed treatments with PHD-1 resulted in a yield increase of 0.67 mt/ha at Ames location where SDS was very severe (Fig. 10.7).

Fig. 10.7 Two-row plots comparison between seed treatments effect of bio-pesticide (Plant Health Defense) on the occurrence of soybean sudden death syndrome and yields, Ames, IA

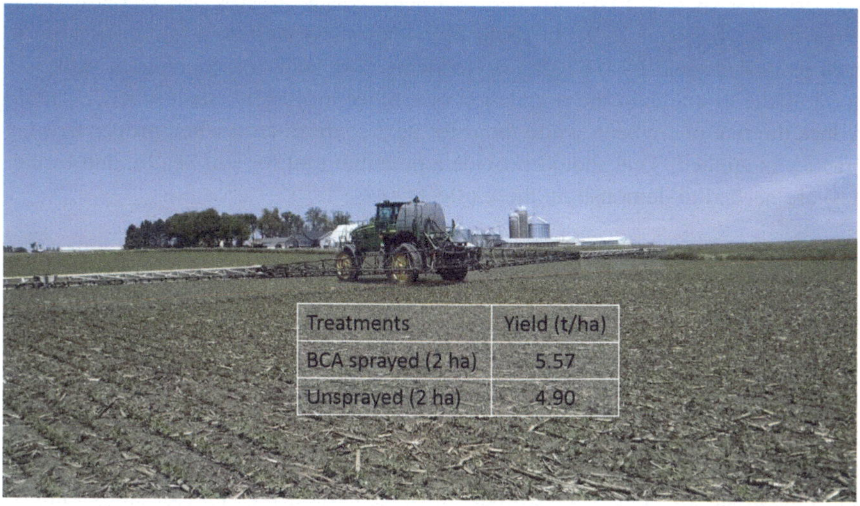

Fig. 10.8 Post emergence application of bio-pesticide (Plant Health Defense) in spring 2016, on soybean white mold and yields, Sioux Center, IA

Fig. 10.9 Effect of bio-pesticide (Plant Health Defense) seed treatments (ST) and foliar/soil application (Spray) on the occurrence of soybean sudden death syndrome and white mold and yields, Nashua, IA

Biological control products are the future of the US North Central Region for disease management where major row crops (soybean, corn, wheat) are in rotation, because the biopesticides we tested reduce various soil-borne pathogens. Biopesticides such as PHD products can help reduce diseases and increase yields (Figs. 10.7, 10.8, and 10.9). In soybean fields rotated with corn, an application in corn increased yield by 0.67 t/ha by reducing corn stalk rot. Since corn stalk is a good media to maintain the population of a biocontrol agent, such practice should reduce the risk of soybean white mold for next soybean crop when proper farming practice is applied. This should provide a potential tool for soil health, and further data are needed to demonstrate these effects.

10.7 Assessment of Seed Treatments of Biocontrol Agents

Taylor and Harman (1990) indicated that seed treatments can provide an inexpensive insurance against soil fungi. Seed treatments with bacteria- and fungi-based BCA are used in protecting against soil-borne pathogens. Though, a comprehensive list of 25 commercial BCA registered under U.S. Environmental Protection Agency is available (Gardener and Fravel 2002; Fravel 2005), only Rootshield and SoilGard are sold with Trichoderma or Gliocladium as their active ingredients. Either most of them are for drench or spray or none of the fungus-based BCA are listed as seed treatments. Harman (1991), indicated that highly effective strain for seed treatments is insufficient to ensure effective biological. Recently, effects of seed treatments and foliar/soil applications assessed in soybean and corn either in research plots and in commercial farming showed encouraging results (Figs. 10.7, 10.8, and 10.9, Navi et al. 2018).

10.7.1 Corn

Seed treatments antagonists reduced corn damping-off caused by species of *Pythium* and *Fusarium* (Mao et al. 1997) and promoted plant growth in corn (Singh and Nautiyal 2012). Effects of seed treatments and foliar applications of BCA assessed against major soil-borne pathogens of corn either in research plots or in commercial farming showed yield advantages in BCA-treated plots compared with synthetic fungicides treatments and untreated controls (Navi and Yang unpublished).

10.7.2 Soybean

Biological agents seed treatments were used to control soybean white mold (Hao et al. 2010). anthracnose/pod blight (Jagtap et al. 2012). Biocontrol agents (species of *Trichoderma*, *Coniothyrium*, and *Talaromyces*) seed treatments in combination with bioAPT microbial carrier tested against SDS and Rhizoctonia root rot showed yield advantages in soybean (Navi et al. 2016). Similarly, there are reports of tests against root rot of okra, sunflower, soybean, and mungbean (Ehteshamul et al. 1990), and charcoal rot (*Macrophomina phaseolina*) using *T. harzianum* (Khaledi and Taheri 2016). Effects of seed treatments and foliar applications of BCA assessed against major soil-borne pathogens of soybean either in research plots or in commercial farming showed yield advantages in BCA-treated plots compared with untreated controls (Figs. 10.7, 10.8, and 10.9, Navi and Yang unpublished).

10.7.3 Cotton and Wheat

Howell (1982, 2002, 2006) and Lewis and Papavizas (1991b) tested *Gliocladium virens* or *Trichoderma virens* against damping-off of cotton seedlings, and, similarly, Sivan and Chet (1986) tested species of *Trichoderma* against *Fusarium* spp. in cotton, wheat, and muskmelon.

10.8 Assessment of BCA Application Methods

Howell (1982) studied the effect of *Gliocladium virens* through in-furrow application method on damping-off of cotton seedlings. *T. harzianum* mixed with greenhouse mix, reduced microsclerotia coverage on soybean root and stems (Barari and Foroutan 2016), and observed reduction in SDS and WM incidences and a yield advantage in foliar/soil applied plots compared with untreated plots. Effects of seed treatments (Navi et al. 2018) and foliar applications of BCA assessed against major soil-borne pathogens of corn either in research plots or in commercial farming showed yield advantages in BCA-treated plots compared with synthetic fungicides treatments and untreated controls (Figs. 10.8 and 10.9, Navi and Yang unpublished).

10.9 Conclusion

Investigations on seed treatment, soil, and foliage applications of biopesticides need further attention. Also, research directions of personnel involved should focus on formulations, large-scale fermentation of biopesticides, and extension efforts to address their conventional adaptation strategies in row crops.

References

Abd-El-Khair H, Khalifa RKM, Haggag KHE (2010) Effect of *Trichoderma* species on damping off diseases incidence, some plant enzymes activity and nutritional status of bean plants. J Am Sci 6:486–497

Allen TW, Bradley CA, Sisson AJ, Byamukama E, Chilvers MI, Coker CM, Collins AA, Damicone JP, Dorrance AE, Dufault NS, Esker PE, Faske TR, Giesler LJ, Grybauskas AP, Hershman DE, Hollier CA, Isakeit T, Jardine DJ, Kelley HM, Kemerait RC, Kleczewski NM, Koenning SR, Kurle JE, Malvick DK, Markell SG, Mehl HL, Mueller DS, Mueller JD, Mulrooney RP, Nelson BD, Newman MA, Osborne L, Overstreet C, Padgett GB, Phipps PM, Price PP, Sikora EJ, Smith DL, Spurlock TN, Tande CA, Tenuta AU, Wise KA, Wrather JA (2017) Soybean yield loss estimates due to diseases in the United States and Ontario, Canada, from 2010 to 2014. Plant Health Prog 18:19–27

Bailey BA, Melnick RL (2013) The endophytic *Trichoderma*. In: Mukherjee PK, Horwitz BA, Singh US, Mukherjee M, Schmoll M (eds) *Trichoderma*: biology and applications, 1st edn. CAB International, London, pp 152–172

Barari H, Foroutan A (2016) Biocontrol of soybean charcoal root rot disease by using *Trichoderma* spp. Cercetări Agronomic Moldova 166:41–51

Broekaert WF, Terras FR, Cammue B, Vanderleyden J (1990) An automated quantitative assay for fungal growth inhibition. FEMS Microbiol Lett 69:55–59

Brozóvá J (2004) Mycoparasitic fungi *Trichoderma* spp. in plant protection. Plant Protect Sci 2:63–74

Celar F (2003) Competition for ammonium and nitrate forms of nitrogen between some phytopathogenic and antagonistic soil fungi. Biol Control 28:19–24

Chaube HS, Mishra DS, Varshney S, Singh US (2003) Biocontrol of plant pathogens by fungal antagonists: historical background, present status and future prospects. Annu Rev Phytopathol 2:1–42

Chet I (1993) Biotechnology in plant disease control. Wiley-Liss Inc., Hoboken, NJ

Chet I, Harman GE, Baker R (1981) *Trichoderma hamatum*: its hyphal interactions with *Rhizoctoniasolani* and *Pythium* spp. Microb Ecol 7:29–38

Dal BGM, Mónaco CI, Cháves AR (1997) Efecto de los metabolitos volátiles de *Trichoderma harzianum* sobre el crecimiento de hongos fitopatógenos procedentes del suelo. Rev Iberoam Mic 14:131–134

Druzhinina IS, Seidl-Seiboth V, Herrera-Estrella A, Horwitz BA, Kenerley CM, Monte E, Mukherjee PK, Zeilinger S, Grigoriev V, Kubicek CP (2011) *Trichoderma*: the genomics of opportunistic success. Nat Rev Microbiol 9:749–759

Dwivedi SK (2014) Role of antagonistic microbes in management of phytopathogenic fungi of some important crops. In: Kharwar RN, Upadhyay R, Dubey N, Raghuwanshi R (eds) Microbial diversity and biotechnology in food security. Springer, India, pp 273–292

Ehteshamul HS, Ghaffar A, Zaki MJ (1990) Biological control of root rot diseases of okra, sunflower, soybean, and mungbean. Pak J Bot 22:121–124

Elad Y, Chet I, Henis Y (1982) Degradation of plant pathogenic fungi by *Trichoderma harzianum*. Can J Microbiol 28:719–725

Etebarian HR (2010) Evaluation of Trichoderma isolates for biological control of charcoal stem rot in melon caused by *Macrophomina phaseolina*. J Agric Sci Technol 8:243–250

Figueroa M, Hammond-Kosack KE, Solomon PS (2018) A review of wheat diseases-a field perspective. Mol Plant Pathol 19:1523–1536

Fravel DR (2005) Commercialization and implementation of biocontrol 1. Annu Rev Phytopathol 43:337–359

García BFJ, Santamarina MP, Roselló J (2005) *Trichoderma:* mecanismos de control. Phytoma 172:106–107

García-Garza JA, Reeleder RD, Paulitz TC (1997) Degradation of sclerotia of *Sclerotinia sclerotiorum* by fungus Gnats *(Bradysia copropila)* and the biocontrol fungi *Trichoderma* spp. Soil Biol Biochem 29:123–129

Gardener BBM, Fravel DR (2002) Biological control of plant pathogens: research, commercialization, and application in the USA. Plant Health Prog 10. https://doi.org/10.1094/PHP-2002-0510-01-RV

Hajieghrari B, Torabi-Giglou M, Mohammadi MR, Davari M (2010) Biological potential of some Iranian *Trichoderma* isolates in the control of soil-borne plant pathogenic fungi. Afr J Biotechnol 7:967–972

Hao J, Wang D, Hammerschmidt R (2010) Using biological agents to control soybean white mold. Michigan Soybean Checkoff. http://www.michigansoybean.org/MSPCSite/Research/FY10ResSum.pdf

Harman GE (1991) Seed treatments for biological control of plant disease. Crop Prot 10:166–171

Harman GE, Taylor AG (1988) Improved seedling performance by integration of biological control agents at favorable pH levels with solid matrix priming. Phytopathology 78:520–525

Harman GE, Chet I, Baker R (1980) *Trichoderma hamatum* effects on seed and seedling disease induced in radish and pea by *Pythium* spp. or *Rhizoctonia solani*. Phytopathology 70:1167–1172

Harman GE, Howell CR, Viterbo A, Chet I, Lorito M (2004) Trichoderma species-opportunistic, avirulent plant symbionts. Nat Rev Microbiol 2:43–56

Hartman GL (2015) Worldwide importance of soybean pathogens and pests. In: Compendium of soybean diseases and pests, 5th edn. The American Phytopathology, St. Paul, MN, pp 4–5, 201pp

Hartman GL, Rupe JC, Sikora EJ, Domier LL, Davis JA, Steffey KL (2015) Soybean diseases in Argentina, Brazil, Canada, China, India and Japan. In: Compendium of soybean diseases and pests, 5th edn. The American Phytopathology, St. Paul, MN, pp 5, 201pp–16

Howell CR (1982) Effect of *Gliocladium virens* on *Pythium ultimum*, *Rhizoctonia solani*, and damping-off of cotton seedlings. Phytopathology 72:496–498

Howell CR (2002) Cotton seedling pre-emergence damping-off induced by *Rhizopus orizae* and *Pythium* spp. and its biological control with *Trichoderma* spp. Phytopathology 92:177–180

Howell CR (2006) Understanding the mechanisms employed by *Trichoderma virens* to effect biological control of cotton diseases. Phytopathology 96:178–180

Jagtap GP, Gavate DS, Dey U (2012) Control of *Colletotrichum truncatum* causing anthracnose/pod blight of soybean by aqueous leaf extracts, biocontrol agents and fungicides. Agric Adv 1:39–52

Jahagirdar S, Kambrekar DN, Navi SS, Kunta M (2019) Plant growth promoting fungi: diversity and classification. In: Jogaiah S, Abdelrahman M (eds) Bioactive molecules in plant immune defense. Springer, Cham, pp 25–34. https://doi.org/10.1007/978-3-030-27165-7_2

Jeffries P (1995) Biology and ecology of mycoparasitism. Can J Bot 73:1284–1290

Kandula DRW, Jones EE, Stewart A, McLean KL, Hampton JG (2015) Trichoderma species for biocontrol of soil-borne plant pathogens of pasture species. Biocontrol Sci Tech 25:1052–1069

Khaledi N, Taheri P (2016) Biocontrol mechanisms of *Trichoderma harzianum* against soybean charcoal rot caused by *Macrophomina phaseolina*. J Plant Protect Res 56:21–31. https://doi.org/10.1515/jppr-2016-0004

Larralde-Corona CP, Santiago-Mena MR, Sifuentes-Rincon AM, Rodríguez-Luna IC, Rodriguez-Perez MA, Shirai K, Narvaez-Zapata JA (2008) Biocontrol potential and polyphasic characterization of novel native *Trichoderma* strains against *Macrophomina phaseolina* isolated from sorghum and common bean. Appl Microbiol Biotechnol 80:167–177

Lawrence K, Hagan A, Nortan R, Hu J, Faske T, Hutmacher R, Mueller J, Small I, Grabau Z, Kemerait B, Overstreet C, Price P, Lawrence G, Allen T, Atwell S, Idowu J, Boman R, Goodson J, Kelly H, Woodward JE, Wheeler TA (2018) Cotton disease loss estimate committee report, 2018. In: Proceedings of the 2018 Beltwide cotton conferences, San Antonio, TX, January 3–5, 2018, pp 161–163

Lewis JA, Larkin RP (1997) Extruded granular formulation with biomass of biocontrol *Gliocladium virens* and *Trichoderma* spp. to reduce damping-off of eggplant caused by *Rhizoctonia solani* and saprophytic growth of the pathogen in soil-less mix. Biocontrol Sci Technol 7:49–60

Lewis JA, Papavizas GC (1991a) Biocontrol of plant diseases: the approach for tomorrow. Crop Prot 10:95–105

Lewis JA, Papavizas GC (1991b) Biocontrol of cotton damping-off caused by *Rhizoctonia solani* in the field with formulations of *Trichoderma* spp. and *Gliocladium virens*. Crop Prot 10:396–402

Lifshitz R, Windham MT, Baker R (1986) Mechanism of biological control of pre-emergence damping-off of pea by seed treatment with *Trichoderma* spp. Phytopathology 76:720–725

Lopes FAC, Steindorff AS, Geraldine AM, Brandão RS, Monteiro VN, Júnior ML, Coelho ASG, Ulhoa CJ, Silva RN (2012) Biochemical and metabolic profiles of Trichoderma strains isolated from common bean crops in the Brazilian Cerrado, and potential antagonism against *Sclerotinia sclerotiorum*. Fungal Biol 116:815–824

Mao W, Lewis JA, Hebbar PK, Lumsden RD (1997) Seed treatment with a fungal or a bacterial antagonist for reducing corn damping-off caused by species of *Pythium* and *Fusarium*. Plant Dis 81:450–454

Mastouri F, Björkman T, Harman GE (2010) Seed treatment with *Trichoderma harzianum* alleviates biotic, abiotic, and physiological stresses in germinating seeds and seedlings. Phytopathology 100:1213–1221

Matroudi S, Zamani MR (2009) Antagonistic effects of three species of *Trichoderma* sp. on *Sclerotinia sclerotiorum*, the causal agent of canola stem rot. Egypt J Biol 11:37–44

Mendoza JLH, Pérez MIS, Prieto JMG, Velásquez JDQ, Olivares JGG, Langarica HRG (2015) Antibiosis of *Trichoderma* spp. strains native to northeastern Mexico against the pathogenic fungus *Macrophomina phaseolina*. Braz J Microbiol 46:1093–1101

Menendez AB, Godeas A (1998) Biological control of *Sclerotinia sclerotiorum* attacking soybean plants. Degradation of the cell walls of this pathogen by *Trichoderma harzianum* (BAFC 742). Mycopathologia 142:153–160

Milanesi PM, Blume E, Antonioli ZI, Muniz MFB, Santos RFD, Finger G, Durigon MR (2013a) Biocontrole de *Fusarium* spp. com *Trichoderma* spp. e promoção de crescimento em plântulas de soja. Revista de Ciências Agrárias 36:347–356

Milanesi PM, Blume E, Muniz MFB, Reiniger LRS, Antoniolli ZI, Junges E, Lupatini M (2013b) Detecção de *Fusarium* spp. e *Trichoderma* spp. e antagonismo de *Trichoderma* sp. em soja sob plantio direto. Semina: Ciências Agrárias 34:3219–3234

Moreno CA, Castillo F, González A, Bernal D, Jaimes Y, Chaparro M, Gonzalez C, Rodriguez F, Restrepo S, Cotes AM (2009) Biological and molecular characterization of the response of tomato plants treated with *Trichoderma koningiopsis*. Physiol Mol Plant Pathol 74:111–120

Mueller D, Wise K, Sisson A (2018) Corn disease loss estimates from the United States and Ontario, Canada – 2017. Corn disease management, CPN-2007-17-W

Navi SS, Bandyopadhyay R (2002) Biological control of fungal plant pathogens. In: Waller JM, Lenné JM, Waller SJ (eds) Plant pathologist's pocketbook. CABI Publishing, Wallingford, UK, pp 354–365

Navi SS, Yang XB (2016) Sudden death syndrome - a growing threat of losses in soybeans. CAB Rev 2016:11. https://doi.org/10.1079/PAVSNNR201611039. No. 039. http://www.cabi.org/cabreviews

Navi SS, Huynh T, Li X, Yang XB (2016) Effects of biocontrol agents in combination with bioAPT microbial carrier as seed treatment on soybean diseases and yields (Abstr). In: Khetarpal RK, Mondal KK, Dubey SC, Rao GP, Celia Chalam V, Singh N et al (eds) IPS 6th international conference on plant, pathogens and people, 23–27 Feb 2016, New Delhi, India, p 64, 676pp

Navi SS, Rajasab AH, Yang XB (2018) Challenges and opportunities in management of soil borne pathogens of soybean using biocontrol agents. In: Singh D, Chakraborty BN, Pandey RN, Sharma P (eds) Biological control of crop diseases: recent advances & perspectives. Today & Tomorrow's Printers and Publishers, New Delhi, pp 531–577

Navi SS, Huynh T, Mayers CG, Yang XB (2019) Diversity of *Pythium* spp. associated with soybean damping-off, and management implications by using foliar fungicides as seed treatments. Phytopathol Res 1:8. https://doi.org/10.1186/s42483-019-0015-9

Nelson EB, Harman GE, Nash GT (1988) Enhancement of *Trichoderma*-induced biological control of *Pythium* seed rot and pre-emergence damping-off of peas. Soil Biol Biochem 20:145–150

Ojaghian MR (2011) Potential of *Trichoderma* spp. and *Talaromyces flavus* for biological control of potato stem rot caused by *Sclerotinia sclerotiorum*. Phytoparasitica 39:185–193

Padder BA, Sharma PN (2011) *In vitro* and *in vivo* antagonism of biocontrol agents against *Colletotrichum lindemuthianum* causing bean anthracnose. Arch Phytopathol Plant Protect 44:961–969

Papavizas GC (1985) *Trichoderma* and *Gliocladium*: biology, ecology, and potential for biocontrol. Annu Rev Phytopathol 23:23–54

Punja ZK, Utkhede RS (2003) Using fungi and yeasts to manage vegetable crop diseases. Trends Biotechnol 21:400–407

Samuels GJ, Hebbar PK (2015) Developing *Trichoderma*-based products for application in agriculture. In: Trichoderma identification and agricultural applications. The American Phytopathology, St. Paul, MN, pp 7–34

Santos A, Garcia M, Cotes AM, Villamizar L (2011) The effect of the formulation on the shelf-life of biopesticides based on two Colombian isolates of *Trichoderma koningiopsis* Th003 and *Trichoderma asperellum* Th034. Rev Iberoam Micol 29:150–156

Sawant IS, Rajguru YR, Salunkhe VP, Wadkar PN (2012) Evaluation and selection of efficient isolates of *Trichoderma* species from diverse locations in India for biological control of anthracnose disease of grapes. J Biol Control 26:50–60

Scheuerell SJ, Sullivan DM, Mahaffee WF (2005) Suppression of seedling damping-off caused by *Pythium ultimum, P. irregulare*, and *Rhizoctonia solani* in container media amended with a diverse range of Pacific northwest compost sources. Phytopathology 95:306–315

Schirmböck M, Lorito M, Wang YL, Hayes CK, Arisan-Atac I, Scala F, Harman GE, Kubicek CP (1994) Parallel formation and synergism of hydrolytic enzymes and peptaibol antibiotics, molecular mechanisms involved in the antagonistic action of *Trichoderma harzianum* against phytopathogenic fungi. Appl Environ Microbiol 60:4364–4370

Singh PC, Nautiyal CS (2012) A novel method to prepare concentrated conidial biomass formulation of Trichoderma harzianum for seed application. J Appl Microbiol 113:1442–1450

Sivan A, Chet I (1986) Biological control of *Fusarium* spp. in cotton, wheat and muskmelon by *Trichoderma harzianum*. J Phytopathol 116:39–47

Smith GS, Carvil ON (1997) Field screening of commercial and experimental soybean cultivars for their reaction to *Macrophomina phaseolina*. Plant Dis 81:363–368

Taylor AG, Harman GE (1990) Concepts and technologies of selected seed treatments. Annu Rev Phytopathol 28:321–339

Templeton GE, Heiny DK (1989) Improvement of fungi to enhance mycoherbicide potential. In: Whipps JM, Lumsden RD (eds) Biotechnology of fungi for improving plant growth. Cambridge University Press, Cambridge, UK, pp 127–152

Verma M, Brar SK, Tyagi R, Surampalli R, Valero J (2007) Antagonistic fungi, *Trichoderma* spp.: panoply of biological control. Biochem Eng J 37:1–20

Weindling R (1932) *Trichoderma lignorum* as a parasite of other soil fungi. Phytopathology 22:837–845

Whipps JM, Budge SP (1990) Screening for sclerotial mycoparasites of *Sclerotinia sclerotiorum*. Mycol Res 94:607–612

Whipps JM, Lumsden RD (2001) Commercial use of fungi as plant disease biological control agents: status and prospects. In: Butt TM, Jackson C, Magan N (eds) Fungi as biocontrol agents: progress, problems and potential. CABI Publishing, Wallingford, UK, pp 9–22

Woo SL, Ruocco M, Vinale F, Nigro M, Marra R, Lombardi N, Pascale A, Lanzuise S, Manganiello G, Lorito M (2014) Trichoderma-based products and their widespread use in agriculture. Open Mycol J 8(1):71–126

Wrather JA, Koenning SR (2009) Effects of diseases on soybean yields in the United States 1996 to 2007. Online. Plant Health Prog 10:24. https://doi.org/10.1094/PHP-2009-0401-01-RS

Wrather JA, Koenning SR (2010) Soybean disease loss estimates for the United States, 1996–2010. Delta Research Center, Agriculture Experiment Station, University of Missouri, College of Agriculture, Food and Natural Resources, Missouri

Zinati GM (2005) Compost in the 20th century: a tool to control plant diseases in nursery and vegetable crops. HortTechnology 15(1):61–66

Chapter 11
Potential of *Trichoderma* spp. for Pest Management and Plant Growth Promotion in NE India

L. C. Bora, Popy Bora, and Monoj Gogoi

Abstract The growing concern about the health and environmental hazards of chemical pesticides worldwide and burgeoning interest on organic or chemical residue-free food have driven plant protection research towards developing a bio-intensive and ecofriendly strategy of pest management. At this juncture, exploring plant-beneficial microbes including the fungal genus *Trichoderma* is considered as the most suitable tactic. *Trichoderma* is known as the most commonly used biocontrol agent, as *Trichoderma* based biopesticides occupy a significant position in the world biopesticide market. The plant-beneficial properties of *Trichoderma* have widened its horizon beyond plant disease management, and the genus is explored for biotic and abiotic stress management as well. *Trichoderma* has shown its potential in plant growth promotion and activating defence mechanism in plants. In the recent decade, the genus has caught researchers' attention in insect pest management as well as bioremediation of organic and inorganic pollutants. This book chapter summarizes the comprehensive information on research findings on *Trichoderma* including plant disease and insect pest management, plant growth promotion, *Trichoderma*-mediated host defence response, role in organic agriculture in NE India and discusses the future line of research works.

Keywords Biocontrol · Plant disease · Pest management · *Trichoderma*

11.1 Introduction

Pest management in agriculture plays a pivotal role in global food security through maintaining crop productivity as well as quality production. The chemical pesticides hugely popularized by green revolution in India once considered as magical cure elements have now been posing serious threat to the environment and human health. Worldwide growing concern about the ill effects of chemocentric management

L. C. Bora (✉) · P. Bora · M. Gogoi
Department of Plant Pathology, Assam Agricultural University, Jorhat, Assam, India

A. K. Sharma, P. Sharma (eds.), *Trichoderma*, Rhizosphere Biology,
https://doi.org/10.1007/978-981-15-3321-1_11

practices and burgeoning interest on organically grown foods have provided a strong impetus for the development of alternative means of sustaining agriculture. At this juncture, further work on development of technology based on natural resources has immense potential for ecofriendly pest management practices. Biological management by exploring plant-beneficial microorganisms against plant diseases as well as insect pests is considered as one such ecofriendly and cost-effective alternative approach. Significant progress has been made worldwide in the research for the identification of efficient and aggressive biocontrol agents and thereby reducing disease incidence in many crops against many economically significant diseases. Some plant-beneficial microbes viz., *Trichoderma* spp., *Aspergillus* spp., *Gliocladium* spp. and fluorescent pseudomonads have been extensively studied and exploited in disease management programmes due to their persistent nature in soil, phyllosphere, and rhizosphere, their high multiplication rate, easiness in handling, and mass multiplication without any adverse effects on non-targeted organisms (Sharma et al. 2001; Singh et al. 2002). Among all the biocontrol agents, voluminous literatures on *Trichoderma* spp. worldwide including India impliy the significance of *Trichoderma* not only in the agricultural sector but also in the industrial sector.

North East India is considered as an organic area by default as a significant portion of land resources are still untouched by conventional chemocentric agricultural practices (Bora et al. 2016b). As of 2017–2018, as per Agricultural and Processed Food Products Export Development Authority (*APEDA*), nearly 90,500 ha of land in the NE region is already under organic cultivation (Anonymous 2017). Considering the central government policies to convert the NE region of India into organic hub of the country, there is a huge demand of quality biopesticides to address diseases and insect pest problem in the region. Microbial bioformulations including *Trichoderma*-based bioformulation can play a significant role here and provide economic advantage to the farmers. Besides use in organic agriculture, *Trichoderma* can play an important role in Integrated Disease Management (IDM) systems as well.

11.2 About Genus *Trichoderma*

The genus Trichoderma has worldwide occurrence in both terrestrial and marine ecosystem. *Trichoderma* spp. are found above and below ground and are commonly associated with root, soil and plant debris, forest humus etc. (Harman et al. 2004; Mukherjee et al. 2013). The diverse metabolic capability and aggressive competitive nature of *Trichoderma* species make the genus a profuse colonizer. Many *Trichoderma* strains are strong opportunistic invaders, fast growing and produce a number of antibiotics (Singh et al. 2009). Friedl and Druzhinina (2012), through a metagenomic approach, showed that relatively small numbers are adapted to soil as habitat and they colonize plant roots, establishing communication with the plant where they systemically alter the expression of several phytogenes. The endophytic association of *Trichoderma* species have also been reported in roots and shoots of many plants and found to have antagonistic and plant growth-promoting properties. Strains of *Trichoderma* are rarely found to be pathogenic to living plants, although the genus is reported to be pathogenic in edible basidiomycetes fungi causing significant economic loss in mushroom cultivation (Samuels 1996).

11.3 *Trichoderma* as Twin Service Provider

Persoon in 1794 described *Trichoderma* as a genus, but the biocontrol potential was established by Weindling in 1930s (Weindling 1934). Among all the successful biocontrol agents available today, *Trichoderma* holds a special position due to its multifarious plant-beneficial properties. The worldwide acceptance of *Trichoderma* as an efficient biocontrol agent against many pathogens has opened up new frontiers in further research, and it has now been established that *Trichoderma* has multifaceted effects that include (1) improvement of plant growth and vigour and solubilisation of nutrients, (2) enhancement of defence mechanism in plants (Harman et al. 2004; Vinale et al. 2008), (3) biodegradation of pesticide residues and bioremediation of heavy metals (Katayama and Matsumara 1993; Harish et al. 2013) and (4) imparting abiotic stress tolerance (Lorito et al. 2010). Besides, *Trichoderma* also produces a wide range of enzymes, including cellulase and hemicellulase, that have widespread industrial application, and this makes the genus a preferred input for bio-industries too (Mukherjee et al. 2013).

11.3.1 Plant Disease Management and Trichoderma

Although the biocontrol potential of *T. lingnorum* (*viride*) on *Rhizoctonia solani*, a soil-borne pathogen by Weindling in 1934 field level demonstration of bioefficacy of *Trichoderma* was shown to the world during 1970s. Ever since, vigorous worldwide research on the genus has established some other *Trichoderma* species as efficient biocontrol agents. *T. harzianum*, *T. hamatum*, *T. ressei*, *T. longibrachiatum*, *T. lingorum* and *T. pseudokoningii* have been proven as bioagents for their efficacy against some major plant pathogens that include *Pythium aphanidermatum*, *Fusarium* spp., *Botrytis cinerea*, *Sclerotinia sclerotiorum*, *Rhizoctonia solani*, *Sclerotium rolfsii* and *Ralstonia solanacearum* (Mukherjee and Raghu 1997; Mishra et al. 2000; Dubey and Suresh 2006; Shanmugam et al. 2012; Bora et al. 2015). *Trichoderma* has now been considered as a successful management option for many soil-borne and aerial pathogenic fungi and bacteria worldwide (Howell 2003; Bora et al. 2016a, b).

Trichoderma employs an array of mechanisms for suppression of plant pathogens that includes hyperparasitism, competition for nutrients, antibiosis and induction of host defence mechanism in plants. Mycoparasitism is a complex mechanism with four sequential steps viz., chemotropism and recognition; attachment and coiling;

cell wall penetration; and digestion of host cell (Chet et al. 1981). Steyaert et al. (2003) demonstrated that *Trichoderma* can detect other fungi growing towards it, coil the hyphae and penetrate the host cell by producing hydrolytic cell wall-degrading enzymes and eventually collapse the host hyphae. The genome analysis of *T. virens* and *T. atroviride* revealed an expanded arsenal of genes encoding cell wall-degrading enzymes of which chitinases, β-1,3-glucanases, α-1,3-glucanases and proteases are of significant importance in multienzymatic system of *Trichoderma* strains (Hermosa et al. 2012; Singh et al. 2014). Kamala and Devi (2012) recorded the high protease and β-1,3glucanase activity of isolates of *Trichoderma* from Manipur, which showed inhibition of *Fusarium* spp. and *Rhizoctonia solani* under in vitro conditions.

Antibiosis is another mechanism of *Trichoderma* spp. that attributes to their biocontrol potential, wherein the efficient strains secrete low molecular weight secondary metabolites detrimental for the growth of plant-pathogenic microbes. *Trichoderma* spp. produce a variety of antibiotics, such as gliotoxin, viridin, gliovirin, koninginins, pyrones and viridol, and peptaibols against many fungal phytopathogens (Howell 2003; Harman et al. 2004). Howell (1999) described gliotoxin and gliovirin as the two most important *secondary* metabolites produced by *Trichoderma* belonging to P and Q group strains, respectively. The 'P' group strains are active against *Pythium ultimum*, but not against *R. solani* and the 'Q' group is active against *R. solani*, but less effective against *P. ultimum.* Viridin is produced by *T. koningii*, *T. viride*, and *T. virens*, which prevents spore germination of *Botrytis allii*, *Colletotrichum lini*, *Fusarium caeruleum*, *Penicillium expansum*, *Aspergillus niger* and *Stachybotrys atra* (Singh et al. 2005).

The biocontrol potential of *Trichoderma* spp. is also attributable to its active colonizing properties. Generally considered as an aggressive competitor against soil-borne fungal pathogens, it grows very fast towards the pathogen and rapidly colonize them (Cuervo-Parra et al. 2011). *Trichoderma* spp. can also exploit the substrates more aggressively than many other plant-pathogenic microbes leading to the death of pathogens growing in their vicinity. It outcompetes pathogenic flora for the nutrients in plant root exudates in the rhizosphere that are rich in amino acid, sugar, vitamins, organic acid, iron etc., making them starved for nutrients. Hence competition is an effective mechanism of biocontrol in *Trichoderma* (Alabouvette et al. 2009).

Besides the direct mechanism in biological control of pathogens, *Trichoderma* enhances defence mechanism in plants and thus protects plants against numerous groups of pathogens. This has been attributed to a *Trichoderma*–plant cross talk that involves an array of bioactive metabolites produced by the antagonist (Woo et al. 2006; Alfano et al. 2007). *Trichoderma* is known to activate salicyclic acid (SA)- and jasmonic acid/ethylene (JA/ET)-mediated induced resistance in plants. This resistance induction may also be attributed by reduced effector triggered susceptibility (ETS), enhanced effector triggered immunity (ETI) and Microbe-associated molecular pattern (MAMP)-triggered immunity (MTI) in plants (Sharma et al. 2017). Seedling treatment of cotton with efficient strain of *T. virens* yielded higher accumulation of defence-related compounds such as terpenoids and peroxidase in the root (Howell 2002). Chowdappa et al. (2013) reported induced systemic resistance in tomato by *T. harzianum* against early blight and late blight diseases. Combined application of *Trichoderma harzianum* and *Pseudomonas* spp. could induce systemic resistance in Arabidopsis and cucumber (Alizadeh et al. 2013).

11.3.2 Trichoderma for Insect Pest Management

A diverse assemblage of microorganisms including entomopathogenic fungi, bacteria and viruses are currently being explored against many insect pests of agricultural crops. Jassim et al. (1990) reported the larvicidal activity of *T. harzianum* against elm bark beetle, *Scolytus* spp. In another study, *T. harzianum* spore suspension resulted in 80% mortality of the African cotton leaf worm *Spodoptera littoralis* larvae when applied at 1×10^8 spore/ml. Recent toxicological, biochemical, histological and ultrastructural studies have revealed the fungus as a potential entomopathogen against several insect pests. Anwar et al. (2016) recorded *T. longibrachiatum* as an entomopathogenic fungi against *Bemisia tabaci* for the first time in Pakistan. In a study made by Khaleil et al. (2016) the biochemical analysis of the *T. hamatum* spore-treated adult cotton aphid revealed different quantitative changes in the total soluble protein, transaminase enzymes and the carbohydrates hydrolyzing enzyme as compared to the untreated ones. Binod et al. (2007) in India evaluated the bioefficacy of *T. harzianum* culture filtrate containing chitinase against Cotton bollworm, *Helicoverpa armigera* where results showed negative effect of *T. harzianum* on growth and metamorphosis of the insect larvae. Ibrahim reported the entomopathogenic activity of *T. longibrachiatum* against cowpea aphid, *Aphis craccivora*. Ghosh and Pal (2016) used the conidial spore suspension of *T. longibrachiatum* against brinjal fruit and shoot borer, *Leucinodes orbonalis*, and recorded maximum inhibition (80%) when applied at a concentration of 1×10^8 spores/ml. The entomopathogenic *Trichoderma* spp. produce antibiotics, proteinases and chitinase enzymes. Besides, *Trichoderma* colonization and development inside the insect haemocoel leads to depletion of nutrients and thereby starvation and subsequent death of insect host (Khaleil et al. 2016).

However, *Trichoderma* has not yet been fully explored as an entomopathogen. Further research on molecular basis of host–pathogen interaction could provide more insight on this issue. Selection of efficient strains that can manage diseases, insect and nematode might prove to be economically viable solutions for management of pest complex.

11.3.3 Trichoderma for Plant Growth Promotion

Use of *Trichoderma* spp. in agriculture has now reached beyond pest management, and the genus is also used as a plant growth promoter. Several workers have reported plant growth-promoting effect of *Trichoderma* in many crops. Studies showed that *Trichoderma* spp. could alter physiological processes in plants viz., photosynthesis rate, stomatal conductance, transpiration, internal CO_2 concentration, water use efficiency and nutrient uptake (Saba et al. 2012). *Trichoderma* has been known to solubilize several plant nutrients and produce plant growth hormones viz., indole acetic acid (IAA), gibberellic acid (GA) and ethylene, which result in enhanced root and shoot growth in plants (Harman and Shoresh 2007). Chowdappa et al. (2013) reported phytohormones produced by *Trichoderma* spp. as the key factor for increased plant height in rice. Seed and/or soil treatment with *Trichoderma* spp. enhances their germination rate, enhances the seedling vigour, accelerates the vegetative and reproductive growth of plants like number of branches, spikes, flowers and fruits per plant. However, the molecular mechanisms behind these beneficial effects are not fully understood (Vinale 2008). Sharma et al. (2012) also demonstrated the potential of *T. harzianum* in plant growth promotion in wheat. Enhanced germination percent had been found in *Trichoderma*-treated okra, maize, beans, mustard, chilli, soyabean, chickpea, tomato etc. (Bjorkman et al. 1998; Lalitha et al. 2012; Kumar et al. 2014; Babychan and Simon 2017). Increased plant height and increase in number of leaves and yield were recorded in *Trichoderma*-inoculated *Bhoot jolokia* (*Capsicum chinense*) plants by Prakash and Saikia (2015). Enhancing plant growth by *Trichoderma* spp. also helps cure the crop in disease. Many seed-invading pathogens such as *Pythium* could not attack the host plants due to faster germination of seed and enhanced seedling vigour (Matsouri et al. 2010).

11.4 Work on *Trichoderma* for Plant Disease Management in NE India

The pioneering work on *Trichoderma* in NE India was made by Roy et al. during the 1970s. Their work showed that the genus can successfully control pathogens like *Pythium* spp., *Rhizoctonia solani*, *Fusarium oxysporum*, and *Sclerotium rolfsii* (Roy 1977). Subsequently the dedicated works of several research groups reported *T. harzianum*, *T. viride*, *T. hamatum* and *Aspergillus* to be successful against many soil-borne pathogens in the region. Since then a number of economically important diseases have been reported to be managed through native *Trichoderma* spp. in N.E India. Sheath blight of rice caused by *Rhizoctonia solani* and root rot caused by *S. rolfsii* and stem and leaf rot caused by *Phytophthora palmivora* in betelvine were also successfully controlled by *T. viride* and *T. harzianum* strains. Research on improving the bioefficacy of the antagonists has also been done by many workers. The efficacy of *Trichoderma* spp. was found to be increased when used in combination with some organic amendments like pressmud in soil (Das 1992). Bora and Bhagabati (1999) reported increased efficacy of *T. harzianum* as seed protectant of rice when used with methylcellulose as a carrier material. Mech

(2004) induced UV-irradiated mutant of *T. harzianum* and recorded its enhanced antagonistic effect on *S. sclerotiourum.*

Management of white mould of French bean caused by *S. sclerotiorum* was successfully attempted by *T. harzianum* and *Gliocladium virens* and both bioagents were also reported to be compatible and showed significant reduction in per cent disease incidence increased plant height, dry weight, root, shoot and crop yield (Bora and Deka 2007). Kamala and Devi (2012) recorded better bioefficacy of native *T. harzianum* strains against damping off and wilting of beans caused by *R. solani* and *Fusarium* spp. respectively as compared to the exotic *T. harzianum* (ITCC No. 6276) strain. Kithan and Daiho (2014) conducted a field study for screening the efficacy of various management practices against *Curvularia lunata* var. *Aeria* causing leaf blight disease of *Etlingera linguiformis*. They found that rhizome treatment and foliar spray of *T. harzianum* could significantly inhibit the mycelial growth of the pathogen. Naglot et al. (2015) isolated native strains of rhizospheric *Trichoderma* spp. from tea gardens of the north eastern states of India. These were found to be antagonistic against two important tea fungal pathogens, namely *Pestalotia theae* and *Fusarium solani*, under in vitro condition. Bora (2017) studied *Trichoderma harzianum* in the management of *Sclerotinia* rot of carrot caused by *S. sclerotiorum* in organically amended soil. The bioformulation of *T. harzianum* added with chitosan (1%), zinc (0.25%) and boron (0.5%) compositions showed increase in the antagonistic ability of *T. harzianum* against *S. sclerotiorum.*

Management of phytobacterial diseases by *Trichoderma* spp. has also been successfully attempted by several workers. *T. harzianum* was found to be successful against wilt causing bacterial pathogen *Ralstonia solanacearum* in ginger (race 4), tomato, brinjal and chilli (Bora et al. 2013; Deuri 2013). Khan et al. (2018) reported successful management of *R. solanacearum* and *F. oxysporum* f. sp. *lactucae* in hydroponically grown lettuce plant through application of *T. viride*.

11.5 Development and Commercialization of *Trichoderma*-Based Biopesticide

Availability of quality biopesticides with aggressive bioagents and their survival during storage are two key factors governing success of disease management through biological control in organic agriculture as well as IDM programmes in conventional agriculture. At Assam Agricultural University, the pioneer in biopesticide research in NE India, a few *Trichoderma*-based bioformulations viz., *Biofor Pf2*, *Biozin PTB*, *Biotime*, *Biozium*, *Bioveer*, *Biogreen*-5 and Biogreen-L, have been developed using native and aggressive strains of *Trichoderma* spp. in combination with other plant growth-promoting microbes. Crop plants are generally subjected to the attack not only by diseases, but also insect pests and nematodes at the same time under field condition. Hence, a single bioformulation carrying different biocontrol agents against diseases as well as insect pests could be a sustainable and cost-effective solution from the farmer's perspective. Keeping this in view, a study was initiated in Assam Agricultural University to assay the compatibility status of *Trichoderma* spp. with other bioagents including entomopathogens and it was found to be compatible with *P fluorescens*, *Bacillus* spp. and *M. anisopleae* (Bora 2007, Deuri 2013, Khan et al. 2018). It was also evident that combination of different bioagents can provide better disease reduction as compared to a single agent.

For successful development of bioformulation selection of substrate is very important, as a good substrate provides nutrients, maintains viable population count at a desired level and promotes synthesis of desired enzymes which helps in biocontrol mechanisms (Singh et al. 2009). Different organic substrates were compared for the preparation of bioformulations with the addition of adhesive carboxymethyl cellulose and osmoticant mannitol. Among all the substrates, the mixture of vermicompost and decomposed mustard oil cake was found most effective for aggressive multiplication of *Trichoderma* and few other microbial bioagents (Bora and Bora 2008a, b).

Trichoderma have also been found successful against some pest and diseases of economically important crops in Assam and NE India, viz., tea (Deb et al. 2017), banana (Dutta et al. 2016), turmeric and ginger (Bora 2008; Deuri 2013; Lalfakawma et al. 2014). Sarkar (2015) used aggressive strains of four different bioagents *Trichoderma viride*, *T. parareesei*, *M. anisopliae* and *B. thuringiensis* against *Ralstonia solanacearum*, which causes bacterial wilt of Bhoot jolokia (*Capsicum chinens*), and recorded significant reduction of wilt incidence (83.87%) with the highest yield (9.84 t/ha). Record of yield enhancement as well as suppression of bacterial wilt incidence in Bhut Jolokia (*Capsicum assamicum*) using consortia of *Trichoderma* along with some PGP microbes was made by Bora et al. (2016b), bacterial wilt in tomato by Nath et al. (2016) and Bora et al. (2016a). Bora et al. (2016c) also recorded *Rhizoctonia* root management in chilli (*Capsicum frutescens*) using *Trichoderma*-based microbial consortia. Khan et al. (2016) used microbial consortia containing *T. viride* against wilt causing pathogen *Fusarium oxysporum* f. sp. *lactucae* in hydroponically grown lettuce (*Lactuca sativa*).

Dutta (2015), while exploring ecofriendly management of Alternaria leaf blight (*Alternaria alternate*) of Mandukaparni (*Centella asiatica* L.), observed that *T. harzianum* could show maximum radial growth inhibition (96.48%) of *A. alternata* as compared to other microbial bioagents. Under field condition, minimum (18.05%) disease incidence with increased (55.36%) disease control was recorded when *T. harzianum* was applied. Besides, *Trichoderma*-based formulations have shown satisfactory results in controlling insect pests under field condition. Deb (2017) recorded significant reduction in insect pest infestation in tea viz., tea mosquito bug, red spider mites, looper caterpillar and aphid along with and grey blight diseases with increased green leaf yield (10.40 q/ha) due to application of bioformulation containing mixture of *T. viride*, *B. thuringiensis*, *P. fluorescens*, *B. Bassiana* and *M. anisopliae*. Gogoi et al. (2017) used bioformulation Biofor-pf-2 containing *T. viride* for the management of root rot disease of patchouli caused by *Fusarium solani*. He recorded that soil application of Biofor-pf-2 @ 100 g/pot showed least disease incidence (44.40%) as compared to inoculated control

(100%). Boruah (2017) observed that seed treatment of vegetable seeds with next-gen nano-bio formulation of *Trichoderma asperellum* and chitosan NP suppressed the growth of *F. oxysporum*, *Sclerotium rolfsii* and *Rhizoctonia solani* as compared to the recommended chemicals.

Recent government regulation on minimizing pesticide use in tea as well as global demand on pesticide-free organic tea has facilitated conversion of many gardens to organic, where *Trichoderma*-based bioformulations have become an integral part of pest and disease management programme.

11.6 Prospects

The advent of envisioned 'Act East Policy' and pronouncement of 'Second Green Revolution' in the country particularly in the North East have reasonably anticipated the life sciences research to provide technologies, products and knowledge based on current understanding that could be directly applied to boost agricultural production and productivity, quality and employment. As per the 'Act East Policy' the North-Eastern region, which shares 98% of its border with South Asian countries, will be turned into a trade and business hub of South East Asia. The border transit would make the North East a gateway to entry into South East Asian market that will be a potential boon for trade of agricultural commodities and processed food products more specifically of organic origin. This inclination creates a scope of the region to grow crops organically and emerge as a key supplier of organic products into the South East Asian organic market.

On the other hand, the North Eastern states are readying itself for the next phase of the Green Revolution by adopting innovative self-sustainable approaches like 'organic farming' by observing the after effects of chemical rich agriculture practiced since 1960s. In 1960s and 1970s the North Eastern states faltered to join the mass agricultural transformation movement of Green Revolution. But, thanks to the talk of 'Second Green Revolution' through organic agriculture in the Eastern states that had started around 2010 is a timely taken effort to tackle the sluggish long-term agricultural growth in this part of India. So it is the most appropriate time to focus on diversification of agricultural production system which can improve farm incomes and accelerated growth, with or without stress on natural resources. We know the fact that, no agriculture system can continue to feed a growing population if it degrades the natural resources, so agriculture needs to be sustainable and the answer to this is seriously planned 'organic farming'.

In Assam Agricultural University, which is the premier institution in the entire N.E. India, work has already been initiated to pave way for 'organic farming' to establish agricultural sustainability in the region. The institute was successful in seeding the organic farming concept among the farming community of Assam and other N.E. states by engaging and activating key stakeholders. Though it was a beginning only, a lot of work needs to be done to make the envisioned 'Act East Policy' and 'Second Green Revolution' successful in this region. Some of the research areas where stress has to be given particularly with aggressive biocontrol and plant growth-promoting agents like *Trichoderma* are as follows:

- Bio-prospecting chemical pollutant-degrading *Trichoderma* to explore and enhance biodegradation and organic management of major pest and diseases. Some of the most common problems associated with crop field, specifically in rice crop, are arsenic contamination and elevated level of methane emission. In Assam, a vast patch of agricultural land of upper Brahmaputra valley zone is already contaminated with arsenic, which later contaminates food and potable water, and appears to be a major concern to human health (Mandal and Suzuki 2002). The conventional arsenic removal method may produce many harmful by-products, and it involves high cost too (Bahar et al. 2012). As an alternative, transformation of As(III) to less-toxic As(V) through a microbe-mediated process has gained significant attention as it edges out conventional treatment methods. *Trichoderma* species may be one of such microbial groups that can fit into the system of microbial transformation of Arsenic contamination in rice ecosystem. Similarly, another serious interrelated fact with rice production is that wetland rice fields emit a large volume of atmospheric methane accounting for 60 Tg/year (6×10^{10} kg/year), with a range of 20–100 Tg/year approximately 20% of the global anthropogenic methane emission sources annually (Anonymous 1966). So it is a major cause of concern and discussion because inter-annual variability in the rate of methane concentration possibly leads to large-scale temperature and precipitation anomalies (Dlugokencky et al. 2001). In recent years there has been a growing interest in the use of microorganisms for environmental restoration. Considering the diverse activities and practical utilities of *Trichoderma* spp., these can be successfully employed in reducing arsenic and methane biodegradation in contaminated rice fields. Though it is too simple to say, practical implementation needs careful planning to materialize the same successfully in rice fields. The major difficulty that may arise is the identification, multiplication and development of a suitable and cost-effective delivery system without adding any extra burden on farmer's cultivation practices. The possible solution of these problems is integration of the methanotrophs with the available *Trichoderma*-based biocontrol technology with simple changes to create a win–win situation without changing the farming practices of cultivation.
- Development of a multi-utility *Trichoderma*-based bioformulation by integrating micronutrient supplements, targeting the defence-related genes expression in crop plants for enhanced crop growth and protection. Nutrients in micro and macro form also enhance soil microfauna that play an important role in crop pest management (Agrios 2005). Micronutrients can enhance disease suppression indirectly by stimulating indigenous populations of antagonistic microorganisms (Huber 1989) and directly by activating systemic acquired resistance (SAR) in host plant systems (Dordas 2007). Simultaneous activation of Induced Systemic Resistance (ISR) through soil *Trichoderma* and micronutrients has been found to provide additive disease control. So, biocontrol with *Trichoderma*-based bioformulation and amending with essential micronutrients can lead to a potential

strategy to target the defence-related gene expression in crop plants for enhanced crop growth and protection against major plant pests.

- Exploration of *Trichoderma* and biofertilizer agents as consortia for bio-intensive crop management practices: The combined applications of microbial antagonists like *Trichoderma* with microbial biofertilizers can significantly reduce crop diseases in field along with enhanced growth and yield (Srivastava et al. 2010). Fungal biofertilizer agent arbuscular mycorrhizal applied along with *T. harzianum*-based formulation could lead to a significant reduction in disease along with enhanced growth and yield of crops like tomato in field (Srivastava et al. 2010). Exploring potential compatible *Trichoderma* and biofertilizer agents can lead to a complete package for biointensive management of many soil-borne plant pathogens along with substantial yield enhancement.
- Risk assessment and evaluation of non-target effect of microbial biopesticides: Scientists around the world are showing concern about the probable risks of biopesticides; also correlating the facts that biopesticides show toxic property against target pests may also transfuse these properties into plant physiology and biochemistry. Therefore, toxicological studies of microbial biopesticides are now essential through accumulation of biological characteristics and chemistry data of the microbes like *Trichoderma* used for preparation of biopesticides. Acute Toxicity Studies (eye irritation, acute oral, dermal, inhalation and skin sensitization) and Repeated Dose Toxicity Studies need to be conducted under the guidance of Medical Toxicology Division of Central Insecticide Board following CIB and OECD test guidelines. Similarly, risk assessment of biopesticide handlers and users requires precise evaluation based on toxicological data.

11.7 Conclusion

Trichoderma has been successfully used to combat plant diseases in organic agriculture as well as conventional agriculture in recent time and is being explored as a plant-beneficial microbe due to its multifaceted effects. Voluminous research worldwide have resulted in better understanding of the mechanism of biocontrol by *Trichoderma*, plant growth promotion and activation of plant defence mechanism at molecular level. The genus has been used in industrial sector too owing to its potential as a source of bio-enzymes. Several studies have shown the genus as an efficient biocontrol agent against some agriculturally important insect pests. However, very limited work has been done to understand the molecular basis of *Trichoderma*–insect pest interaction. Also, the non-target effect of entomopathogenic *Trichoderma* on beneficial insects is poorly understood. Exploring the multifaceted effects of *Trichoderma* spp. for bioremediation of organic pollutants and combining the genus with existing biofertlizers are some other areas that need special attention. A multi-utility *Trichoderma* formulation fortified with micronutrients or microbial fertilizers can not only be used for disease management but also can enhance soil fertility without adding additional financial burden to the farming community.

References

Agrios G (2005) Plant pathology, 5th edn. Elsevier, London

Alabouvette C, Olivain C, Migheli Q, Steinberg C (2009) Microbiological control of soil-borne phytopathogenic fungi with special emphasis on wilt inducing *Fusarium oxysporum*. New Phytol 184:529–544

Alfano G, Ivey MLL, Cakir C, Bos JIB, Miller SA, Madden LV, Kamoun S, Hoitink HAJ (2007) Systemic modulation of gene expression in tomato by *Trichoderma hamatum*. Phytopathology 97:429–437

Alizadeh H, Behboudi K, Ahmadzadeh M, Javan-Nikkhah M, Zamioudis C, Pieterse CMJ, Bakker PAHM (2013) Induced systemic resistance in cucumber and *Arabidopsis thaliana* by the combination of *Trichoderma harzianum* Tr6 and *Pseudomonas sp.* Ps14. Biol Control 65:14–23

Anonymous (1966). https://www.ipcc-nggip.iges.or.jp/public/gl/guidelin/ch4ref5.pdf

Anonymous (2017). http://agriexchange.apeda.gov.in/news/newssearch

Anwar W, Haider M, Shahid A, Mushtaq Rehman H, Hameed ZMU, Javed S (2016) First record of *Trichoderma longibrachiatum* as entomopathogenic fungi against *Bemisia tabaci* in Pakistan. Pak J Phytopathol 28(2):287–294

Babychan M, Simon S (2017) Efficacy of *Trichoderma* spp. against *Fusarium oxysporum* f. sp. *lycopersici* (FOL) infecting pre- and post-seedling of tomato. J Pharmacogn Phytochem 6:616–619

Bahar MM, Mallavarapu M, Naidu R (2012) Arsenic bioremediation potential of a new Arsenite-oxidizing bacterium *Stenotrophomonas sp* MM-7 isolated from soil. Biodegradation 23 (6):803–812

Binod P, Sukumaran RK, Shirke SV, Rajput JC, Pandey A (2007) Evaluation of fungal culture filtrate containing chitinase as a biocontrol agent against *Helicoverpa armigera*. J Appl Microbiol 103:1845–1852

Bjorkman T, Blanchard LM, Harman GE (1998) Growth enhancement of shrunken-2 sweet corn when colonized with *Trichoderma harzianum* 1295-22 effect of environmental stress. J Am Soc Hort Sci 123:35–40

Bora M (2007) An integration of bioagents and fungicides for the management of white mold of French bean. MSc (Agri) Thesis, Assam Agric University, Jorhat, Assam

Bora LC (2008) Use of *Pseudomonas fluorescens* as biopesticide for management of wilt disease caused by *Ralstonia solanacearum*. Trop Agric 85(2):57–61

Bora M (2017) Aggrieved *Trichoderma harzianum* in management of *Sclerotinia* rot of carrot in organically amended soil. PhD (Agri) Thesis, Assam Agric University, Jorhat, Assam

Bora LC, Bhagabati, KN (1999) Quantitative aspects of assessing virulence of *Xanthomonas campestris* pv. citri (Hasse) Dye, the incitant of citrus canker in Assam. Intl Symp Citriculture, National Res Centre for Citrus, Nagpur, Maharashtra

Bora LC, Bora P (2008a) Biological control strategies for management of bacterial wilt of brinjal (*Solanum melongena*). J Mycol Plant Pathol 38(3):512–517

Bora LC, Bora P (2008b) Mass culture of saprophytic antagonists and their application for management of bacterial wilt and rhizome rot of ginger. J Mycol Plant Pathol 38(1):245–251

Bora LC, Deka SN (2007) Wilt disease suppression and yield enhancement in tomato (*Lycopersicon esculentum*) by application of *Pseudomonas fluorescens* based bio pesticide (Biofor-Pf) in Assam. Indian J Agric Sci 77(8):490–494

Bora LC, Sarkar, R, Kataky, L (2013) Genomic characterization of microbial antagonists, their interactive effects and utility in management of bacterial wilt of *Bhut Jolokia* (*Capsicum chinense* Jacq). In Acta Phytopathologica Sinica, 10th International Congress of Plant Pathology ICPP 2013. 20–30 Aug 2013, Beijing, China

Bora LC, Kataki L, Talukdar K, Nath BC, Sarkar R (2015) Molecular characterizations of microbial antagonists and development of bioformulations for management of bacterial wilt of Naga chilli (*Capsicum chinens* Jacq.) in Assam. J Exp Biol Agric Sci 3(2):109–122

Bora LC, Kataki, L, Talukdar, K (2016a) Yield enhancement and *Rhizoctonia* root disease management of chilli (*Capsicum frutescens*) using microbial consortia in Assam. In National seminar on Integrated Development of Horticulture in Sub Tropical & Hill Region. pp 117–127

Bora LC, Kataki L, Talukdar K, Khan P, Dutta J (2016b) Yield enhancement and bacterial wilt suppression in Bhut Jolokia (*Capsicum assamicum*) using consortia of microbial antagonists. Biopest Int 12(2):165

Boruah S (2017) Development of next-gen nano-bioformulation of *Trichoderma* for seed treatment of vegetable crops. PhD (Agri) thesis, Assam Agric University, Jorhat, Assam

Chet I, Harman GE, Baker R (1981) *Trichoderma hamatum*: its hyphal interactions with *Rhizoctonia solani* and *Pythium spp*. Microb Ecol 7:29–38

Chowdappa P, Mohan Kumar SP, Lakshmi MJ, Upretti KK (2013) Growth stimulation and induction of systemic resistance in tomato against early and late blight by *Bacillus subtilis* OTPB1 or *Trichoderma harzianum* OTPB3. Biol Control 65(1):109–117

Cuervo-Parra JA, Ramirez SM, Sanchez LV, Ramirez LM (2011) Antagonistic effect of *Trichoderma harzianum* VLS291on phytopathogenic fungi isolated from cocoa (*Theobroma cacao*). Afr J Biotechnol 10(52):10657–10663

Das BC (1992) Antagonism of *Aspergillus terreus* on *Rhizoctonia solani* and its effect on the incidence of sheath blight of rice. PhD (Agri) Thesis, Assam Agric University, Jorhat, Assam

Deb R (2017) PGPM formulation mediated biochemical changes and defense response of tea against major diseases and pests. MSc (Agri) Thesis, Assam Agric University, Jorhat, Assam

Deb R, Bora LC, Das P (2017) Microbial bioformulations for suppression of major insect pests and diseases and enhanced biochemical properties of tea crop. Intl J Curr Microbiol Appl Sci 6 (6):1872–1879

Deuri D (2013) Bio-intensive approach for management of bacterial wilt of ginger (*Zinziber officinale*). MSc (Agri) Thesis, Assam Agric University, Jorhat, Assam

Dlugokencky EJ, Walter BP, Masarie KA, Lang PM, Kasischke ES (2001) Measurements of an anomalous global methane increase during 1998. Geophys Res Lett 28:499–502. https://doi.org/10.1029/2000GL012119

Dordas C (2007) Nitrogen fertilization influences yield, physiology and N use efficiency in flax (*Linum usitatissimum* L.). In 15th N Workshop, 28–30 May 2007, Lleida, Spain

Dubey SC, Suresh M (2006) Randomly amplified polymorphic DNA markers for *Trichoderma* species and antagonism against *Fusarium oxysporum* fsp Ciceris causing chickpea wilt. J Phytopathol 154:663–669

Dutta R (2015) Eco-friendly management of *Alternaria* leaf blight of Mandukaparni (*Centella asiatica* L). MSc(Agri) thesis, Assam Agric University, Jorhat, Assam

Dutta J, Bora LC, Singh S (2016) Suppression of *Colletotrichum musae* associated with micro-propagated banana plantlets using microbial bio agents. NE zonal meeting, 2016 Indian Phytopathological Society. p 90

Friedl MA, Druzhinina IS (2012) Taxon-specific metagenomics of *Trichoderma* reveals a narrow community of opportunistic species that regulate each other's development. Microbiology 158 (1):60–83

Ghosh SK, Pal S (2016) Entomopathogenic potential of *Trichoderma longibrachiatum* and its comparative evaluation with malathion against the insect pest *Leucinodes orbonalis*. Environ Monit Asses 188(1):37

Gogoi M, Sarmah DK, Bora LC, Barua IC (2017) Management of root rot disease of patchouli caused *Fusarium solani* (Mart) Sacc through fungicide, bioagent and oilcake in field. Int J Curr Microbiol App Sci 6(9):3385–3397

Harish R, Kumar SN, Chauhan JB (2013) Biodegradation of organophosphate pesticide by soil fungi. Adv BioTech 12(9):2319–6750

Harman GE (2006) Overview of mechanism and uses of Trichoderma spp. Phytopathology 96:190–194

Harman GE, Shoresh M (2007) The mechanism and application of symbiotic opportunistic plant symbionts. In: Vurro M, Gressel J (eds) Novel biotechnologies for biocontrol agent enhancement and management. NATO security through science series. Springer, Dordrecht, pp 131–155

Harman GE, Howell CR, Viterbo A, Chet I, Lorito M (2004) *Trichoderma* species opportunistic, avirulent plant symbionts. Nat Rev Microbiol 2(1):43–56

Hermosa R, Viterbo A, Chet I, Monte E (2012) Plant beneficial effects of *Trichoderma* and of its genes. Microbiology 158:17–25

Howell CR (1999) Selective isolation from soil and separation *in vitro* of P and Q strains of *Trichoderma virens* with differential media. Mycologia 91:930–934

Howell CR (2002) Cotton seedling preemergence damping-off incited by Rhizopus oryzae and Pythium spp. and its biological control with Trichoderma spp. Phytopathology 92:177–180

Howell CR (2003) Mechanisms employed by *Trichoderma* species in the biological control of plant diseases: the history and evolution of current concepts. Plant Dis 87:4–10

Huber DM (1989) Introduction. In: Engelhard AW (ed) Soil-borne plant pathogens: management of diseases with macro and microelements. APS Press, St Paul, MN, pp 169–206

Jassim HK, Foster H, Foster HA, Fairhurst CP (1990) Biological control of Dutch elm disease: larvicidal activity of *Trichoderma harzianum T. polysporum* and *Scytalidium lignicola* in *Scolytus scolytus* and *S. multistriatus* reared in artificial culture. Ann Appl Biol 117:187–196

Kamala T, Devi SI (2012) Biocontrol properties of indigenous *Trichoderma* isolates from North-East India against *Fusarium oxysporum* and *Rhizoctonia solani.* Afr J Biotechnol 11 (34):8491–8499

Katayama A, Matsumara F (1993) Degradation of organochlorine pesticides, particularly endosulfan, by *Trichoderma harzianum*. Environ Toxicol Chem 12(6):1059–1065

Khaleil M, El-Mougith A, Hashem H, Lokma N (2016) Biocontrol potential of entomopathogenic fungus, *Trichoderma hamatum* against the cotton aphid, Aphis Gossypii. J Environ Sci Toxicol Food Technol 10(5):11–20

Khan P, Borah PK, Bora LC (2016) *Fusarium oxysporum* f.sp. *lactucae* (FoL) a potential threat to hydroponically grown lettuce (*Lactuca sativa*) and its management using microbial consortia. NE zonal meeting, 2016, Indian Phytopathological Society. p 35

Khan P, Bora LC, Bora P, Talukdar K, Kataky L (2018) Efficacy of microbial consortia against bacterial wilt caused by Ralstonia solanacearum in hydroponically grown lettuce plant. Int J Curr Microbiol App Sci 7(6):3046–3055

Kithan C, Daiho L (2014) *In vitro* evaluation of botanicals, bio-agents and fungicides against leaf blight of *Etlingera linguiformis* caused by *Curvularia lunata* var Aeria. J Plant Pathol Microbiol 5:3

Kumar V, Shahid M, Singh A, Srivastava M, Mishra A, Srivastava YK, Pandey S, Sharma A (2014) Effect of biopriming with biocontrol agents *Trichoderma harzianum* (Th Azad) and *Trichoderma viride* on chickpea genotype (Radhey). J Plant Pathol Microbiol 5:1

Lalfakawma C, Nath BC, Bora LC, Srivastava S, Singh JP (2014) Integrated disease management of *Zingiber officinale* Rosc rhizome rot. Bioscan 9(1):265–269

Lalitha P, Srujana, Arunalakshmi K (2012) Effect of *Trichoderma viride* on germination of mustard and survival of mustard seedlings. Int J Life Sci Biotechnol Pharm Res 1(1):137–140

Lorito M, Woo SL, Harman GE, Monte E (2010) Translational research on *Trichoderma*: from 'Omics to the field. Annu Rev Phytopathol 48:395–341

Mandal BK, Suzuki KT (2002) Arsenic round the world: a review. Talanta 58(1):201–235

Matsouri F, Björkman T, Harman GE (2010) Seed treatment with *Trichoderma harzianum* alleviates biotic, abiotic and physiological stresses in germinating seeds and seedlings. Phytopathology 100:1213–1221

Mech S (2004) UV-radiated mutant of *Trichoderma harzianum* Rifai and its carrier-based formulation for management of white mold of French bean. MSc (Agri) Thesis, Assam Agric University, Jorhat, Assam

Mishra RC, Singh R, Singh HB, Dixit A (2000) *In situ* efficacy of *Trichoderma harzianum* as mycoparasite on *Sclerotium rolfsi* and *Rhizoctonia solani*. TropAgric 77(3):205–206

Mukherjee PK, Raghu K (1997) *Trichoderma* sp. as a microbial suppressive agent of *Sclerotium rolfsii* on vegetables. World J Microbiol Biotechnol 13(5):497–499

Mukherjee PK, Horwitz BA, Herrera EA, Schmoll M, Kenerley CM (2013) *Trichoderma* research in the genome era. Annu Rev Phytopathol 51:105–129

Naglot A, Goswami S, Rahman I, Shrimali DD, Yadav K, Gupta VK, Rabha AJ, Gogoi HK, Veer V (2015) Antagonistic potential of native *Trichoderma viride* strain against potent tea fungal pathogens in north East India. Plant Pathol J 31(3):278–289

Nath BC, Bora LC, Kataki L, Talukdar K, Sharma P, Dutta J, Khan P (2016) Plant growth promoting microbes, their compatibility analysis and utility in biointensive management of bacterial wilt of tomato. Int J Curr Microbiol App Sci 5(6):1007–1016

Persoon CH (1794) Disposita methodical fungorum. Römers Neues Mag Bot 1:81–128

Prakash V, Saikia AJ (2015) Production and multiplication of native compost fungal activator by using different substrates and its influence on growth and development of *Capsicum chinensis* Jacq "Bhut Jolokia". Biotechnol Res Int 2015:1. https://doi.org/10.1155/2015/481363

Roy AK (1977) Parasitic activity of *Trichoderma viride* on the sheath blight fungus of rice (*Cortium sasakii*). J Plant Dis Prot 84:675–683

Saba H, Vibhash D, Manisha M, Prashant KS, Farhanm H, Tauseef A (2012) *Trichoderma*-a promising plant growth stimulator and biocontrol agent. Mycosphere 3(4):524–531

Samuels GJ (1996) *Trichoderma*: a review of biology and systematics of the genus. Mycol Res 100:923–935

Sarkar R (2015) Genomic analysis of few bioactive microorganisims, their interactive effects and utility as consortia in management of bacterial wilt of Bhoot Jalakia (*Capsicum chinense* Jacq). MSc (Agri) thesis, Assam Agric University, Jorhat, Assam

Shanmugam V, Gupta S, Dohroo NP (2012) Selection of compatible biocontrol strain mixture based on co cultivation for control of rhizome rot of ginger. Crop Prot 43:119–127

Sharma P, Singh L, Adlakha D (2001) Antagonistic potential of *Trichoderma* and *Aspergillus* species on *Sclerotinia sclerotiorum* (Lib) de Barry causing rots in cabbage and cauliflower. Pestic Info 2:41–44

Sharma P, Patel AN, Saini MK, Deep S (2012) Field demonstration of *Trichoderma harzianum* as a plant growth promoter in wheat (*Triticum aestivum* L). J Agric Sci 4(8):65–73

Singh HB, Singh A, Nautiyal CS (2002) Commercialization of biocontrol agents: problems and prospects. In: Rao GP et al (eds) Frontiers of fungal diversity in Indian subcontinent. International Book Distribution Company, Lucknow, India, pp 847–861

Singh S, Dureja P, Tanwar RS (2005) Production and antifungal activity of secondary metabolites of *Trichoderma virens*. Pestic Res J 17:26–29

Singh V, Lal RJ, Awasthi SK, Verma MR (2009) Managing red rot of sugarcane by Trichoderma harzianum. Indian Sugar 59(4):25–30

Singh A, Sarma BK, Singh HB, Upadhyay RS (2014) *Trichoderma*: a silent worker of plant rhizosphere. In: Gupta VG et al (eds) Biotechnology and biology of *Trichoderma*. Elsevier, Amsterdam, pp 543–549

Srivastava R, Khalid A, Singh U, Sharma AK (2010) Evaluation of arbuscular mycorrhizal fungus, *Fluorescent Pseudomonas* and *Trichoderma harzianum* formulation against *Fusarium oxysporum* f.sp. *lycopersici* for the management of tomato wilt. Biol Control 53:24–31

Steyaert JM, Ridgway HJ, Elad Y, Stewart A (2003) Genetic basis of mycoparasitism: a mechanism of biological control by species of Trichoderma. N Z J Crop Hort Sci 31:281–291

Vinale F, Sivasithamparam K, Ghisalberti EL, Marra R, Woo SL, Lorito M (2008) Trichoderma-plant-pathogen interactions. Soil Biol Biochem 40:1–10

Weindling R (1934) Studies on a lethal principle effective in the parasite action of *Trichoderma lignorom*, *Rhizoctonia solani* and other soil fungi. Phytopathology 24:1153–1179

Woo SL, Scala F, Ruocco M (2006) The molecular biology of the interactions between *Trichoderma spp.*, phytopathogenic fungi, and plants. Phytopathology 96:181–185

Chapter 12
Deployment of *Trichoderma* for the Management of Tea Diseases

Pranab Dutta, R. P. Bhuyan, and Pratibha Sharma

Abstract Tea, *Camellia sinensis* (L.) O. Kuntze, is considered as one of the important plantation crops that has a great role in the GDP of the nation. Tea leaves contain more than 700 chemicals, among which flavonoids, amino acids, vitamins, caffeine and polysaccharides are beneficial to human health. They also improve the beneficial intestinal microflora, provide immunity against intestinal disorders and protect cell membranes from oxidative damage. The crop is infected by various diseases at different stages of the crop. Many of the diseases can be curbed through biological approaches, more particularly using *Trichoderma* spp. In this chapter we have discussed the works on potential use of *Trichoderma* spp. for the management of tea diseases, mode of action of *Trichoderma* against the pathogens causing tea diseases and the role of different biotic and abiotic factors on *Trichoderma* spp. leading to a successful result.

Keywords *Trichoderma* · Tea · Disease

Tea, *Camellia sinensis* (L.) O. Kuntze, is one of the important plantation crops. Tea determines the economy of many countries, and it contributes annually US $1500 million to the gross national products of India, which includes US $480 million through export earnings. Tea is cultivated right from the high peaks of the Himalayas in North India up to the Nilgiri hills of South India, and after China, India is the second largest tea producer in the world (Anonymous 2014). Tea is also the 'state drink' of Assam (Anonymous 2012). In October 2012, Assam officially declared tea

P. Dutta (✉)
School of Crop Protection, College of Post Graduate Studies in Agricultural Sciences, Central Agricultural University (Imphal), Umiam, Meghalaya, India

R. P. Bhuyan
Department of Tea Husbandry and Technology, Assam Agricultural University, Jorhat, Assam, India

P. Sharma
Division of Plant Pathology, ICAR-Indian Agricultural Research Institute, New Delhi, India

A. K. Sharma, P. Sharma (eds.), *Trichoderma*, Rhizosphere Biology,
https://doi.org/10.1007/978-981-15-3321-1_12

as the state drink (Anonymous 2012). The deputy chairman of Niti Aayog, Montek Singh Ahluwalia, planned to recognise tea (Anonymous 2013b) as the "National Drink" in 2013 (Anonymous 2013c). ASSOCHAM reported India as the world's largest tea consumer that uses approximately 30% of the global output, and it is also the second largest exporter after China (Anonymous 2013a).

Asia is the largest tea producing continent in the world (81.4%), followed by Africa (15.5%), and then by Latin America (2.0%) (FAO 1999). Tea cultivation is done best in deep and well-drained acidic soils, with a pH 4.5–5.0, well distributed rainfall of 1500–2000 mm/year, average temperatures of 18–20 °C and sufficient sunshine. These ecological conditions are prevalent in the tropics at higher altitudes of about 800–2000 m (1400–2100 near equator). According to Carr (1972) the minimum rainfall for commercial yields is 1150–1400 mm/year, depending on how it is distributed. Temperature should not fall below 13 °C and not exceed a mean maximum of 30 °C. In the subtropics, tea has a dormancy period during winter and also in dry periods. The optimum sunlight intensity for tea is 50–80% from full radiation (Eden 1976; Purseglove 1968). Suppression of weeds is essential. Tea is sometimes interplanted in a nurse crop or with green manure. Deep cultivation, which would damage the surface roots, should not be done. The plucking interval depends on the length of time needed for the production of new shoot (flushing) and, depending on the altitude, varies from 70 to 90 days.

Ayurveda has a long-standing tradition of using herbal teas as medicine. The medicinal property of different herbs and spices like holy basil, cardamom, pepper, liquorice and mint has long been utilized in Indian kitchen. Teas traditionally made with these plant leaves or spices have been in use for centuries against maladies. Tea leaves contain more than 700 chemicals, among which flavonoids, amino acids, vitamin C, vitamin E and vitamin K, caffeine and polysaccharides are beneficial to human health. They also improve the beneficial intestinal microflora, provide immunity against intestinal disorders and protect cell membranes from oxidative damage (Bhagat et al. 2010).

Many a times, tea is also mixed with a number of herbs. The sweet and milky taste of chai helps to give a new appearance like with some medicinal additives that make it, the stronger and more bitter flavours, and may with more pleasant flavours such as cardamom, clove and ginger add amusing flavour and bouquet to the tea along with many health benefits.

The first ever clear documentation of consumption of tea in India was in the Ramayana (750–500 BCE). But for the next 1000 years there is no documentation related to tea in India. After that, with stories of the Bodhidharma and Gan Lu (the two Buddhist monks), and their involvement with tea, the records re-emerged during the first century CE. In India, tea became a commercial crop after the arrival of the British East India Company, and now India has become of the largest tea producers in the world, in which 70% of its produce is being consumed within the country (Fig. 12.1). Assam and Darjeeling teas are the renowned varieties of teas that grow exclusively in the country. In India, the production, certification and exportation of tea is controlled by the Tea Board of India.

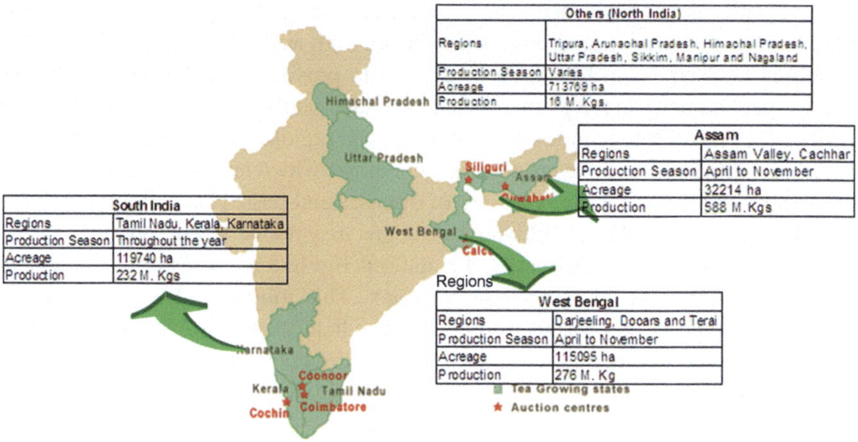

Fig. 12.1 Tea growing states of India with acreage and production. Source: ITA, Home page

Tea industry is one of the most important agroindustries in the country. It provides more remunerative employment to large number of workers and rural people, particularly women, especially for harvest.

Being perennial, tea provides permanent support for food and shelter to various insect pests. The monoculture nature of its cultivation also provides a congenial atmosphere to different insect pests attacking all the parts of the tea plant viz., root, stem, leaf, flower and seed, causing 10–15% loss in yield. As such tea eco-system supports a rich faunal diversity of several species of insect pests, nematodes and mites, and distribution of such insects largely depends on the climate and weather of the tea-growing areas. Pest incidence in tea is far more complex and more damaging in the Asian region, including India, and it has adverse consequences in tea production. Pest infestation also adversely affects the quality of processed tea. Use of chemical pesticides has been the most dominating practice for minimizing the pest population on tea crop in almost all tea-growing areas for nearly three decades. The overuse of pesticides in tea crop resulted in the rejection of India's export consignments due to toxic residues. In this context, the biological control methods are being used as an alternative to synthetic pesticides under the overall package of IPM, which recommends the judicious use of pesticides only when the pest reaches an economic threshold level (ETL).

Currently, the main challenge for tea sector is the production of pesticide residue-free tea, which has a high demand in the highly industrialised countries and from which a producing country can earn huge foreign money. If tea can be produced without the use of pesticides in some area or region which will not only retain the quality of tea but also fulfill the demand of the producer for high yield with low labour input. Reduction in labour started in the 1970s due to the use of herbicides over manual weeding. Risk of herbicide residues in the plucked tea has become a new problem for the growers. Though the elimination of shade trees increased the yield, it also increased the growth of weeds, the use of synthetic fertilizers and also the pests and disease incidence. The economic losses incurred on tea due to biotic stresses such as diseases caused by pathogen are high as compared to the losses due to animal pests (pests). The diseases and pests pressures on tea depend on the management strategy in the plantation area and also on the agro-ecological environment, which is different in specific tea-growing regions. Therefore, adoption of disease escaping method like growing in a condition which less favourable for biotic stresses may help in excessive pressure of diseases and pest. This will minimize the pesticides application during cultivation. But this may not be always possible. Other alternatives will be addressed below. Plantation crop like tea is grown as monoculture on the same site for more than 50 years. This might be desirable from an economic viewpoint, but not so from a phytomedical viewpoint. A strong equilibrium with the environment is essential for a healthy tea bush to survive for a long period, which otherwise is not favourable for a highly efficient plantation.

12.1 Diseases of Tea

Tea is grown as a monoculture for several decades. Crops under monoculture are cultivated by two methods: (a) annuals cultivated for short term in succession in the same area year after year or season after season (e.g. paddy) and (b) perennials grown for long term like 50–100 years or more. Monocultures provide ambient atmosphere for pests and diseases, and if allowed to spread it may lead to an epidemic. Diseases are the greatest barriers for the production of tea. More than 400 pathogens have been recorded to cause various diseases in tea (Thoudam and Dutta 2012). These pathogens cause infection in different plant parts viz., foliage, stem and root resulting in heavy losses. Diseases of tea are categorized into primary and secondary diseases. Both the types attack all the plant parts like leaf, stem and roots.

12.2 Leaf and Stem Diseases

12.2.1 Blister Blight

The most important tea leaf disease i.e., blister blight is caused by *Exobasidium vexans* Massee (Basidiomycotina, Ustomycetes, Exobasidiales). The disease is distributed only in Asia. It attacks the upper leaf surface and initially appears as a pale-yellow translucent spot, then a concave depression and lower surface of the infected leaf showing concave depression becomes a convex wart with white velvety or brown growth where white spore grows during moist weather. Later it becomes a circular brown spot. The affected young stems initially bend, then show distortion and may break off or die.

12.2.2 Black Rot or Black Blight or Thread Blight

Black rot or black blight or thread blight is caused by *Corticium invisum Corticium theae syn., Pellicularia koleroga* (Basidiomycotina, Hymenomycetes, Aphyllophorales, Corticiaceae). The disease is distributed worldwide. Symptoms develop as small circular brownish warts on the leaf surface. Infected leaves and twigs turn brown. Brownish circular small warts develop on infected leaf surface. Leaf apex and margins of the infected leaves turn black. In the rainy season the infected leaf becomes sodden. Under humid conditions in the tropics dead leaves hang on thin threads from the branches and the remaining two or more leaves are attached to the infected branches with gummy exudation.

12.2.3 Grey Blight

The leaf disease grey blight is caused by *Pestalotiopsis theae* (Deuteromycotina, Coelomycetes), which is distributed worldwide. Symptoms develop as round, irregular, grey and necrotic leaf lesions. In the centre of older spots black fruiting bodies (acervuli) develop. The pathogen mostly attacks the weak, damaged or unnourished tea bushes.

12.2.4 Red Rust

The disease is caused by *Cephaleurus parasiticus* Karst, *C. virescens* (Algae, Heterokontophyta [=Chrysophyta]) and is distributed worldwide. Symptom develops as red film covering leaves and branches. On the upper surface of infected leaves, orange yellow, circular patches appear and in green twigs reddish hair-like growth appears. On later stage, the spots turn brown and dry up. Infected stems show circular patches, and stems and twigs harden prematurely forming scaly patches and later crack.

12.2.5 Brown Blight, Grey Blight

Colletotrichum sp., *Pestalotiopsis* sp. etc. are the two main pathogens that cause brown blight or grey blight disease of tea. These are weak fungal pathogens that usually infect plants that are weakened due to poor management or adverse abiotic stresses. Symptoms first develop as small, oval, light yellow-green spots on young leaves. Later on, spots are found to be surrounded by a narrow, yellow halo. Further, the spots grow and turn brown or grey, with scattered concentric rings. Tiny black dots appear on the mature spots and eventually the infected dried tissue falls off. The organism can infect leaves of any age.

12.2.6 Pink Disease

The disease is caused by *Corticium salmonicolor* Berk & Broome (Basidiomycotina, Hymenomycetes, Aphyllophorales, Corticiaceae) and is distributed worldwide. Symptoms arise as pink crust on twig and branches. The portion above the branches dies off and cause dieback. The disease is prevalent in humid conditions in the tropics.

12.2.7 Stem Canker or Twig Die Back

Macrophomina theicola (Petch) (Deuteromycotina, Coelomycetes) causes the disease and is mostly prevalent in Sri Lanka. The first symptoms develop as browning and drooping of affected leaves and at a later stage, the infection spreads to the shoots, and the infected shoots become dry and die from the tip downward. Shallow cankers and slowly spreading lesions appear on the dried branches which are surrounded by a thick area of bark.

12.2.8 Thorny Stem Blight

The disease is caused by *Tunstallia aculeate* Petch (Ascomycotina, Sphaeriales) and is mostly prevalent in India. The pathogen is a weak wound parasite and causes symptoms such as dying of branches and fruiting bodies (perithecia) projecting from the bark of dead branches giving a thorny appearance. Small, black, thorn-like, pointed projections appear either singly or in group. These are observed at the collar region of the tea bushes. White strands of mycelium are seen on the surface underneath the bark of infected tea plants.

Aglaospora aculeata is another fungi that also causes thorny blight. This organism produces similar symptoms as mentioned for *T. aculeate*.

12.2.9 Brown Blight

Colletotrichum camelliae Mass. is the causal organism of brown blight of tea. Symptoms mainly appear on the old leaves as circular irregular spots with red and grey mycelium. On young leaves, yellowish green spots develop.

12.2.10 Grey Blight

The disease is caused by *Pestalozzia theae* Sawada (Deuteromycotina, Coelomycetes). The disease is distributed worldwide. The pathogen mostly attacks weak or damaged tea bushes. Symptoms develop as light or dark brown, circular- or oval-shaped patches on the foliage. Round, irregular, grey spots or patches develop and cause necrotic leaf lesions. In the centre of older spots black fruiting bodies (acervuli) appear. The mycelium of the pathogen appears as concentric rings, which are distinctly visible. Like the mycelium, fruiting bodies are also visible with a concentric pattern on the dorsal surface of leaves.

12.3 Root Diseases

12.3.1 Poria Disease or Red Root Disease

The disease is caused by *Poria hypobrunnea*. Due to the disease, on the collar region of the infected tea bushes, yellowish encrustation is seen, which changes to dull grey colour and appears corky. A thin film of yellowish-brown mycelium is present below the bark. At the base of the big branches, fruiting bodies are found. But if the infection is caused by *Poria hypolateritia*, roots with encrustation of soil, sand, stone particles and white to chocolate red mycelium are seen. Sometimes, black cords or sheets are also seen. The crusts that are seen in the infected tea bushes can be easily removed by rubbing or washing, and white films of mycelium are seen to present.

Another organism named *Ganoderma pseudoferreum* (Wakef.) van Overh.et Steinm; *G. philippi* (Basidiomycotina, Hymenomycetes, Aphyllophorales, Ganodermataceae) also causes red root diseases. This pathogen causes wilting of the whole bush and branches die from tip downwards. Roots are seen with white surface mycelium which later turns red, to which soil adheres. Root pale brown and hard, later pale buff and either dry or spongy.

12.3.2 Nectria

The disease is caused by *Nectria cinnabarina*. Symptom develops as dying of stems from tip downward. Newly emerged shoots which arise lowers down the progressed infection are generally thin and weak, and infected bush looks like at the point of death.

12.3.3 Root Black Root Rot

Rosellinia arcuata Petch; *R. bunodes* (Berk. et Broome) Sacc. (Ascomycotina, Pyrenomycetes, Sphaeriales, Xylariaceae) is the causal agent of root black root rot of tea. The pathogen causes rotting of roots and base of stem and discolouration of woods. In this disease, only the mycelial growth is seen but without any encrustation. Sometimes wooly, greyish stocking of mycelium is noticed extending to the base of the branches. Presence of white mycelia is also seen on the roots, later it becomes grey to black. Star-shaped, small, white to black strands of mycelium accompanied by irregular dots and dashes are present beneath the bark.

12.3.4 Brown Root Rot

Fomes lamaoensis is the pathogen of the brown root rot disease of tea. Symptom develops at the collar region of the bush without any encrustation, but brown soft mycelial growth is seen extending up to the main stem to several inches. Roots of infected tea bushes are found be with encrustation of soil, sand, stone particles etc. In this disease, the crusts that form on the infected plant parts are difficult to remove by washing or rubbing, and a thin layer of light to dark brown mycelium is present beneath the bark.

12.3.5 Charcoal Stump Rot or Charcoal Root Rot

The disease is caused by *Ustulina deusta* (Hoffm.) Lind; *U. zonata* (Lév.) Sacc. (Ascomycotina, Pyrenomycetes, Sphaeriales) and is distributed throughout the world. Symptom develops as white fan-like patches under the bark of the roots with no surface mycelium. Wood at the base of the stem is seen with irregular double lines. No surface mycelium is seen on roots. The collar region of the infected bush is white at first, then turns greenish grey and finally black, but without any encrustation. The surface of the infected parts is rough, brittle and if broken looks like charcoal. The root surface of infected bush is without any mycelial strands, and white or black cushion or lumps are seen on the root. Silky, off white and fan-like mycelial film are present beneath the peeled bark.

12.3.6 Diplodia Root Rot

Lasiodiplodia theobromae (Pat.) Griffon et Maubl. (syn. *Botryodiplodia theobromae* Pat.) (Deuteromycotina, Coelomycetes) is the causal organism of this disease, and it spreads in pantropically. Dieback is the common symptom of this disease. Besides root and stem-base, rots are also seen in the infected tea bushes. Infected bushes have blackened vascular system and black discoloured wood.

12.3.7 *Armillaria* Root Rot or Root Splitting Disease

The disease is caused by *Armillariella mellea* (Vahl:Fr.) (syn. *Armillaria mellea*) (Basidiomycotina, Hymenomycetes, Agaricales). The disease is severe in Africa and Indonesia, and in India it is rare. The disease causes sudden browning of leaves and splitting of roots. Diseased plants have white mycelial mat under the bark of stem base and roots with shoestring rhizomorph.

12.3.8 White Root Rot

White root rot disease of tea is caused by *Rigidoporus lignosus* (Klotzsch) Imazeki (Basidiomycotina, Hymenomycetes, Aphyllophorales, Polyporaceae). Dieback, presence of white mycelium (rhizomorphs) on root and rotting of roots are common symptoms of the diseases.

12.3.9 Brown Root Disease

The disease is caused by *Phellinus noxius* (Corner) Cunn. (syn. *Fomes noxius* Murr.) (Basidiomycotina, Aphyllophorales, Hymenochaetaceae). The disease is prevalent in Africa, SE Asia and Australia. Symptoms appear as adherence of a crust of earth and gravel around the entire root, white or brown mycelium under the bark and pale dry rot of wood.

12.3.10 *Rhizoctonia* Seedling Blight

The disease is caused by *Rhizoctonia bataticola* (Taub.) Butl. (Deuteromycotina, Agono-host range. Soil-inhabiting fungi mycetes); teleomorph probably *Tanatephorus cucumeris* (Basidiomycotina, Hymenomycetes). It has wide host range. Symptom of the disease develops as wilt of seedlings.

12.3.11 Tarry Root Rot

It is caused by *Hypoxylona sarcodes*. The pathogen attacks the collar region of the bush, and symptoms such as dark coloured, tough encrustation and smooth surface crust with dried black point develop. The wood of infected plant parts develops dark bands and small reticulations of light brown thin lines.

12.3.12 Purple Root Rot

This disease is caused by *Helicobasidium compactum*. Due to the infection of the pathogen, the collar region of the bush is covered with thick velvety pad mycelial strands. The collar region is surrounded with purplish brown mycelium. Root surfaces are covered with purplish brown highly branched mycelial strands. The woods are pink to pinkish brown in colour without any mycelial strands.

12.3.13 Ganoderma

Ganoderma applanatum and *Ganoderma lucidum* are the main causal agents of this disease. If the organism attacks a tea-cultivated area, a pronounced bracket-like growth is seen in the collar region of tea bushes. Upper surface of the brackets is thick, dull, marked with grey to brown concentric zones and the lower surfaces are dull white in colour.

12.3.14 Violet Root Rot

The causal organism of this disease is *Sphaerostilbe repens*. Symptoms develop as light violet-coloured roots. Premature leaf falling occurs. At the collar region, fruiting bodies develop in clusters and they resemble tiny flower buds. In water-logged areas, white to orange-purplish mycelial strands, which are irregular, thick, flat, are present beneath the peeled bark.

12.4 Disease Management

Several options are available to combat the diseases in a given situation, which include physical, cultural, agrochemical, biotechnological and biological means. Over the decades, chemical measures have become so popular that they are being treated as indispensable, because of their quick response to diversified pests and diseases. Use of pesticides has multiple roles such as to get good production against probable losses by diseases and pests, to enhance food quality and quantity, to reduce vector-borne diseases in mammals etc. Nobel Laureate Norman Borlaug said, "Complete ban on agrochemicals use in agriculture might result in 50% reduction in global food production and 4 to 5 times increase in food prices".

Adoption of agrochemicals like pesticides (insecticides, fungicides, herbicides, rodenticides, acardicides, molluscaesides etc.) can reduce the pest problem to a considerable level. But we rarely think about what happens to the agrochemicals after their application and the direct or indirect effect of chemicals and their breakdown products on the non-targets and the environment.

Indiscriminate use of chemicals pesticides causes problems of resistance, resurgence and residues. Besides these, they also cause elimination of natural enemies, disturbance of the ecological balance, environmental degradation, biomagnification and bioaccumulation (Bues et al. 2004; Chen et al. 2001; Gerhardson 2002; Punja and Utkhede 2003). Moreover, the trends of pest attack are increasing consistently due to the misuse and/or overuse of synthetic pesticides (Dhaliwal and Koul 2010). Build-up of pest resistance to these chemicals is also another aspect of concern (Gerhardson 2002; Hagn et al. 2002; Hewitt 1998; Pittendrigh and Gaffney 2001).

By realizing the ill effect of indiscriminate use of synthetic pesticides both the producers and enusers prompt them for an alternative mangement approaches. This may be a new form of pesticides, safe to the environment with low or no residual toxicity and host specific in action.

Biological control is one such alternative option fulfilling the requirement of the producers and end users as well as the environmentalists. Work on biological control for soil-borne plant pathogens was initiated in the 1930s (Weindling 1932). Later, during 1963, the first international meeting on soil-borne plant pathogens was held in Berkeley, California, and the book *Ecology of Soil-Borne Plant Pathogens—Prelude to Biological Control* (Baker and Snyder 1965) was the result of the meeting. This is regarded as one of the benchmarks in plant pathology history.

Biological control can be achieved through the use of living microorganisms such as bacteria, fungi, nematodes, protozoa, parasitoids and predators that control the pests by biological not toxic means. These are broadly classified as entomopthogenic fungi, fungal antagonist, bacterial antagonist, entomopathogenic bacteria, nematopathogenic bacteria, viral entomopathogens, parasites, predators, etc. These biological control agents have some specific characteristics like shelf life, cost-effectiveness, easy production, safety and acceptability, ease of application and effectiveness against target pests.

The biological control methods using biocontrol agents offer an advantage over chemical methods in different ways. By the use of living organisms, the risk to air, water and soil can be reduced (Hagn et al. 2002; Melero-Vara et al. 2000). There are less chances for development of resistance in the targeted pathogen and short persistence compared to synthetic pesticides, so there is no build-up of pollutants in the soil. Many biological control agents are compatible with sub-lethal and/or lethal dose of commonly used pesticides, some facilitate growth through the crop period and have high degree of specificity.

Microorganisms that are used as biocontrol agents are generally isolated by selective screening methods from natural sources like, soil, air, plant parts etc. (Knudsen et al. 1997). Because of this, the microorganisms do not need to be introduced exogenously and the pest management may be possible through the enhancement of organisms that are present naturally, which involves low cost (Gerhardson 2002) and renewable alternative to synthetic pesticides. A number of microorganisms have so far been identified as potential biological control agents (Killham 1994) with a great potential to control plant pathogens and insect pests through different direct and indirect mechanisms such as antibiotic secretion, mycoparasitism, nutrient competition, enhancement of plant defence mechanism etc. (Gams and Bissett 1998; Klein and Eveleigh 1998; Knudsen et al. 1997; Papavizas 1985; Paulitz 2000; Punja and Utkhede 2003).

12.5 *Trichoderma* as Biological Control Agents

Among the antagonistic microorganisms, fungal antagonists, and especially *Trichoderma* species (Deutermycetes, with no known sexual stage), which are cosmopolitan saprophytic soil-inhabiting fungus (Gams and Bissett 1998), are heavily researched for their various mechanisms that contribute to biological control potentiality. Based on the branching pattern of conidiophores and the morphology of the conidium, Rifai (1969) distinguished nine species of *Trichoderma* namely, *Trichoderma aureoviride*, *T. hamatum*, *T. harzianum*, *T. koningii*, *T. longibrachiatum*, *T. piluliferum*, *T. polysporum*, *T. pseudokoningii* and *T. viride*.

Bissett (1991a) proposed a sectional classification of *Trichoderma* recognizing the five different sections; section *Trichoderma*, *Longibrachiatum*, *Saturnisporum*, *Pachybasium* and *Hypocreanum*. He assigned 20 species to *Trichoderma* section *Pachybasium* and explained and classified them on the basis of morphology of conidiophore and conidium (Bissett 1991b). Persoon (1794) characterized *Trichoderma* section by narrow and flexuous conidiophores with branches and phialides uncrowded and frequently paired.

Approximately 10–27% of research papers described the mechanism of antagonism and disease suppression. Areas like host–pathogen interaction, tritrophic interaction, improvement of strains and different commercial aspects need more attention in future. In this chapter, the status of *Trichoderma* for management of tea diseases is discussed.

Out of the different biocontrol agents, the *Trichoderma* species (*Trichoderma harzianum* Rifai, *Trichoderma hamatum* (Bon.) Bain., or *Trichoderma virens* (J. Miller, Giddens & Foster) von Arx.) have been identified as active agents against many plant pathogens (Weindling 1932; Gams and Bissett 1998; Gracia-Garza et al. 1997; Papavizas 1985; Punja and Utkhede 2003; Samuels 1996; Steyaert et al. 2003). The biological control research group at the Assam Agricultural University,

Jorhat, has also developed commercial products from *Trichoderma* spp. to control soil-borne plant pathogens of tea and vegetable crops. Commercial bioformulation has been done with different species of *Trichoderma* for crop protection and enhancement of plant growth in several countries (Mcspadden Gardner and Fravel 2002).

Despite the above-mentioned advantages, the test for persistence and virulence of artificially introduced *Trichoderma* species in the field showed erratic results (Papavizas 1981). On the other hand, little is known on the effect of indegenious *Trichoderma* species. For successful integration of *Trichoderma* with other crop health management approaches, it is necessary to understand the ecology and dynamics of *Trichoderma* populations in the field (Chet 1990; Gerhardson 2002; Killham 1994; Paulitz 2000). But the biocontrol workers tend to study the suppressive qualities of *Trichoderma* species on plant pathogens (Gracia-Garza et al. 1997; Kay and Stewart 1994; Lewis and Lumsden 2001; Schoeman et al. 1996).

12.6 Ecology of *Trichoderma*

Trichoderma species are ubiquitous soil fungi and can grow well from 0 °C for *T. polysporum* and as high as 40 °C for *T. koningii* (Domsch et al. 1980; Tronsmo and Dennis 1978). Out of the different factors, temperature plays an important role not only in the growth and development of *Trichoderma* species but also in the metabolic activity, enzyme production ability, ability to produce volatile compounds and many more (Tronsmo and Dennis 1978).

Trichoderma species are widely distributed throughout the world (Domsch et al. 1980). They can grow in all types of soils i.e., forest humus layer (Wardle et al. 1993), agricultural orchard, forest, pasture and desert soils (Roiger et al. 1991; Hagn et al. 2003) and in soil that contains organic matter (Papavizas 1985). They can be found on the surface of plants roots and on decaying bark, especially when it is damaged by other fungi; and on resting structure of soil-borne fungi like sclerotia or other fungal propagules (Papavizas 1985). Besides, they can be recovered in extreme environments like mangrove swamps, salt marshes and estuarine sediments (Domsch et al. 1980) where adverse osmotic potential conditions pose a real challenge for fungal survival. Such environmental conditions are colonized well by *T. viride* Pers. ex Gray (Domsch et al. 1980). Species, like *T. harzianum*, are mostly present in rhizoispheric soil (Parkinson et al. 1963) of fields, horticulture farms and forest crops (Gochenaur and Backus 1967).

Trichoderma species can produce chlamydospores on natural substrates such as oat kernels. These structures may play an important role in the survival of the *Trichoderma* species in the soil (Henis and Papavizas 1983). Hypha also survives in soil (Papavizas et al. 1984). Caldwell (1958) was the first to observe that chlamydospores survive in soil better than conidia. Papavizas (1982) found that if the conidia are applied in the soil as inoculum of *Trichoderma harzianum*, it can persist from 110 to 130 days even without any amendments. This persistence varies from isolate to isolate.

Davet (1979) determined that conidial population applied to soil decreased initially and then stabilized for 24 months to about 1/10th of the original population added to the soil.

Lewis and Papavizas (1984) demonstrated the potentiality of different *Trichoderma* species that form resting structures like chlamydospores in natural soil either in aggregates or in fragments after its application in the soil. Mass multiplication of *Trichoderma* species is enhanced by the low pH, in the former study indicating that acidification of soil could induce suppressiveness by *Trichoderma* (Chet and Baker 1980).

Soil moisture or water potential enhances the *Trichoderma* conidia to survive longer than in soil with low moisture (Lui and Baker 1980). Water potential has a great influence on the growth and development of *Trichoderma* species, which has relationship with soil temperature. Water potential becomes more negative with increase in temperature (Dix and Webster 1995). Inoculum like hypha or mycelium, production and germination of spore and biocontrol potential of *Trichoderma* species are generally negatively affected by highly negative soil water potentials (Clarkson et al. 2004; Eastburn and Butler 1991). Danielson and Davey (1973) reported that for most of the *Trichoderma* species, a low soil water potential is required for potential growth. They are found more abundantly in soil with high moisture, more particularly in humid litter. But *T. viride* can grow well at soil water potentials of −240 MPa (Domsch et al. 1980).

As mentioned earlier, soil temperature has a great influence on the persistence of *Trichoderma* species, which are found abundantly in a wide range of soil and distributed throughout the world. Some species of *Trichoderma* prefer warm tropical soil (*T. harzianum*), whereas some prefer cool temperature (*T. polysporum* and *T. viride*) (Danielson and Davey 1973; Klein and Eveleigh 1998). One of the greatest advantage of *Trichoderma* is that it can grow in a wide range of temperatures i.e., as low as low as 0 °C for *T. polysporum* and as high as 40 °C for *T. koningii* (Domsch et al. 1980; Tronsmo and Dennis 1978). Optimum temperature plays an important role not only in the growth and development of *Trichoderma* but also on their metabolic activity, production of volatile antibiotics (Tronsmo and Dennis 1978) and enzymes.

Papavizas (1981) reported that *T. harzianum* did not survive well in the rhizosphere of bean and pea seedling, where seeds are coated with the conidia of the fungus. In addition, he found that the number of colony-forming units recovered per gram of rhizosphere soil including the roots and decaying seed coats and cotyledons was less than the number of conidia added per individual seed. The minimal effective amount of *Trichoderma* was found by Chet and Baker (1980) to be around 1×10^6 cfu/g soil.

12.7 Modes of Action of *Trichoderma*

Trichoderma species act against fungal plant pathogens through multiple ways. They may act directly and/or indirectly by mycoparasitism, antibiosis, competition for space, providing food and nutrients, enhancement of plant growth and development, SAR, solubilization of nutrients and sequestration of inorganic nutrients, inactivation of the pathogens' enzymes (Samuels 1996), enhancement of the nutrient-uptake ability of plants etc.

12.7.1 Mycoparasitism

Mycoparasitism, also known as direct antagonism, is the direct means of action of one fungus on another (Dix and Webster 1995). This is considered as an important mode of action of *Trichoderma* spp. and they parasitize the pathogens with lytic enzymes like β-1,3-gluconase and proteases (Haran et al. 1996a). Besides, chitinolytic enzymes like endochitinases, exochitinases and 1, 4-β-*N*-acetylglucosaminidases that are induced during growth in liquid medium containing chitin as carbon source (Haran et al. 1996b) also help in mycoparasitism.

Mycoparasitism is a complex process of parasitization that comprises four sequential steps. In the first step of mycoparasitism, the secretion of chemicals stimulus by the target fungus attracts the antagonistic fungus towards it and this step is known as chemotrophic growth. That is, during the initial interaction between the biocontrol agents and the pathogen, the hyphae of the fungal biocontrol agents grow directly towards their host (Chet et al. 1981).

In the second step, the mycoparasite reaches the host. In this step, the antagonistic fungus recognizes the cell surface of the pathogen. This step is also known as specific recognition. In the third step, hypha of the antagonist coils around the pathogen hypha or attaches to it by forming a hook-like structure or simply grows along with the hypha of the pathogen. In the fourth step, in many cases, the hypha of the fungal biocontrol agent penetrates the host mycelium, by degrading its cell wall partially (Elad et al. 1983) with different types of lytic enzymes like β-glucanase, chitinase and proteinases (Chet and Baker 1981). Lytic enzymes and antibiotics of biocontrol agents act synergistically, which enhances the ability of *Trichoderma* species to suppress the growth of pathogens (Kay and Stewart 1994; Steyaert et al. 2003).

12.7.2 Antibiosis

The word antibiosis means production of antibiotics by microorganisms like *Trichoderma* and many microorganism causes breakdown of target host cell.

Antibiosis is considered as an indirect mode of action of *Trichoderma* species where there is no need for hyphal interaction or contact (Dix and Webster 1995). *Trichoderma* strains can produce volatile and non-volatile antibiotics such as gliovirin and also produces 43 substances with antibiotic activity (alkyl pyrones, isonitriles, polyketides, peptaibols, dikeyopiperazines, sesquiterpenes, steroids etc.) other than many enzymes (Sivasithamparam and Ghisalbert 1998; Howell 1998). In the case of antibiosis, synergy occurs along with mycoparasitism (Schirmbock et al. 1994), where the hydrolytic enzymes allow the antibiotics to penetrate the host cells and antibiotics suppress the synthesis of cell wall, thereby enhancing the action of the hydrolytic enzymes (Lorito et al. 1996). Antibiotics, which are species-specific, act on the target organism through growth inhibition, production of metabolites and disturbing the nutrient uptake and suppressing the production of inoculum (Howell 1998; Wilcox et al. 1992). Different species of *Trichoderma* have different suppressing mechanisms against the same pathogen. This is also true at the strain level, that is, within the same *Trichoderma* species, different strains exhibit different biocontrol activities (Ghisalberti et al. 1990; Howell et al. 1993).

12.8 Competition and Rhizosphere Competence

Competition is also another mechanism of biological control. Competition occurs for space or nutrients (primarily carbon, nitrogen and iron). Competition is regarded as one of the classical biocontrol mechanisms of *Trichoderma* species. (Elad et al. 1999).

Out of the different biocontrol agents, *Trichoderma* species are the most aggressive competitors and have the ability to compete highly as compared to other and the competitive nature is totally species-dependent (Wardle et al. 1993). In recent years, competition through rhizosphere competence has gained momentum (Howell 2003). *Trichoderma* species, which have wider colonizing ability in different substrates, are considered to be the most efficient soil colonizers. They can easily displace the other less-aggressive substrate colonizers (Papavizas 1985). Environmental factors like soil pH, temperature and water potential greatly influence the colonizing ability of *Trichoderma* species (Carreiro and Koske 1992; Danielson and Davey 1973; Domsch et al. 1980; Eastburn and Butler 1988).

12.9 Induced Resistance and Enhancement of Plant Growth

Induction of resistance in the host plant is another mechanism in biological control (Howell 2003). De Meyer et al. (1998) reported that a number of morphological and biochemical changes are observed in *Trichoderma*-treated plants because of the colonization and penetration of plant root tissues, which ultimately leads to induced systemic resistance (ISR) in the treated plant (De Meyer et al. 1998). Among the resistance inducers, xylanase and other elicitors produced by *Trichoderma* species are considered to play an important role in biological control (Anderson et al. 1993), and currently, researchers are showing interest in resistance elicitation in plants by *Trichoderma*.

Trichoderma species enhance plant growth by stimulating the nutrient uptake ability of treated plants (Kleifeld and Chet 1992; Ousley et al. 1994) or by releasing hormones or secreting plant growth regulators (Windham et al. 1986) in an area from where the plant can absorb easily. Stimulation of plant growth indirectly inhibits the growth of pathogens, thereby avoiding disturbance to the metabolic system of plants (Elad et al. 1987). Yet, the significance of such results has to be regarded with caution, as most trial results reported in the literature have been conducted in the glasshouse and often do not reflect the field situation. In the field, the potential of *Trichoderma* species to stimulate plant growth can be overshadowed by more efficient plant growth promoting rhizobacteria (PGPR) such as *Pseudomonas* species (Schroth and Hancock 1982).

12.10 Solubilization and Sequestration of Inorganic Plant Nutrients

Plant roots play an important role that influences their accessibility and absorption are influenced by microorganisms (Altomare et al. 1999). In an in vitro study, Altomare et al. (1999) showed that a strain of *Trichoderma harzianum* produces a number of metabolites to solubilize rock phosphate, Zn, Mn^{4+}, Fe^{3+} and Cu^{2+} and increases iron availability and enhances iron uptake. But, till date, the direct role of *Trichoderma* in solubilizing and chelating nutrients has not been demonstrated, but literature is available on solubilization of iron and making it available to plants (Harman 2000).

12.11 Inactivation of the Pathogen Enzymes

The rigidity of fungal cell wall is determined by polysaccharides, chitin and β-glucans. As mentioned earlier, a number of enzymes like chitinases and/or glucanases of the pathogen degrade the polysaccharides that are responsible for maintaining the cell wall integrity (Howell 2003).

Enzymes like proteases can degrade the cell wall of the pathogen and reduce the virulence of the pathogen. They also showed that protease may be directly toxic to the germination of the pathogen and also may inactivate its enzymes (Elad and Kapat 1999).

12.12 Factors Affecting the Antagonistic Activity

A variety of environmental factors are shown to have interesting effects on the potentiality of biocontrol agents against insect pest/pathogens. The main influencing parameters for the success of biocontrol agents are solar radiation, temperature, relative humidity, rainfall and soil factors.

12.12.1 Solar Radiation

Out of the different parameters affecting the persistence of conidia, hyphal bodies and hyphae, solar radiation plays an important role. All the fungal propagules are highly sensitive to solar radiation, and there is a possibility of damage by solar radiation, particularly, the UV-B portion of the solar spectrum (285–315 nm). However, the susceptibility to irradiation varies according to the type of strain used for biological control. For example, the conidia of *Metarhizium anisopliae* var. acridum were found to be the most resistant to artificial sunlight followed by conidia of *Beauveria bassiana.* So, UV-resistant genotypes should be considered for IPM programmes. Some fungi that produce pigmented propagules like melanin have more UV resistance than non-pigmented fungi.

12.12.2 Temperature

Like solar radiation, temperature also plays an important role in the success of *Trichoderma* species as biocontrol agents. The *Trichoderma* species *T. viride* and *T. polysporum* are cold tolerant and show better results at low field temperatures than *T. harzianum* that cannot tolerate cold temperature (Tronsmo and Dennis 1978). The virulence of *T. viride* is reduced greatly at temperatures less than 10 °C (Clarkson et al. 2004). The action of the antibiotics produced by *Trichoderma* spp. is also influenced by temperature. The efficacy of gliotoxin on plant pathogens is more at 32 °C than at 3 °C (Lumsden et al. 1992; Weindling 1941). Hence, it is evident that antagonistic fungi perform efficiently if both the pathogen and the biocontrol agent have similar temperature requirements (Hjeljord and Tronsmo 1998).

12.12.3 Relative Humidity

Relative humidity of the environments can influence the persistence and efficacy of fungal biocontrol agents in several ways. Evaporation of spray droplets is dependent on the relative humidity along with temperature. This can result in the loss of small particles, which adversely affects targeting. For most fungi, the conidia produced by fungal biocontrol agents exhibit more stability in cool and dry environmental condition. In contrast, more conidial persistence is recorded at moderate temperature with high relative humidity. Water is very much essential for the germination of fungal propagules and to cause infection. The pre-requisite for moisture during germination of inoculum has led to the belief that moisture is essential for the effective use of entomopathogenic fungi. Literature reveals that dry conditions during or immediately after application of fungal inoculum are less detrimental than thought earlier.

12.12.4 Rainfall

Rainfall helps in the dislodging and dispersal of conidia from substrate, and it aids in the dispersion of fungal propagules. Thereby, the propagules can travel a good distance and disperse in and around their source.

12.12.5 Soil Factors

Many biocontrol agents are soil borne in nature and found to have potential effect against soil pests. But, the factors like soil type (i.e. texture, CEC, organic matter content, pH, MBC etc.), soil moisture and population of viable soil microflora play an important role in the persistence and/or efficacy of biocontrol agents. In an experiment, Storey and Gardner (1987) observed that the water infiltration values, soil texture and organic matter content of the soil determines the ability of vertical movement of biocontrol agents. They observed that in most of the cases, higher loss occurs in sandy soil as compared to fine-textured soils. But, conidia remained at the soil surface in the sandy soils and soils with low organic matter content retain lower propagules than dry-textured and organic soils. The mechanism of conidial retention in such soils is not known, but many relate to their high cation exchange capacity and/or their reduced pore sizes. Besides these, the vertical movement of propagules is also dependent on their quality and quantity. The role of other soil parameters like pH and ionic conductivity is not understood. Lockwood (1977) reported that the fungicidal potential of soil-borne fungal pathogens is higher in neutral or slightly acidic soils.

12.12.6 Substrates Quality

The antagonistic ability of *Trichoderma* spp. also depends on the quality of substrate and availability of nutrient in the substrate (Howell 1998). *Trichoderma* species produce higher amount of secondary metabolites in substrate containing high carbon-to-nitrogen ratio (C:N). Since antagonistic ability is dependent on the quality of substrates, substrates with organic content should be used for the field delivery of *Trichoderma* species. Lewis and Lumsden (2001) and McLean and Stewart (2000) demonstrated that the use of wheat bran formulation of *Trichoderma* species allows better proliferation and antagonistic activity against soil-borne plant pathogens.

12.12.7 Biotic Factors

Among the biotic factors, the complex nature of rhizosphere plays an important role in the success of biological control agents. Though *Trichoderma* species are fast-growing, aggressive colonizers, they still have to engage in competition with other rhizospheric microbes. The presence of opportunistic microbes and their ability to grow fast in the vicinity of *Trichoderma* can reduce the availability of substrate.

12.12.8 Benefit of Using Trichoderma

Besides the disease controllability of *Trichoderma* against various pathogens like Fusarium, Rhizoctonia, Sclerotinia, Sclerotium, Phytophthora and Colletotrichum, they can also help in promoting the plant growth. *Trichoderma* strains have the ability to solubilize phosphates and a number of micronutrients. They can enhance root growth, more particularly deep roots, thereby increasing the plant's ability to resist drought. *Trichoderma* strains are known to induce plant resistance by inducing the secondary metabolism of plants like ethylene production, hypersensitive responses and other defense-related reactions of host plants. Besides these, the selection of chitinase gene of *Trichoderma* species and its insertion into host plants like tobacco and potato plants has increased their resistance to fungal growth. Selected transgenic lines were found highly tolerant to *Alternaria alternata*, *A. solani*, *Botrytis cirerea* and *Rhizoctonia* spp. Further, *Trichoderma* strains can be used as agents for the bioremediation of soils that are loaded with inorganic pesticides. *Trichoderma* has the ability to degrade inorganic insecticides like organochlorines, organophosphates and carbonates.

12.13 Management of Tea Diseases with *Trichoderma* spp.

Trichoderma is the most widely used biocontrol agent. Among the various isolates of *Trichoderma*, *T. viride*, *T. asperellum*, *T. harzianum*, *T. virens* and *T. hamatum* are used in the management of various diseases of crop plants. It has many advantages as a bio-control agent:

(a) High rhizosphere competence
(b) High competitive saprophytic ability
(c) Enhances plant growth
(d) Great arsenal of inducible polysaccharide-degrading enzymes
(e) Produces secondary metabolites/antibiotics of volatile and non volatile nature
(f) Ease for mass multiplication
(g) Broad spectrum of action against various pathogens
(h) Provides excellent and reliable control
(i) Environmentally safe (Jeyarajan and Nakkeeran 2000)

Use of *Trichoderma* species either as a single management component or as a component of an integrated disease management tool showed a significantly positive result. Foliar spray of native strain of *Trichoderma harzianum* at 5 ml/l of water + 0.02% carboxy methyl cellulose (CMC) immediately after the appearance of first visible symptom followed second spray after 15 days of the first spray and another spray after 21 days of the second spray resulted reduction of blister blight incidence significantly. Similarly, *T. harzianum* and/or *Gliocladium virens* was found to be effective for the management of the twig die back or stem canker of tea. The biocontrol agents should be applied at 1500 g/bushes as enriched compost containing 1×10^6 cfu/g. Besides, when the disease first appears in the plantation area, foliar spraying either of the biological control agents should be done at 5 ml/l of water during the afternoon hours.

Tea bushes are pruned every year, and due to this practice, the pruned end gets damaged and paves way for a number of invading pathogens. So, *T. harzianum* can be applied as a paste in the pruned end immediately after pruning. In the tea garden, the use of *T. harzianum* as paste or paint immediately after pruning alone or in combination with copper fungicides showed encouraging results in protecting the pruned end from the infection of Poria. Further, enhanced plant growth with early bud break (7 days ahead) was observed in *T. harzianum* applied gardens.

In a study conducted by Sarkar et al. (2015) in Bangladesh, *Trichoderma* spp. were collected and studied for their cultural, morphological and antagonistic potentials against tea grey blight pathogen *Pestalozzia theae* and maximum inhibition of $84.45 \pm 0.77\%$ after 72 h of inoculation under in vitro condition. This was followed by $76.02 \pm 3.50\%$ after 24 h of inoculation. They concluded that there is a possibility to use these fungi as a component of integrated disease management of grey blight disease in tea.

Earlier attempts on biological control of blister blight were made with popular antagonists like *Trichoderma harzianum*, *Gliocladium virens*, *Serratia marcescens*, *Pseudomonas fluorescens* and *Bacillus subtilis* (Premkumar and Baby 2005). These were applied as formulations based on talc/vermicompost, but the results were not encouraging. When *Pseudomonas fluorescens* and *B. subtilis* were applied as liquid culture, addition of nutrients like $(NH_4)_2$, SO_4 and salicylic acid improved their bio-efficacy, especially when the disease incidence was low.

Balasuriya (1998) reported that inoculation of *T. viride* on wood caused wood decay, which was marginally high as compared to inoculation of wood rotting fungus, *Nemania diffusa*, alone. Application of *Trichoderma* as paste on the pruned end and the application of soil around the bush (1 kg/bush) was found to be effective in protecting from stem disease of tea including *Hypoxylon serpens* (Premkumar and Baby 2005).

Thoudam and Dutta (2012) reported that heavy spray of aqueous solution of *T. atroviride and T. citrin* collected as aerosopre in tea ecosystem can be used as possible biocontrol agent against black rot disease of tea caused by *Corticium theae*.

An in vitro study on the interaction between collar canker pathogen *Phomopsis theae* and fungal antagonists like *T. harzianum* and *G. virens* revealed that antibiosis and hyperparasitism are mechanisms that prevent the establishment of pathogens (Ponmurugan and Baby 2003). In the case of pathogens grown on tea stem bits, pre-inoculation and simultaneous inoculation of antagonists prevented their complete development, whereas antagonists could not inhibit the growth of pathogens already established due to the normal development of pycnidia followed by exudation of cirrhus (spore mass).

Field efficacy of talc-based formulation of *T. harzianum* and *G. virens* as paste and dressed ion pruned cut end along with soil application of either biocontrol agents (2 kg/bush) or fungicide solution (0.5%) in controlling collar canker was reported by Ponmurugan et al. (2002). The results showed that wound healing (reduction in canker size) was high in the case of biocontrol treatments applied both on wound and in soil. They also reported that biocontrol agents survived well in soil and also on pruned cuts.

A wide number of antagonistic microbes have been tried under gnotobiotic conditions against major root pathogens of tea. They were *Trichoderma* spp. (Baby and Chandra Mouli 1996; Hazarika et al. 2000; Balasuriya 2001), *Gliocladium virens* (Baby and Chandra Mouli 1996; Hazarika et al. 2000), *Bacillus pumilis*, *Bacillus megaterium*, *Serratia marcescens* and *Ochrobactrum anthropic* (Chakrabarty et al. 2013), strains of fluorescent *Pseudomonas* (Hazarika et al. 2000; Dileep Kumar et al. 2005) and non-fluorescent *Pseudomonas* (Dileep Kumar et al. 2005), *B. subtilis* (Hazarika et al. 2000) and strains of actinomycetes are mentionable. Field experiments conducted against red root disease showed that coir pith-based bioformulation of *T. harzianum* and *Gliocladium virens* at 4 tonnes/ha in furrow application or planting pit application at 100 g/pit were effective in controlling the disease in new planting area as there was no casualty in young plants (Baby et al. 2004). There was a mild incidence of casualty among new plants in chemical-treated plots, and interestingly in chemical treatments where biocontrol agents were given as sub-treatments no casualty was noticed. Similar observations were reported from Sri Lanka, with *T. harzianum* (5 g/l), which gave good control of Poria root disease (Balasuriya 2005).

Hazarika et al. (2000), reported that spore suspensions (10^8 ml^{-1}) of antagonistic microbes like *T. harzianum*, *T. viride* and *G. virens* can give good control of disease caused by *Ustulina zonata* on potted tea plants, which can be compared to carbendazim treatment. Soil drenching was found superior compared to seedling root dip in reducing the mortality of the plants. In another experiment with *U. zonata* (Hazarika and Phookan 2003), a consortium of *Psuedomonas*, *Trichoderma* and

VAM fungi *Glomus fasciculatum* was used and the level of disease control achieved was significantly superior to the application of individual components.

In an eco-friendly management approach to tea disease caused by *Colletotrichum camelliae*, nine different indigenous antagonistic microbial agents i.e., *Trichoderma viride*, *Trichoderma harzianum*, *Trichoderma virens*, *Trichoderma pseudokoningii, Trichoderma asperellum, Pseudomonas fluorescens, Paeicilomyces lilacinus, Beauveria bassiana* and *Metarhizium anisopliae* are used by in vitro dual culture method. After 165 h of incubation, the percentage of inhibition over control was analysed by the following statistical procedure. It was observed that in the in vitro antagonistic test there were considerable differences between the bioagents in their ability to inhibit the radial growth of *C. camelliae* (Table 12.1). However, the highest growth inhibition (83.44%) was observed in *T. harzianurn*, which was followed by *T. viride* (77.66%) and *T. pseudokoningii* (77.22%). However, statistically all the bioagents were at par in their efficacies. The lowest growth inhibition (52.44%) was observed in *M. anisopliae* which proved to be less effective (Dutta et al. 2016). The present results confirm that *T. harzianum* and *T. viride* inhibit the growth of *C. camelliae* causing die back in tea and therefore can be suggested for the management of die back disease in field condition.

Similarly in another experiment conducted against the tea disease caused by *Poria hypobrunea,* seven biocontrol agents viz., *Beauveria bassiana*, *Metarhizium anisopliae*, *Trichoderma pseudokoningii*, *Paecilomyces lilacinus*, *Trichoderma asperellum*, *T. harzianum* and *Pseudomonas fluorescens* were tested. *T. harzianum* was found to be the highly effective biocontrol agent in inhibiting the mycelial growth of *P. hypobrunnea* with a mycelial growth inhibition of 52.66% followed by *T. asperellum* (Table 12.2). Out of all these tested biocontrol agents, *P. fluorescence* and *M. anisopliae* could not inhibit the mycelial growth of *P. hypobrunea*.

In an another experiment, *Trichoderma pseudokoningii* was tested for its compatibility with 11 insecticides, 7 fungicides, 2 herbicides, 1 sticker cum spreader, 4 different inorganic fertilizers commonly applied in tea ecosystem and agricultural field at two different concentrations i.e., recommended dose (RD) and half of the RD, and four different concentrations were tested for inorganic fertilizer and organic sticker cum spreader. All the pesticides significantly inhibited the mycelial growth of the fungus *T. pseudokoningii* except Thiamethonaus 25% WG at 0.125%. Ritha was found to be compatible with the biocontrol agent in all the tested doses. Urea and muriate of potash were found compatible with *T. pseudokoningii*, whereas single super phosphate and calcium ammonium nitrate inhibited its growth (Dutta et al. 2017). The observed variations in the inhibitory potential could be due to the inherent variability of the chemical insecticide to the biocontrol agent, *T. pseudokoningii*. CAN inhibited the colony growth of *T. pseudokoningii* in all the tested concentrations. Present investigations showed varying effects of pesticides, sticker and inorganic fertilizers on the fungal biocontrol agents, but their actual effects at the cellular and field level need to be investigated to understand if the effects are permanent or temporary (Dutta et al. 2017). In case of temporary arrest of fungus activity, it may recover after degradation of toxicant and such chemicals can be employed in combination with the fungus under field conditions. The results

Table 12.1 Mycelial growth inhibition of *Colletotrichum camelliae* by different biocontrol agents

Treatments	Radial growth (cm) after 168 h of incubation	% Mycelial growth inhibition at 168 h of incubation
Colletotrichum camelliae + *Beauveria bassiana*	3.67[c]	59.22 (36.77)[f]
C. camelliae + *Metarhizium anisopliae*	4.28[b]	52.44 (31.62)[g]
C. camelliae + *Trichoderma pseudokoningii*	2.05[g]	77.22 (52.81)[b]
C. camelliae + *Paeicilomyces lilacianus*	1.49[h]	83.44 (59.88)[a]
C. camelliae + *T. asperellum*	2.12[g]	76.44 (52.04)[b]
C. camelliae + *T. harzianum*	2.01[g]	77.66 (53.23)[b]
C. camelliae + *Pseudomonas fluorescence*	3.1[e]	65.55 (42.23)[d]
C. camelliae + *Verticilium lecani*	2.36[f]	73.77 (49.38)[c]
C. camelliae + *T. viride*	3.42[d]	62.00 (38.95)[e]
Control	9.00[a]	–
$CD_{0.05}$	**0.15**	**1.46**

Source: Final project report submitted by Dutta, 2014
Means within columns separated by Duncan's multiple range test $P = 0.05$; Figures followed by same letters are statistically not different; Figures in the parentheses are angular transformation value

are of permanent importance for both the development of protocols for fermentation technology and field applications of the application and efficacy of biocontrol agents. *Trichoderma* is also reported to be compatible with biofertilizers like *Rhizobium*, *Azospirillum*, *Bacillus subtilis* and *Phosphobacteria*. From the result of this work, it can be concluded that the compatible components can be mixed as tank mix at the same time. This will reduce the cost of spraying. Further, it will give the synergistic effect of the different components. This means that, if we use the compatible *Trichoderma* with biofertilizer this will not only help to protect the crop from plant diseases by the former but also help to supply the required nutrient by the later. Similarly, if we combine *Trichoderma* with entomopathogens, then besides curbing tea diseases the insect pest will be taken care of by the entomopathogen. Further, if *Trichoderma* is mixed with sublethal or lethal dose of pesticides, the use of chemical will help to weaken the pathogen, and a weaker pathogen is easily attacked and killed by biocontrol agents.

Due care should be taken for application of biocontrol agents like *Trichoderma*, i.e., use of non-compatible chemical pesticides should be avoided at least for 4–5 days, *Trichoderma* should be applied in dry soil as moisture is an essential factor for its growth and survivability, foliar spray should be done during afternoon hours as UV rays during bright sunshine hours may harm the inoculum, for soil

Table 12.2 Mycelial growth inhibition of *Poria hypobrunnea* by different biocontrol agent

Treatment	Mycelial growth at different hours of incubation (cm)			% Mycelial growth inhibition over control
	48 h	72 h	96 h	
P. hypobrunnea + *Beauveria bassiana*	4.50[cd]	5.74[b]	5.96[bc]	33.77
P. hypobrunnea + *Metarhizium anisopliae*	4.12[c]	7.16[cd]	9.00[d]	0.00
P. hypobrunnea + *Trichoderma pseudokoningii*	4.01[c]	5.24[b]	6.66[c]	26.00
P. hypobrunnea + *Paecilomyces lilacinus*	4.78[de]	5.99[bc]	6.48[c]	28.00
P. hypobrunnea + *T. asperellum*	3.22[b]	5.33[b]	5.49[b]	39.00
P. hypobrunnea + *T. harzianum*	1.14[a]	2.79[a]	4.26[a]	52.66
P. hypobrunnea + *Pseudomonas fluorescens*	5.61[f]	7.81[d]	9.00[d]	0.00
Control	5.16[ef]	8.01[d]	9.00[d]	0.00
CD (0.005)	**0.612**	**1.175**	**0.900**	

Source: Final project report submitted by Dutta, 2014
Data are mean of five replications. Data followed by same letter are statistically at par with each other

application of *Trichoderma*, it should be mass multiplied in organic compost with good quantity of nutrients that support the growth of *Trichoderma*.

The success of various experiments on the effectiveness of *Trichoderma* as a biological control agent vis-à-vis its production and application technology has motivated manufacturers to produce these potential biological control agents on a large scale. However, legal and commercial considerations are equally important to its further development. Research workers on biocontrol developed *Trichoderma*-based large-scale production technology with higher shelf life and virulence in field conditions and available in the market under different trade name. Org-Trichojal is one such kind of product developed at the Assam Agricultural University, Jorhat, Assam. This technology is developed with native isolates of *Trichoderma harzianum* with good field efficacy in tea sectors. Such a product could be used as one of the components of integrated disease management in tea.

Further research approaches on *Trichoderma* should be concentrated on genomics, transcriptomics, proteomics and metabolomics for developing next-generation products with higher shelf life and without any erratic field results.

12.14 Conclusion

The literature on use of *Trichoderma* spp. for management of tea diseases showed the potentiality of the biocontrol agent in tea plantation. But further studies are needed to determine the environmental effects on the fate of the bio-agents. New technologies such as micro-encapsulation of biocontrol agents may be of high priority in enhancing their potential. Moreover, works on integration of *Trichoderma* spp. with botanical pesticides are few, so more work on this aspect may pave way for searching more effective management approaches for tea disease.

References

Altomare C, Norvell WA, Björkman T, Harman GC (1999) Solubilization of phosphates and micronutrients by the plant-growth promoting and biocontrol fungus *Trichoderma harzianum* Rifai 1295-22. Appl Environ Microbiol 65:2926–2933

Anderson RD, Baily BA, Taylor R, Sharon A, Avni A, Mattoo AK, Fuchs Y (1993) Fungal xylanase elicits ethylene biosynthesis and other defense responses in tobacco. In: Pech JC, Latche A, Balague C (eds) Cellular and molecular aspects of the plant hormone ethylene. Kluwer Academic, Dordrecht, Netherlands, pp 197–204

Anonymous (2012) Reporter, B S (17 October 2012). Assam cabinet's nod to tea as 'state drink' of Assam. Business Standard India. Accessed 12 Feb 2019

Anonymous (2013a) Archived copy. Archived from the original on 29 Oct 2013. Accessed 25 Feb 2013

Anonymous (2013b) India to declare tea national drink in 2013. The Times of India. Archived from the original on 30 May 2013

Anonymous (2013c) Tea to get hotter with national drink tag? - Times of India. The Times of India

Anonymous (2014) Tea Board of India. Indian Tea culture. Govt. of India. Retrieved 6 Oct 2014

Baby UI, Chandra Mouli B (1996) Biological antagonism of Trichoderma and Gliocladium spp. against certain primary root pathogens of tea. J Plantation Crops 24(Suppl):249–255

Baby UI, Premkumar R, Ajay D, Sanjay R, Sasidhar R (2004) Chemical and biological control of red root disease. Newsl UPASI Tea Res Foundation 14(2):1–2

Baker KF, Snyder WC (1965) Ecology of soil-borne plant pathogens: prelude to biological control, vol 100. University of California Press, Berkeley, CA, p 224

Balasuriya A (1998) Report of the plant pathology division. Annual Report of Tea Research Institute, Sri Lanka. pp 88–89

Balasuriya A (2001) Report of the plant pathology division. Annual Report of Tea Research Institute, Sri Lanka. p 128

Balasuriya A (2005) Report of the plant pathology division, Annual Report of Tea Research Institute, Sri Lanka. p 177

Bhagat RM, Baruah DR, Safique S (2010) Climate and tea, *Camellia sinensis* (L) O. Kuntze production with special reference to north east India: a review. J Environ Res Dev 4 (4):1017–1028

Bissett J (1991a) A revision of the genus *Trichoderma*. II. Infrageneric classification. Can J Bot 69:2357–2372

Bissett J (1991b) A revision of the genud *Trichoderma*. III. Section *Pachybasium*. Can J Bot 69:2373–2417

Bues R, Bussieres P, Dadomo M, Dumas Y, Garcia-Pomar MI, Lyannaz JP (2004) Assessing the environmental impacts of pesticides used on processing tomato crops. Agric Eco Environ 102:155–162

Caldwell R (1958) Fate of spores of *Trichoderma viride* Pers. Ex Fr. introduced into soil. Nature 181:1144–1145

Carr MKV (1972) The climatic requirements of the tea plant: a review. Exp Agric 8:1–14

Carreiro MM, Koske RE (1992) Effect of temperature on decomposition and development of microfungal communities in leaf litter microcosms. Can J Bot 70:2177–2183

Chakrabarty U, Chakrabarty BN, Chakrabarty AP, Sunar K, Dey PL (2013) Plant growth promoting rhizobacteria mediated improvement of health status of tea plants. Ind J Biotechnol 12:20–31

Chet I, Baker R (1980) Induction of suppressiveness to *Rhizoctonia solani* in soil. Phytopathology 70:994–998

Chen SK, Edwards CA, Subler S (2001) A microcosm approach for evaluating the effects of the fungicides benomyl and captan on soil ecological processes and plant growth. Appl Soil Ecol 18:69–82

Chet I (1990) Biological control of soilborne pathogens with fungal antagonists in combination with soil treatment. In: Hornby D, Cook RJ, Henis Y, Ko WH, Rovira AD, Schippers B, Scott PR (eds) Biological control of soilborne pathogens. CAB Publishing House, New York, pp 15–25

Chet I, Baker R (1981) Isolation and biocontrol potential of *Trichoderma hamatum* from soil naturally suppressive to *Rhizoctonia solani*. Phytopathology 71:286–290

Chet I, Harman GE, Baker R (1981) *Trichoderma hamatum*: its hyphal interactions with *Rhizoctonia solani* and *Pythium* spp. Microbiol Ecol 7:29–38

Clarkson JP, Mead A, Payne T, Whipps JM (2004) Effect of environmental factors and *Sclerotium cepivorum* isolate on sclerotial degradation and biological control of white rot by *Trichoderma*. Plant Pathol 53:353–362

Danielson RM, Davey CB (1973) Non nutritional factors affecting the growth of *Trichoderma* in culture. Soil Biol Biochem 5:495–504

Davet P (1979) Technique pour l'analyse des populations de *Trichoderma* et de *Gliocladium virens* dans le sol. Ann Phytopathol 11:524–533

De Meyer G, Bigirimana J, Elad Y, Hofte M (1998) Induced systemic resistance in *Trichoderma harzianum* biocontrol of *Botrytis cinerea*. Eur J Plant Pathol 104:279–286

Dhaliwal GS, Koul O (2010) Quest for pest management: from green revolution to gene revolution. Kalyani Publishers, New Delhi

Dileep Kumar BS, Mishra AK, Dutta S (2005) Role of plant growth promoting rhizobacteria in suppression of root diseases of tea. International Symposium on Integrated Management of Fungal Diseases. Section-II. p 23

Dix N, Webster J (1995) Fungal ecology. Chapman and Hall, London, p 549

Domsch KH, Gams W, Anderson TH (1980) Compendium of soil fungi, vol 1. Academic, New York

Dutta P, Kaushik H, Puzari KC, Bhuyan RP (2016) Ecofriendly management of tea diseases in current scenario of climate change. In: Dynamics in crop protection and climate change. Publ. Studera Press, New Delhi, pp 325–340

Dutta P, Kakati N, Das A, Kaushik H, Boruah S, Bhowmick P, Kaman P, Bhuyan RP, Hazarika GN (2017) *Trichoderma pseudokoningii* showed compatibility with certain commonly used inorganic pesticides, fertilizer and sticker cum spreaders. Int J Curr Microbiol App Sci 6(2):140–146

Eastburn DM, Butler EE (1988) Microhabitat characterization of *Trichoderma harzianum* in natural soil: evaluation of factors affecting population density. Soil Biol Biochem 20:541–545

Eastburn DM, Butler EE (1991) Effects of soil moisture and temperature on the saprophytic ability of *Trichoderma harzianum*. Mycologia 83:257–263

Eden T (1976) Tea, 93rd edn. Longman Group, London, p 236

Elad Y, Kapat A (1999) The role of *Trichoderma harzianum* protease in the biocontrol of *Botrytis cinerea*. Eur J Plant Pathol 105:177–189

Elad Y, Chet I, Henis Y (1983) Parasitism of *Trichoderma* spp. on *Rhizoctonia solani* and *Sclerotium rolfsii*. Scanning electron microscopy and fluorescence microscopy. Phytopathology 73:85–88

Elad Y, Chet I, Baker R (1987) Increased growth response of plants induced by rhizobacteria antagonistic to soilborne pathogenic fungi. Plant Soil 98:325–330

Elad Y, David DR, Levi T, Kapat A, Kirshner B, Guvrin E, Levine A (1999) *Trichoderma harzianum* T-39 mechanisms of biocontrol of foliar pathogens. In: Lyr H (ed) Modern fungicides and antifungal compounds II. Intercept Ltd., Andover, Hampshire

FAO (1999) FAO yearbook of production Vol 52, 1998. FAO Statistic Series No. 148. Food and Agriculture Organization of the United Nations, Rome, p 170

Gams W, Bissett J (1998) Morphology and identification of *Trichoderma*. In: Kubicek CP, Harman GE (eds) *Trichoderma and Gliocladium*. Taylor & Francis, London, pp 3–31

Gerhardson B (2002) Biological substitutes for pesticides. Trends Biotechnol 20:338–343

Ghisalberti EL, Narbey MJ, Dewan MM, Sivasithaparam K (1990) Variability among strains of *Trichoderma harzianum* in their ability to reduce take all and to produce pyrones. Plant Soil 121 (20):281–291

Gochenaur SE, Backus MP (1967) Mycoecology of willow and cotton-wood lowland communities in southern Wisconsin. II. Soil microfungi in the sandbar willow stands. Mycologia 59:893–901

Gracia-Garza JA, Reeleder RD, Paulitz TC (1997) Degradation of sclerotia of *Sclerotinia sclerotiorum* by fungus gnats (*Bradysia coprophila*) and the biocontrol fungi *Trichoderma* spp. Soil Biol Biochem 29:123–129

Hagn A, Geue H, Pritsch K, Schloter M (2002) Assessment of fungal diversity and community structure in agricultural used soils. Research Signpost, Kerala, India

Hagn A, Pritsch K, Schloter M, Munch JC (2003) Fungal diversity in agricultural soil under different farming management systems, with special reference to biocontrol strains of *Trichoderma* spp. Biol Fertil Soils 38:236–244

Haran S, Schickler H, Oppenheim A, Chet I (1996a) Molecular mechanisms of lytic enzymes involved in the biocontrol activity of *Trichoderma harzianum*. Microbiology 142:2321–2331

Haran S, Schickler H, Oppenheim A, Chet I (1996b) Differential expression of *Trichoderma harzianum* chitinases during mycoparasitism. Phytopathology 86:980–985

Harman GE (2000) Myths and dogmas of biocontrol: changes in perceptions derived from research on *Trichoderma harzianum* T-22. Plant Dis 84:377–393

Hazarika DK, Phookan A (2003) Combination of *Glomus fasciculatum* with *Pseudomonas fluorescens* and *Trichoderma harzianum*: effect on biocontrol potential and growth promotion in tea seedling. Sixth International Workshop on PGPR, IISR Calicut. pp 289–294

Hazarika DK, Phookan AK, Saikia GK, Borthakur BK, Sarma D (2000) Management of charcoal stump rot of tea with biocontrol agents. J Plant Crop 28(2):149–153

Henis Y, Papavizas GC (1983) Factors affecting germinability and susceptibility to attack of sclerotia of *Sclerotium rolfsii* by *Trichoderma harzianum* in field soil. Phytopathology 73:1469–1474

Hewitt HG (1998) Fungicides in crop protection, 1st edn. C.A.B International, Wallingford, Oxon

Hjeljord L, Tronsmo A (1998) *Trichoderma* and *Gliocladium* in biological control: an overview. In: Kubicek CP, Harman GE (eds) *Trichoderma and Gliocladium*. Taylor & Francis, London, pp 131–152

Howell CR (1998) The role of antibiosis in biocontrol. In: Harman GE, Kubicek CP (eds) Trichoderma and Glioladium, vol 2. Taylor and Francis, London, pp 173–184

Howell CR (2003) Mechanisms employed by *Trichoderma* species in the biological control of plant diseases: the history and evolution of current concepts. Plant Dis 87:4–10

Howell CR, Stipanovic RD, Lumsden RD (1993) Antibiotic production by strains of *Gliocladium virens* and its relation to the biocontrol of cotton seedling diseases. Biocontrol Sci Tech 3:435–441

Jeyarajan R, Nakeeran S (2000) Exploitation of microorganisms and virus as biocontrol agents for crop disease management. In: Upadhyay RK et al (eds) Biocontrol potential and its exploitation in sustainable agriculture. Springer Science + Business Media, New York, pp 95–116

Kay SJ, Stewart A (1994) Evaluation of fungal antagonists for control of onion white rot in soil box trials. Plant Pathol 43:371–377

Killham K (1994) Soil ecology, 1st edn. Cambridge University Press, Cambridge

Kleifeld O, Chet I (1992) *Trichoderma harzianum* – interaction with plants and effect on growth response. Plant Soil 144:267–272

Klein D, Eveleigh E (1998) Ecology of *Trichoderma*. In: Kubicek CP, Harman GE (eds) *Trichoderma and Gliocladium*. Taylor & Francis, London, pp 57–69

Knudsen IMB, Hockenhull J, Jensen DF, Gerhardson B, Hokeberg M, Tahvonen R, Teperi E, Sundheim L, Henriksen B (1997) Selection of biological control agents for controlling soil and seed-borne diseases in the field. Eur J Plant Pathol 103:775–784

Lewis JA, Lumsden RD (2001) Biocontrol of damping-off of greenhouse-grown crops caused by *Rhizoctonia solani* with a formulation of *Trichoderma* spp. Crop Prot 20:49–56

Lewis JA, Papavizas GC (1984) A new approach to stimulate population proliferation of *Trichoderma* species and other potential biocontrol fungi introduced into natural soils. Phytopathology 74:1240–1244

Lockwood JL (1977) Fungistasis in soils. Biol Rev 52:1–43

Lorito M, Farkas V, Rebuffat S, Bodo B, Kubicek CP (1996) Cell wall synthesis is a major target of mycoparasitic antagonism by *Trichoderma harzianum*. J Bacteriol 178:6382–6385

Lui S, Baker R (1980) Mechanism of biological control in soil suppressive to *Rhizoctonia solani*. Phytopathology 70:404–412

Lumsden RD, Locke JC, Adkins ST, Walter JF, Ridout C (1992) Isolation and localization of the antibiotic gliotoxin produced by *Gliocladium virens* from alginate prill in soil and soilless media. Phytopathology 82:230–235

McLean KL, Stewart A (2000) Application strategies for control of onion white rot by fungal antagonists. N Z J Crop Hort Sci 28:115–122

Mcspadden Gardner BB, Fravel DR (2002) Biological control of plant pathogens: research, commercialization, and application in the USA. Plant Health Prog. https://doi.org/10.1049/PHP-2002-0510-01-RV

Melero-Vara JM, Prados-Ligero AM, Basallote-Ureba MJ (2000) Comparison of physical, chemical and biological methods of controlling garlic white rot. Eur J Plant Pathol 106:581–588

Ousley MA, Lynch JM, Whipps JM (1994) Potential of *Trichoderma* spp. as consistent plant growth stimulators. Biol Fertil Soils 17:85–90

Papavizas GC (1981) Survival of *Trichoderma harzianum* in soil and in pea and bean rhizosphere. Phytopathology 71:121–125

Papavizas GC (1982) Survival of *Trichoderma harzianum* in soil and in pea and bean rhizosphere. Phytopathology 72:121–125

Papavizas GC (1985) *Trichoderma* and *Gliocladium*: biology, ecology, and potential for biocontrol. Annu Rev Phytopathol 23:23–54

Papavizas GC, Dunn MT, Lewis JA, Beagle-Ristaino J (1984) Liquid fermentation technology for experimental production of biocontrol fungi. Phytopathology 74:1171–1175

Parkinson D, Taylor GS, Pearson R (1963) Studies on fungi in the root region. I. The development of fungi on young roots. Plant Soil 19:322–349

Paulitz TC (2000) Population dynamics of biocontrol agents and pathogens in soils and rhizospheres. Eur J Plant Pathol 106:401–413

Persoon CH (1794) Disposition methodica fungorum. Römer's neues. Magazin Botanische 1:81–128

Pittendrigh BR, Gaffney PJ (2001) Pesticide resistance: can we make it a renewable resource? J Theor Biol 211:365–375

Ponmurugan P, Baby UI (2003) In vitro interaction of fungal antagonists with *Phomopsis theae*, the causal agent of collar canker disease in tea. Indian J Microbiol 43:41–44

Ponmurugan P, Baby UI, Premkuma R, Radhakrishnan B (2002) Integrated control of Phomopsis Canker disease of tea. Proceedings of Placrosym XV. pp 599–602
Premkumar R, Baby UI (2005) Recommendations on the control of root and stem diseases of tea. Handbook of Tea Culture. Section 15/16
Punja ZK, Utkhede RS (2003) Using fungi and yeasts to manage vegetable crop diseases. Trends Biotechnol 21:400–407
Purseglove JW (1968) Tropical crops: dicotyledons 2. Longmans, London, pp 599–612
Rifai MA (1969) A revision of the genus Trichoderma. Mycol Pap 116:1–56
Roiger DJ, Jeffer SN, Caldwell RW (1991) Occurrence of *Trichoderma* species in apple orchard and woodland soil. Soil Biol Biochem 23:353–359
Samuels GJ (1996) *Trichoderma*: a review of biology and systematic of the genus. Mycol Res 100:923–935
Sarkar A, Das RC, Rashid MH, Ali M, Islam MS, Asaed T, Wang Q (2015) Biocontrol potentiality of isolates of Trichoderma spp. against Pestalozzia theae Saw. in tea. Acta Phytopathol Entomol Hungarica 50(2):179–186
Schirmbock M, Lorito M, Wang Hayes YL (1994) Parallel formation and synergism of hydrolytic enzymes and peptaibol antibiotics, molecular mechanisms involved in the antagonism action of *Trichoderma harzianum* against phytopathogenic fungi. Appl Environ Microbiol 60 (12):4364–4370
Schoeman MW, Webber JF, Dickinson DJ (1996) The effect of diffusible metabolites of *Trichoderma harzianum* on in vitro interactions between basidiomycete isolates at two different temperature regimes. Mycol Res 100:1454–1458
Schroth MN, Hancock JG (1982) Disease-suppressive and root colonizing bacteria. Science 216:1376–1381
Sivasithamparam K, Ghisalbert EL (1998) Secondary metabolism in *Trichoderma* and *Glioladium*. In: Kubicek CP, Harman GE (eds) Trichoderma and Glioladium, vol 1. Taylor and Francis, London, pp 139–191
Steyaert JM, Ridgway HJ, Elad Y, Stewart A (2003) Genetic basis of mycoparasitism: a mechanism of biological control by species of *Trichoderma*. N Z J Crop Hort Sci 31:281–291
Storey GK, Gardner WA (1987) Vertical movement of commercially formulated conidia of the entomogenous fungus, *Beauveraibassiana* in four Georgia soil types. Environ Entomol 16:17–181
Thoudam R, Dutta B (2012) Control of black rot disease of tea (*Camellia sinensis I* (L.) O Kuntze) with the mycoflora isolated from tea environment and phyllosphere. J Mycopathol Res 50:205–210
Tronsmo A, Dennis C (1978) Effect of temperature on antagonistic properties of *Trichoderma* species. Trans Br Mycol Soc 71:469–474
Wardle A, Parkinson D, Waller JE (1993) Interspecific competitive interaction between pairs of fungal species in natural substrates. Ecologia 94:165–172
Weindling R (1932) *Trichoderma lignorum*as a parasite of other fungi. Phytopathology 22:837–845
Weindling R (1941) Experimental consideration of the mold toxins of *Gliocladium* and *Trichoderma*. Phytopathology 31:991–1003
Wilcox WF, Harman GE, Di Pietro A (1992) Effect of gliotoxin on growth, sporulation, and zoospores motility of seven *Phytophthora* spp *in vitro*. Phytopathology 82:1121
Windham MT, Elad Y, Baker R (1986) A mechanism for increased plant growth induced by Trichoderma spp. Phytopathology 76(5):518–521

Chapter 13
Multipartite Interaction of *Trichoderma harzianum* (MTCC 5179) as Endophyte and a Growth Promoter of Black Pepper (*Piper nigrum* L.)

M. Anandaraj and P. Umadevi

Abstract Species of *Trichoderma* are used to control soilborne diseases as this versatile organism can adapt itself to varied ecological niche and utilize available nutrients. When it colonizes the plant root, entire microbial communities are altered, and the microbial community helps the plant by solubilizing the nutrients, secreting growth hormones, and preventing harmful effects of pathogens. Though all these effects were attributed to *Trichoderma*, recent metagenomic studies have revealed the complex multipartite interactions in the rhizosphere and inside host cell. *Trichoderma* species are now acknowledged as endophytes, and the mechanisms involved in suppressing the host defenses are documented.

The comparative genomics of *Trichoderma* species revealed the presence of genes in these mycoparasites to attack other pathogenic fungi and interact with plants. The advent of new-generation sequencing (NGS) techniques has opened new ways to analyze and find out the community pattern and functions in the soil rhizosphere. Most *Trichoderma* species function almost similar to the mycorrhiza. When a root is colonized by mycorrhiza, the microbial community changes around the rhizosphere, hence referred to as "mycorrhizosphere effect." Similarly, metagenomics studies have found the microbial community changes in the rhizosphere of *Trichoderma harzianum*-applied roots leading to a "trichorhizosphere effect." Multipartite interaction studies have shown the mechanisms of *Trichoderma* to maintain the plant metabolism and suppress the plant immunity for its own establishment. Genes mediating beneficial interactions with associated microbes can be manipulated to increase the efficiency of the rhizosphere by the gene editing techniques.

Keywords Multipartite interactions · Microbial community · Trichorhizosphere · Immunity suppression · Nutrient mobilization

---

M. Anandaraj (✉) · P. Umadevi
ICAR—Indian Institute of Spices Research, Kozhikode, Kerala, India

A. K. Sharma, P. Sharma (eds.), *Trichoderma*, Rhizosphere Biology,
https://doi.org/10.1007/978-981-15-3321-1_13

13.1 Introduction

Trichoderma spp. are free-living, spore-producing ascomycetous fungi and are considered as opportunistic, avirulent plant symbionts (Harman et al. 2004; Mukherjee et al. 2012). They live in soil and show close association with the rhizosphere of many crop plants and crop residues, capable of utilizing various polysaccharides as carbon source. *Trichoderma* was introduced to farmers initially as a "biofungicide" as it was known to control soilborne diseases (Mukhopadhyay 1987). This versatile organism can adapt itself to varied ecological niche and utilize available plant debris as carbon source. When it colonizes the plant root, the entire microbial communities are triggered to grow to help the plant by solubilizing the nutrients, secreting growth hormones, and preventing harmful effects of pathogens (Harman 2000; Harman et al. 2004; Mukherjee et al. 2012). Species of *Trichoderma* are known for their ubiquitous distribution and dominance over other soil fungal flora. Their nutritional requirement is minimal and has the mechanism to utilize the cell wall components such as cellulose and chitin as carbon source. They are capable of producing abundant propagules, viz., mycelium, conidia, and quiescent resting spores. To survive in varied habitats such as decaying wood, cotton fabrics, and aerial and subterranean plant parts, they are equipped with increased competitive nature and possess an arsenal of cellulolytic enzymes. The various volatile and nonvolatile metabolites produced make them efficient antagonists for many pathogens of agricultural crop plants (Papavizas 1985). The capacity to grow on inexpensive substrates makes *Trichoderma* most amenable fungi for mass production, formulations, and delivery systems. *Trichoderma harzianum* as a group have adaptation to survive varying pH conditions and temperatures ranging from 25 to 35 °C. These characters make *Trichoderma* species successful biocontrol agents.

13.2 Genomics

Omics studies have great contribution to the development of translational research in *Trichoderma*. The first member of the genus to be whole genome sequenced was *T. reesei*/*H. jecorina* (Martinez et al. 2008), which is an industrially important producer of the enzyme cellulases but not a biocontrol agent. In the recent past, the genomes of biocontrol species, viz., *T. harzianum* and *T. asperellum*, have been sequenced (Kubicek et al. 2011). The comparative genomics of these two species explained the presence of genes in these mycoparasites to attack other fungi and interact with plants (Kubicek et al. 2011). Now more than seven *Trichoderma* spp. have their whole genomes sequenced (Baroncelli et al. 2016; Srivastava et al. 2014). Genomics on *Trichoderma* spp. is regarded as a successful case of translational research, in which the genomic data are quickly applied to develop new active principles/molecules for commercial products, new types of formulations, and development of safer application protocols (Lorito et al. 2010).

13.3 Metagenomics and Its Impact in Community Analysis

Lorenz Hiltner, in the year 1904, coined the term "rhizosphere" and emphasized that plant nutrition is considerably influenced by the composition of the rhizosphere microbes. He suggested that in the rhizosphere both beneficial and pathogenic microbes which are capable of adapting to the specific root exudates are attracted. He relied on the hypothesis that the plant's resistance toward pathogenesis and the quality of plant products are very much dependent on the microbial composition of the rhizosphere (Hartmann et al. 2008). The portion/volume of the soil in which the microbe-mediated processes are under the influence of plant roots is termed as the rhizosphere. The microbial composition of rhizosphere soil differs from the bulk soil due to the selective recruitment of the organism by the root exudates that are specific to the plants. This brings major impact on plant fitness and productivity. Hence, the manipulation of rhizosphere microbes is a viable opportunity in improving the plant health and to restore the ecosystem function under biotic/abiotic stress and poor land management practices. Identification and manipulation of rhizosphere soil microbiomes were difficult before as majority of the soil microbes have not yet been isolated and molecular details underlying their functions were largely unknown. But the advent of new-generation sequencing (NGS) techniques has opened the way to analyze and find out the community pattern and functions in the soil rhizosphere. The techniques such as partial community analysis and whole community metagenome analysis are the useful technologies for the study of rhizosphere microbiome. "Metagenome" is the collective genome of microorganisms present in specific environment that provides the information on the diversity, ecology, and functionality of the microbes.

The soil metagenomics can be of whole genome sequencing (shotgun) and amplicon sequencing (16S-rRNA/gene targeted metagenomics). Due to the difference in the data obtained from each method, the bioinformatics tools in data analysis would differ. The application of whole genome sequencing (WGS) is a major development in metagenomics. Unlike the targeted 16S rRNA approach, which specifically studies a single gene or individual genomes, WGS approach brings a global view of the community, levels of phylogenetic diversity, and intraspecies polymorphism. This technique also has the potential to discover new genes, to identify the full genes and metabolic pathways, and to construct draft genome sequences.

The high-throughput sequencing methods, viz., Roche 454, Illumina HiSeq, SOLiD, Ion Torrent, and PacBio RSII, and integrated bioinformatics platforms like MEGAN and MG-RAST enable identification of community patterns, abundance, interaction network, and functional dynamics of microbes in the soil sample and thereby provide information on community ecology. MG-RAST (Meyer et al. 2008) is one of the prominent public domain platforms which support deposition and analysis of metagenomic datasets.

Though there are plenty of sequencing and bioinformatics platforms available, the attempt on soil whole genome metagenomics is very limited. The shotgun sequencing using Roche 454 and Illumina was done from forest, desert, grassland, Arctic, and mangrove sediment, and 33 metagenome datasets were generated and analyzed using integrated bioinformatics tools. Apart from microbial community profile, the indicator metagenomic biomarkers were developed for differentiating the soil communities (Xu et al. 2014). The selective microbial enrichment for biogeochemical cycling of nutrients was demonstrated toward the bioremediation at Loktak Lake, the largest freshwater lake of India. The freshwater rhizospheric microbial community metagenomes were found to be different from the other four aquatic habitats that are varying from pristine to highly polluted eutrophic habitats (Puranik et al. 2016).

In another study, *Ascomycota* was found to be the dominant phylum (76–85%) in fungi using MG-RAST in the rhizosphere metagenome of gray mangroves of the Red Sea along with several commercially used fungi (Simoes et al. 2015). The pathways on metabolism of aromatic compounds, potassium, and osmolytes were identified from the rhizosphere microbiome metagenomics of gray mangroves in the Red Sea using 454 GS FLX Titanium technology coupled with MG-RAST (Alzubaidy et al. 2016). The comparative metagenomics of the vineyards in Central Chile and the native forest soil before the establishment of the vineyard showed the bacterial abundance than fungi in both types of habitats (Castañeda and Barbosa 2016). Functional analysis revealed the genes for metabolism of potassium in vineyard soils and genes for metabolism of amino acids, fatty acids, nucleotides, and secondary metabolism in forest soils.

13.4 Colonization Patterns of Plant Roots by *Trichoderma* spp.

Though the members of the genus *Trichoderma* are saprophytes, the root colonization by certain strains is common. Some rhizosphere competent strains of *Trichoderma* also colonize the intercellular region in roots (Yedidia et al. 1999). Colonization studies using light microscopy and SEM (De Souza et al. 2008) showed that *T. stromaticum* isolates, a biocontrol agent of witches' broom diseases that were used in cacao plantations, were able to make endophytic colonization. The endophytic colonization was found better in seedlings under sterile condition than under unsterile condition. The conidia of *T. harzianum* CECT 2413 on tomato roots showed profuse adhesion of hyphae and colonization of the epidermis as well as the cortex. The confocal microscopy of *Trichoderma* transformant with GFP tag showed hyphal growth in intercellular region and the papilla-like hyphal tip formation (Chacon et al. 2007). Glandular trichomes were demonstrated to be the entry point for four *Trichoderma* species (*T. ovalisporum*, *T. hamatum*, *T. koningiopsis*, and *T. harzianum*) into the cacao stem that leads to systemic colonization of the tissue (Bailey et al. 2009). In black pepper roots, colonization of *T. harzianum* could be noticed within few hours of inoculation (Anandaraj and Umadevi 2017; Umadevi et al. 2018).

13.5 *Trichoderma*-Arbuscular Mycorrhizal Fungi Interaction

Studies have demonstrated that certain *Trichoderma* strains influence the AMF activity. The addition of *T. harzianum* significantly increased the root colonization of AMF species *G. intraradices*, *G. constrictum*, and *G. claroideum* compared to the colonization of the most efficient *G. mosseae* inoculated alone (Martinez et al. 2004) in melon plants. Camprubi et al. (1995) observed nil effect by *Trichoderma* on AMF in citrus. Antagonistic effect was observed by Martinez et al. (2004). Green et al. (1999) observed that the growth and P uptake by external mycelium of AMF were not inhibited by *Trichoderma*; rather the adverse effect of AMF on growth of *Trichoderma* was recorded. Mycoparasitic action of *Trichoderma* on AMF mycelium has been demonstrated (De Jaeger et al. 2010). The presence of *Trichoderma* was found to affect the P translocation into host plant by the *Glomus* sp. (De Jaeger et al. 2011). *T. harzianum* stimulated the spore germination and hyphal growth of *G. mosseae* by producing volatile compounds (Calvet et al. 1992). In black pepper the native AMF colonized the black pepper roots pre-inoculated with *T. harzianum* (Anandaraj and Umadevi 2017; Umadevi et al. 2017).

13.6 Plant-*Trichoderma* Interaction

In general the induced systemic resistance (ISR) of plants is initiated not only by the infecting pathogens but also by the avirulent microbes interacting with them. The colonization of the plant growth-promoting rhizobacteria (PGPR) at the plant rhizosphere results in increased state of resistance to subsequent pathogen attack which is known as systemic ISR. PGPR-mediated ISR can occur in many plant species and were found to be effective against a wide range of pathogens (Van Loon et al. 1998). In contrast to SA-dependent SAR, ISR depends primarily on ethylene (ET) and jasmonic acid (JA) which was evidenced from the extensive work with *Arabidopsis* (Ton et al. 2002). It is also that there is a cross talk and certain degree of overlap between SA and JA (Dong 2001; Glazebrook et al. 2003; Umadevi and Anandaraj 2019). In addition to PGPR, certain group of root-colonizing beneficial fungi also induces/enhances the plant resistance to pathogens. They belong to the anamorphic stages of fungi, commonly found in most soils throughout the world, viz., *Trichoderma* spp., *Fusarium* spp., binucleate *Rhizoctonia*, and *Pythium oligandrum* (Harman et al. 2004). Among these fungi, it has been proved that the biocontrol potential of *Trichoderma* spp. resides on their ability to induce both local and systemic resistance responses.

Howell et al. (2000) and Yedidia et al. (2003) have reported the occurrence of both local (roots) and systemic (leaves) accumulation of antimicrobial compounds as a result of root colonization by *Trichoderma* spp. The ET/JA signaling from plant and a mitogen-activated protein kinase signaling pathway from the fungus were identified as the major players for the *Trichoderma*-mediated ISR in *Cucumis sativus* (Shoresh et al. 2006; Viterbo et al. 2005). Cellulase and xylanase were described as proteinaceous elicitors in *Trichoderma* spp. (Calderon et al. 1993). Many reports indicate that the *Trichoderma* is capable of producing the secondary which elicits the plant defense (Hanson and Howell 2004; Harman et al. 2004). Djonovic et al. (2006) characterized Sm1, a proteinaceous nonenzymatic elicitor from *Trichoderma virens*. The purified Sm1 effectively induced the plant defense responses and systemic resistance against a foliar pathogen of *Gossypium hirsutum*, and the Sm1 response was associated with the accumulation of phenolic compounds, reactive oxygen species, and genes involved in the biosynthesis of sesquiterpenoid phytoalexins and upregulation of genes from SA and JA/ET pathway. The induction of Ca^{2+}-mediated signal perception as an early step during the interaction of soybean cells with *Trichoderma* metabolites was demonstrated by Navazio et al. (2007). The increase in total phenol, total protein, and major three fatty acid contents and the induction of jasmonic acid (JA) were found after treatment of mustard with powder form of *Trichoderma* (Dave et al. 2013). In general, the molecular mechanisms to the induced resistance mediated by *Trichoderma* have been studied in dicot plants to a certain extent (Shoresh et al. 2005; Viterbo et al. 2005; Djonovic et al. 2006) compared to monocots where it is very limited. But many lines of similarity in various aspects of induced resistance that occur between monocots and dicots had been observed (Dong 2004; Chem et al. 2005).

Root colonization (Shoresh and Harman 2008a, b) of *Trichoderma harzianum* {Rifai strain 22 (T22)} induced changes in the shoot proteome of 7-day-old *Zea mays* seedlings, though T22 is infected only on the roots. Two-dimensional gel electrophoresis followed by matrix-assisted laser desorption/ionization tandem time-of-flight mass spectrometry and nano-spray ion trap tandem mass spectrometry was used to identify the differentially expressed proteins in response to colonization of maize plants with T22 and identified that 91 spots out of 114 were upregulated and 30 out of 50 were downregulated proteins. The analysis showed the involvement of a large number of upregulated proteins from carbohydrate metabolism, photosynthesis, and up- and downregulated proteins from stress and defense responses for root colonization. The stress responsive proteins were of glutathione *S*-transferase (upregulated spot), peroxidase (upregulated spot), glutathione-dependent formaldehyde dehydrogenase (upregulated spot), and heat shock proteins (two up- and three downregulated spots). The nucleotide-binding site (NBS)/Leu-rich repeat (LRR) resistance protein-like proteins, such as phenylalanine ammonia lyase (PAL) protein, were found to be upregulated.

The MAPK is activated by inoculation of the roots with *T. asperellum* (Shoresh et al. 2006) in cucumber. The activation of this gene was found to be important for the plant's *Trichoderma*-conferred defense against bacterial pathogen, and silencing of the MAPK had completely eliminated this protection (TIPK). The level of expression of TIPK was higher and more prolonged when plants were inoculated with *Trichoderma* prior to pathogen challenge. Comparing monocots and dicots, the research attempt and the knowledge on perennial crop-*Trichoderma interaction* are rudimentary. The analysis of grapevine leaves in response to T39 treatment and *P. viticola* inoculation using the iTRAQ platform (Palmieri et al. 2012) showed the proteins involved in stress response with increased abundance in T39-treated plants. Increase in abundance of two receptors (a leucine-rich repeat receptor-like protein kinase and a receptor-like protein kinase precursor), resistance RPP protein and two TMV resistance proteins N, GTPase-activating protein, proteins of abscisic acid and auxin signaling and metabolism, thioredoxin and a ferredoxin-thioredoxin reductase, proteins involved in signal transduction (a *Pseudomonas syringae*-resistant protein, a Rabgap/TBC domain-containing protein, and two disease resistance proteins), and redox balance (aglutaredoxin, a copper/zinc superoxide dismutase, and a glutathione reductase) were induced in T39-treated plants after *P. viticola* inoculation.

13.7 *Trichoderma* in Black Pepper System

Anandaraj and Sarma (2003) demonstrated that the probiotic application of *T. harzianum* (MTCC 5179) enhanced growth of black pepper with increased number of nodes and has been incorporated in the integrated disease management in black pepper (Anandaraj 2000; Rajan et al. 2002). Sibi (2013) reported the positive influence of *T. harzianum* (MTCC 5179) on both fresh and dry root and shoot weight of black pepper. Treatment with *T. harzianum* (MTCC 5179) alone reflected in better growth promotion and disease suppression when comparing the effect of a consortium with plant growth-promoting rhizobacteria alone or in combination with *T. harzianum* (MTCC 5179). The effect of probiotic application of *T. harzianum* on growth of black pepper was at par with *T. harzianum* along with AMF. The treatments with AMF alone and in combination with *Pseudomonas* sp. failed to enhance the growth (Sibi 2013). Most *Trichoderma* species function almost similar to the mycorrhiza. When a root is colonized by mycorrhiza, the microbial community changes around the rhizosphere, hence known as "mycorrhizosphere effect" (Linderman 1998). Working with spices, especially black pepper, the author with his coworkers using modern tools like metagenomics have found the microbial community changes in the rhizosphere of *Trichoderma harzianum*-applied roots (Anandaraj and Umadevi 2017) and explained the "trichorhizosphere effect" (Umadevi et al. 2018). The mechanism of enhanced plant growth is not only by the introduced *Trichoderma* but by the combined activity of native soil organisms including mycorrhiza inducted by the inoculated *Trichoderma*.

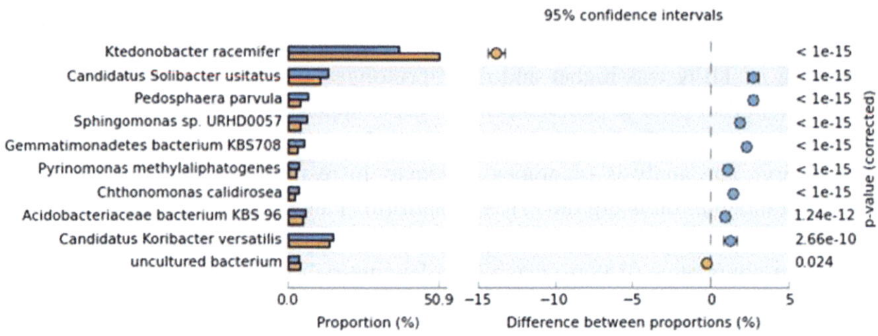

Fig. 13.1 Species-level extended error bar chart for top ten bacteria using STAMP tool analysis. *T. harzianum* treatment is denoted as blue bar and control by orange bar. The differential abundance between the samples was calculated with G-test (w/Yates') + Fisher's test for two-sample analysis. (Source: Umadevi et al. 2018)

The microbial community dynamics at the black pepper rhizosphere upon inoculation of *Trichoderma harzianum* was studied using Illumina HiSeq whole genome metagenomics of rhizosphere soil DNA. The results showed that *T. harzianum* influenced rhizosphere microbial community with selective abundance of beneficial fungi and bacteria which resulted in growth promotion in black pepper. The increased bacterial chemotaxis, virulence, disease, and defense and ion metabolism were the impact at functional level. The study on whole genome metagenomics of rhizosphere soil from control and treatment revealed that the population abundance of bacteria, archaea, and eukaryote were found to be less in *Trichoderma*-treated than in control. In general, it showed selective abundance of certain bacteria, viz., *Candidatus Koribacter versatilis* and *Acidobacteriaceae bacterium*—out of top ten bacterial species, viz., *Acidobacteriaceae bacterium* KBS 96, *Ktedonobacter racemifer*, *Pedosphaera parvula*, *Candidatus Koribacter versatilis*, *Candidatus Solibacter usitatus*, *Sphingomonas* sp. URHD0057, *Pyrinomonas methylaliphatogenes*, *Gemmatimonadetes bacterium*, *Chthonomonas calidirosea*, and uncultured bacteria from the STAMP analysis on population abundance. This selective abundance of these beneficial bacteria in the *T. harzianum*-treated metagenome was the major impact for the growth promotion activity. The active breakdown of complex organic compounds by these recruited organisms created microclimate for the colonization of other microbes in the roots and subsequent interaction (Fig. 13.1).

The metagenome of *Trichoderma*-treated rhizosphere showed abundant reads of the beneficial fungi, viz., *Fusarium oxysporum*, *Talaromyces* sp., *Pestalotiopsis* sp., and *T. harzianum*, with expected beneficial activities when compared to the control. The high abundance of *Talaromyces* species and *Fusarium* in the treatment indicated the significance on their population abundance imparted by the addition of *T. harzianum* toward the fitness of black pepper (Fig. 13.2).

SEM analysis revealed swellings resembling appressoria on the glandular trichome tips in cacao. The anatomical studies in black pepper during the colonization

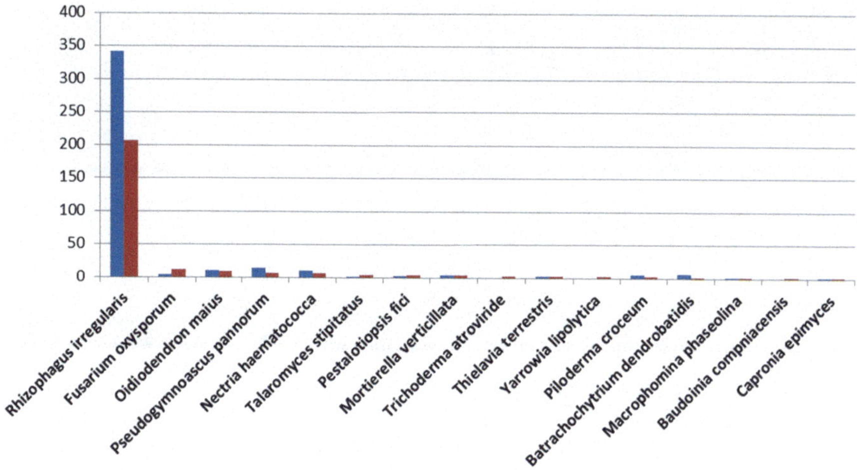

Fig. 13.2 Fungal species population abundance (blue bar, control; maroon bar, *Trichoderma* treated). (Source: Umadevi et al. 2018)

of *Trichoderma* showed the nature of colonization. The abundant bacterial chemotaxis in treatment than in control showed that the probiotic addition of *T. harzianum* in black pepper enabled the active interaction of the selectively recruited bacterial community in the roots. The root sample of the control showed weak adherence of bacteria on the surface, while the treated roots showed abundant adherence on the surface upon imaging with SEM (Umadevi et al. 2017). The external colonization of *T. harzianum* occurred on the roots of the in vitro-derived pepper plants as surface adherence at 12 h of co-cultivation; it continued as the time increases (at 24 h and at 48 h) with profusely growing mycelia bearing chlamydospores. The intercellular colonization was found at 24 h incubation. *T. harzianum* established endophytic colonization inside the cell at 48 h with luxurious intracellular mycelia. Root clearings of plants inoculated with *T. harzianum* showed fungal mycelium in the intercellular spaces at 24 h, and hyphal tips as dark blue granules inside the cell and intracellular chlamydospores were observed at 48 h. Analysis by SEM showed intact cell structures in control roots, but *T. harzianum*-inoculated samples taken at 24 and 48 h showed an indication of the interaction with the root cells and hyphal growth on the surface at 24 h. At 12 h, the enlargement of hyphal tip as papillae showed its interaction for intracellular colonization. Massive colonization was observed on root surface at 48 h. We observed the appearance of papillae on root surfaces at 24 h of inoculation, suggesting its quick interaction with the black pepper root system (Fig. 13.3).

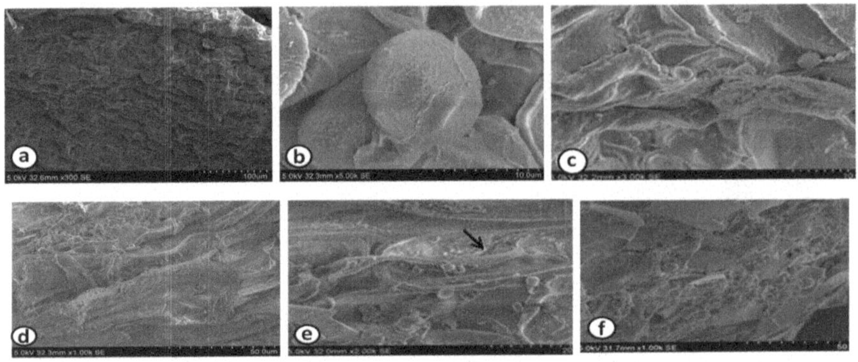

Fig. 13.3 SEM micrographs taken after co-cultivation of *T. harzianum* 5179 with in vitro-derived black pepper plant roots; samples taken at 12, 24, and 48 h of inoculation fixed with methanol. (**a**) Control (without *T. harzianum*); (**b**) spore on cell surface; (**c**) surface attachment at 12 h; (**d**) hyphal growth at 24 h; (**e**) enlargement of hyphal tip (arrow—papilla); and (**f**) dense colonization at 48 h. (Source: Umadevi 2018)

13.7.1 Nature of Colonization

In the treatment with *T. harzianum*, the endophytic colonization of *T. harzianum* accompanied by AMF was luxuriant. The maturation zone of the black pepper roots showed moniliod hyphae along with structurally differentiated AMF mycelium with prominent septation. Microsclerotia were also observed inside the cells, and AMF colonization was found as arum type. The region of maturation showed vesicles with different shapes. *T. harzianum* mycelium and conidia along with the AMF mycelium and vesicles were noticed—suggesting co-colonization inside the roots of black pepper. Since the experiment was set up in field soil, the observation that the roots inoculated with *Trichoderma* had abundant VAM colonization as indicated by the presence of both vesicles and arbuscules shows that the native mycorrhiza colonize pepper roots without any hindrance by inoculated *Trichoderma harzianum* MTCC 1579 suggesting the inoculated *Trichoderma* has facilitated mycorrhizal colonization, whereas in the control soil, though there was AMF colonization, it was sparse. The native beneficial microbes like mycorrhiza had equal opportunity of colonizing black pepper roots; the fact that in *Trichoderma harzianum*-inoculated roots, there are more AMF present suggests its active role in helping mycorrhizal colonization (Fig. 13.4).

13.7.2 **Trichoderma** *and AMF on Growth Promotion in Black Pepper*

Reports on growth promotion effect of VAM on black pepper show that *G. fasciculatum* incorporation as infective propagules (cultured on Rhodes grass)

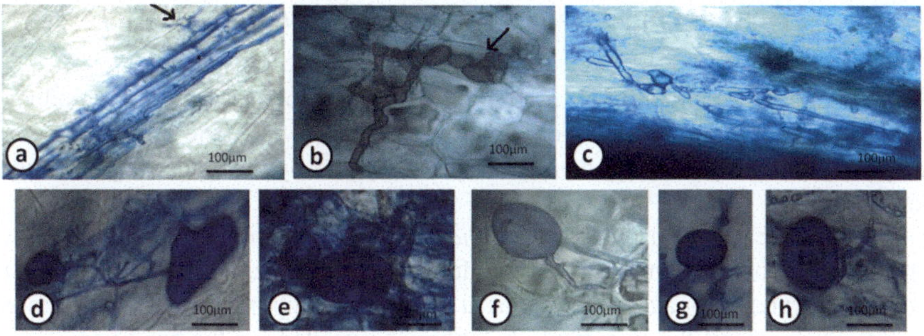

Fig. 13.4 Maturation zone of black pepper roots showing different AMF structures during long-term colonization (4 months) of *T. harzianum*. (**a**) Tree-like intracellular arbuscules (arrow showing arum type growth of AMF mycelium); (**b**) intracellular microsclerotia; (**c**) moniliod hyphae of AMF; and (**d–h**) different shapes of AMF vesicles at the maturation zone of roots. (Source: Umadevi 2018)

increased the rooting of black pepper at nursery condition (Anandaraj and Sarma 1994). The authors showed more than 80% colonization of *G. fasciculatum* in black pepper roots than control. Detailed study on the effect of biocontrol agent on hardening of tissue-cultured black pepper with VAM demonstrated that the treatments wherein *G. fasciculatum* and *T. harzianum* were inoculated showed higher root and shoot mass compared to control (Sibi 2013). The growth promotion effect on plantlets inoculated with species of *Pseudomonas*, *Rhizobium*, and *Trichoderma* was checked in the presence or absence of VAM, and it was found that the overall growth with VAM was higher than that without VAM (Sibi 2013). Compared to the effect of *Trichoderma* in black pepper, the treatments inoculated with *Pseudomonas* or *Rhizobium* alone with or without VAM recorded low profile on growth promotion, which further indicates the synergistic effect of this fungus. The comparison of *T. harzianum* treatments with or without VAM recorded higher growth promotion in *T. harzianum* (alone) without VAM, which suggests the principal action of *T. harzianum* in helping the native VAM fungi present in the soil to colonize the black pepper plants. This was evident from the results with microscopy, wherein we showed the endophytic colonization (100% colonization frequency) of AMF along with *T. harzianum* mycelium in *T. harzianum*-inoculated soil, compared to control (Umadevi et al. 2017). This study demonstrated the localization and endophytic colonization of *T. harzianum* MTCC 5179 in black pepper. The enhanced AMF root colonization by the inoculation of *Trichoderma* in black pepper indicates that *T. harzianum* acts as helper organism in the root ecosystem of black pepper for colonizing AMF on the plant. Moreover, the native microbes that were selectively recruited by black pepper rhizosphere system by the *T. harzianum* would have helped to mobilize nutrients and enhanced the growth. Further, detailed studies on *T. harzianum*, AMF, and native microflora with the host in a multipartite interaction would help in developing targeted biocontrol strategy to overcome soilborne pathogens.

13.7.3 Trichoderma-*Induced Systemic Resistance (T-ISR)*

Even though the *Trichoderma* is used in black pepper production system for a long time, the molecular aspects on induced systemic resistance mediated by *Trichoderma* against *Phytophthora* were demonstrated by the author's group very recently.

The bipartite study (black pepper-*Trichoderma harzianum*) using proteomics has profiled the proteome under expression during 24, 48, 72, and 96 hai for the first time. The expression pattern of proteins at all time of sampling showed the absence of SA involvement, the important phytohormone needed for the synthesis of PR proteins. In all samples collected after inoculation, the PR proteins were found to be downregulated to a greater degree so SA involvement can be ruled out. The other phytohormone, ethylene, was also not having a role as the important proteins (methionine synthase, alanine:glyoxylate aminotransferase) needed for ethylene biosynthesis are downregulated. The upregulated and new protein induced group had PTI related receptors and R genes which sends the clue that the pattern recognition step itself the *Trichoderma* might be shifting the plant activities favorable toward its own establishment by suppressing the plant immunity. Some important downregulated proteins were found to be the marker proteins for the suppression of immunity in black pepper by *T. harzianum*. At 24 hai the SOD, CAT, Glyceraldehyde-3 PO_4; at 48 hai Rap guanine nucleotide exchange factor 2, Histone 2A; at 72 hai, the germin-like protein and subtilisin marker proteins for plant defense; at 96 hai, the SAR marker ascorbate peroxidase and RAB GTPase which is immunity associated programmed cell death were the identified marker proteins. Thus, *T. harzianum* inoculation on the roots alters the host physiology so much at the farther end of leaves. These results add information on how the fungi could maintain the plant metabolism and suppress the plant immunity for its own establishment.

The label-free proteomics on *T. harzianum* mediated early in black pepper showed many marker proteins for ISR. The defense response in leaves in *T. harzianum*-primed plant roots which are also infected with *P. capsici* was identified. The interaction patterns were studied as black pepper × *T. harzianum* (two-way), black pepper × *P. capsici* (two-way), and black pepper × *T. harzianum* × *P. capsici* (three-way). The proteins induced only in the three-way interaction were identified as *Trichoderma*-induced resistance proteins (T-ISR). Many T-ISR proteins from the ROS-scavenging proteins group, the defense-related proteins group, and other important proteins group served as marker proteins of the tripartite interaction to prove that the *Trichoderma* induces systemic resistance in black pepper against *Phytophthora capsici*. The major players of ISR were found. The strong ROS-related activity suggested that ROS-mediated signaling was a major component of the defense, with other defense-related proteins in T-ISR. The ethylene synthesis and lignin synthesis and the isoflavanoid pathway were enhanced in T-ISR. *Trichoderma*-induced systemic resistance is thus a major component of the defense readiness of black pepper against *Phytophthora capsici* and sets in early (72–96 hai) (Umadevi and Anandaraj 2019).

13.8 Conclusions

Modifying growing conditions through cultural practices is an important crop improvement platform in agriculture, to get the desired performance/plant fitness. The manipulation of rhizosphere microbiome known as "rhizosphere engineering" is an important component for reshaping the soil rhizosphere microbial community. Reshaping the rhizosphere microflora by the introduction of beneficial microorganisms will safeguard the host plant against pathogen infections. The cost-effective sequencing technologies and use of integrated bioinformatics databases make the metagenomics of non-model crop-associated microbiomes a possible and efficient research method. This is evident from the work shown here in the black pepper system with *Trichoderma*. The soil rhizosphere metagenomics would reveal the major microbiome in the microclimate, and that could be used to engineer the rhizosphere by probiotic application of major microbes either alone or as consortia. Genes mediating beneficial interactions with associated microbes can be manipulated in the elite germplasm to increase the efficiency of the rhizosphere by the gene editing techniques. Study on the quantitative trait loci (QTL) governing the responses of plant to beneficial rhizosphere microbes will help in marker-assisted breeding toward developing the varieties with improved fitness against the biotic and abiotic stresses.

References

Alzubaidy H, Essack M, Malas TB, Bokhari A, Motwalli O, Kamanu FK, Jamhor SA, Mokhtar NA, Antunes A, Simoes MF, Alam I, Bougouffa S, Lafi FF, Bajic VB, Archer JAC (2016) Rhizosphere microbiome metagenomics of gray mangroves (*Avicennia marina*) in the Red Sea. Gene 576:626–636

Anandaraj M (2000) Diseases of black pepper. In: Ravindran PN (ed) Black pepper *Piper nigrum*, Medicinal and aromatic plants – industrial profiles. Harwood Academic Publishers, Amsterdam, pp 239–267

Anandaraj M, Sarma YR (1994) Effect of vesicular arbuscular mycorrhizae on rooting of black pepper (*Piper nigrum* L.). J Spices Arom Crops 3:39–42

Anandaraj M, Sarma YR (2003) The potential of PGPRs in disease management of Spice crop. In: 6th International PGPR workshop, 5–10 Oct 2003, Calicut, India

Anandaraj M, Umadevi P (2017) Multipartite interaction of introduced biocontrol agents in the rhizosphere. In: Fifth National conference on biological control: integrating recent advances in pest and disease management, ICAR-National Bureau of Agricultural Insect Resources, Bengaluru, India

Bailey BA, Strem MD, Wood D (2009) *Trichoderma* species form endophytic associations within *Theobroma cacao* trichomes. Mycol Res 113:1365–1376

Baroncelli R, Zapparata A, Piaggeschi G, Sarrocco S, Vannacci G (2016) Draft whole-genome sequence of *Trichoderma gamsii* T6085, a promising biocontrol agent of *Fusarium* head blight on wheat. Genome Announc 4(1):e01747–e01715

Calderon AA, Zapata JM, Munoz R, Pedreno MA, Barcelo AR (1993) Resveratrol production as a part of the hypersensitive-like response of grapevine cells to an elicitor from *Trichoderma viride*. New Phytol 124:455–463

Calvet C, Estaun V, Camprubi A (1992) Germination, early mycelia growth and infectivity of a vesicular-arbuscular mycorrhizal fungus in organic substrates. Symbiosis 14:405–411

Camprubi A, Calvet C, Estaun V (1995) Growth enhancement of *Citrus reshni* after inoculation with *Glomus intraradices* and *Trichoderma aureoviridae* and associated effects on microbial population and enzyme activity in potting mixer. Plant Soil 173:233–238

Castañeda LE, Barbosa O (2016) Metagenomic analysis exploring taxonomic and functional diversity of soil microbial communities in Chilean vineyards and surrounding native forests. Peer J 5:e3098. https://doi.org/10.7717/peerj.3098

Chacon MR, Rodriguez-Galan O, Benitez T, D'Sousa S, Benitez T, Sousa S, Rey M, Llobell A, Delgado-Jarana J (2007) Microscopic and transcriptome analyses of early colonization of tomato roots by *Trichoderma harzianum*. Int Microbiol 10:19–27

Chem M, Fitzgerald HA, Canlas PE, Navarre DA, Ronald PC (2005) Over expression of a rice NPR1 homolog leads to constitutive activation of defense response and hypersensitivity to light. Mol Plant Microbe Interact 18:511–520

Dave N, Prajapati K, Patel A, Patel Z, Nandini D, Bariya H (2013) *Trichoderma harzianum* elicits defense response in *Brassica juncea* plantlets. Int Res J Biol Sci 2(11):1–10

De Jaeger N, Declerck S, de la Providencia IV (2010) Mycoparasitism of arbuscular mycorrhizal fungi: a pathway for the entry of saprotrophic fungi into roots. FEMS Microbiol Ecol 73:312–322

De Jaeger N, de la Providencia I, Dupre de Boulois H, Declerck S (2011) *Trichoderma harzianum* might impact phosphorus transport by arbuscular mycorrhizal fungi. FEMS Microbiol Ecol 77:558–567

De Souza JT, Bailey BA, Pomella AWV, Erbe EF, Murphy CA, Bae H, Hebbar KP (2008) Colonization of cacao seedlings by *Trichoderma stromaticum* a mycoparasite of the witches' broom pathogen, and its influence on plant growth and resistance. Biol Control 46:36–45

Djonovic S, Pozo MJ, Dangott LJ, Howell CR, Kenerley CM (2006) Sm1, a proteinaceous elicitor secreted by the biocontrol fungus *Trichoderma virens* induces plant defense responses and systemic resistance. Mol Plant Microbe Interact 19:838–853

Dong X (2001) Genetic dissection of systemic acquired resistance. Curr Opin Plant Biol 4 (4):309–314

Dong X (2004) NPR1, all things considered. Curr Opin Plant Biol 7(5):547–552

Glazebrook J, Chen WJ, Estes B, Chang HS, Nawrath C, Metraux JP, Zhu T, Katagiri F (2003) Topology of the network integrating salicylate and jasmonate signal transduction derived from global expression phenotyping. Plant J 34:217–228

Green H, Larsen J, Olsson PA, Jensen DF, Jakobsen I (1999) Suppression of the biocontrol agent *Trichoderma harzianum* by mycelium of the arbuscular mycorrhizal fungus *Glomus intraradices* in root-free soil. Appl Environ Microbiol 65:1428–1413

Hanson LE, Howell CR (2004) Elicitors of plant defense responses from biocontrol strains of *Trichoderma virens*. Phytopathology 94:171–176

Harman GE (2000) Myths and dogmas of biocontrol: changes in perception derived from research on *Trichoderma harzianum* T-22. Plant Dis 84:377–393

Harman GE, Howell CR, Viterbo A, Chet I, Lorito M (2004) Trichoderma species-opportunistic, avirulent plant symbionts. Nat Rev Microbiol 2:43–56

Hartmann A, Rotballer M, Schimid M (2008) Loenz Hiltner, a pioneer in rhizosphere microbial ecology and soil bacteriology research. Plant Soil 312:7–14

Howell CR, Hanson LE, Stipanovic RD, Puckhaber LS (2000) Induction of terpenoid synthesis in cotton roots and control of *Rhizoctonia solani* by seed treatment with *Trichoderma virens*. Phytopathology 90:248–252

Kubicek CP, Herrera-Estrella A, Seidl-Seiboth V et al (2011) Comparative genome sequence analysis underscores mycoparasitism as the ancestral life style of *Trichoderma*. Genome Biol 12:R40. https://doi.org/10.1186/gb-2011-12-4-r40

Linderman RG (1998) Mycorrhizal interactions with the rhizosphere microflora: the mycorrhizosphere effect. Phytopathology 78:366–371

Lorito M, Woo SL, Harman GE, Monte E (2010) Translational research on *Trichoderma*: from 'Omics to the field. Annu Rev Phytopathol 48:395–417

Martinez A, Obertello M, Pardo A, Ocampo JA, Godeas A (2004) Interactions between *Trichoderma pseudokoningii* strains and the arbuscular mycorrhizal fungi *Glomus mosseae* and *Gigaspora rosea*. Mycorrhiza 14:79–84

Martinez D, Berka RM, Henrissat B, Saloheimo M, Arvas M, Baker SE (2008) Genome sequencing and analysis of the biomass-degrading fungus *Trichoderma reesei* (syn. *Hypocrea jecorina*). Nat Biotechnol 26(5):553–560

Meyer F, Parman D, D'Souza M et al (2008) The metagenomics RAST server - a public resource for the automatic phylogenetic and functional analysis of metagenomes. BMC Bioinformatics 9. https://doi.org/10.1186/1471-2105-9-386

Mukherjee PK, Horwitz BA, Kenerley CM (2012) Secondary metabolism in Trichoderma–a genomic perspective. Microbiology 158:35–45

Mukhopadhyay AN (1987) Biological control of soil-borne plant pathogens by *Trichoderma* spp. Indian J Mycol Plant Pathol 17:1–10

Navazio L, Baldan B, Moscatiello R, Zuppini A, Woo SL, Mariani P, Lorito M (2007) Calcium-mediated perception and defense responses activated in plant cells by metabolite mixtures secreted by the biocontrol fungus *Trichoderma atroviride*. BMC Plant Biol 7:41. https://doi.org/10.1186/1471-2229-7-41

Palmieri MC, Perazzolli M, Matafora V, Moretto M, Bachi A, Pertot I (2012) Proteomic analysis of grapevine resistance induced by *Trichoderma harzianum* T39 reveals specific defence pathways activated against downy mildew. J Exp Bot 63:6237–6251

Papavizas GC (1985) *Trichoderma and Gliocladium:* biology, ecology and potential for biocontrol. Annu Rev Phytopathol 23:23–54

Puranik S, Pal RR, More RP, Purohit HJ (2016) Metagenomic approach to characterize soil microbial diversity of Phumdi at Loktak Lake. Water Sci Technol 74(9):2075–2086

Rajan PP, Sarma YR, Anandaraj M (2002) Management of foot rot disease of black pepper with *Trichoderma* spp. Indian Phytopathol 55:34–38

Shoresh M, Harman GE (2008a) The relationship between increased growth and resistance induced in plants by root colonizing microbes. Plant Signal Behav 3:737–739

Shoresh M, Harman GE (2008b) The molecular basis of shoot responses of maize seedlings to *Trichoderma harzianum* T22 inoculation of the root: a proteomic approach. Plant Physiol 147:2147–2163

Shoresh M, Yedidia I, Chet I (2005) Involvement of jasmonic acid/ethylene signaling pathway in the systemic resistance induced in cucumber by *Trichoderma asperellum* T203. Phytopathology 95:76–84

Shoresh M, Gal-On A, Leibman D, Chet I (2006) Characterization of a mitogen-activated protein kinase gene from cucumber required for *Trichoderma*-conferred plant resistance. Plant Physiol 142:1169–1179

Sibi MC (2013) Development of biocontrol consortia for tissue cultured black pepper (*Piper nigrum* L.,) plants. Dissertation, Mangalore University

Simoes MF, Antunes A, Cristiane A, Ottoni CA, Amini MS, Alam I, Alzubaidy H, Mokhtar N, John AC, Archer JAC, Bajic VB (2015) Soil and rhizosphere associated fungi in gray mangroves (*Avicennia marina*) from the Red Sea - a metagenomic approach. Genom Proteom Bioinf 13:310–320

Srivastava M, Shahid M, Pandey S, Singh A, Kumar V, Gupta S, Maurya M (2014) Trichoderma genome to genomics: a review. Data Mining Genom Proteom 5:3. https://doi.org/10.4172/2153-0602.1000162

Ton J, Van Pelt JA, Van Loon LC, Pieterse CMJ (2002) Differential effectiveness of salicylate-dependent and jasmonate/ethylene-dependent induced resistance in *Arabidopsis*. Mol Plant Microbe Interact 15:27–34

Umadevi P (2018) Microbial community dynamics and modulation of defense responses in black pepper by *Trichoderma harzianum*. Dissertation, University of Calicut

Umadevi P, Anandaraj M (2019) Proteomics analysis of the tripartite interaction between black pepper, *Trichoderma harzianum* and *Phytophthora capsici* provides insights into induced systemic resistance mediated by *Trichoderma* spp. Eur J Plant Pathol 154:1–14

Umadevi P, Anandaraj M, Benjamin S (2017) Endophytic interactions of *Trichoderma harzianum* in a tropical perennial rhizo – ecosystem. Res J Biotechnol 12(3):22–30

Umadevi P, Anandaraj M, Srivastav V, Benjamin S (2018) *Trichoderma harzianum* MTCC 5179 impacts the population and functional dynamics of microbial community in the rhizosphere of black pepper (*Piper nigrum* L.). Braz J Microbiol 49(3):463–470

Van Loon LC, Bakker PA, Pieterse CM (1998) Systemic resistance induced by rhizosphere bacteria. Annu Rev Phytopathol 36:453–483

Viterbo A, Harel M, Horwitz BA, Che I, Mukherjee PK (2005) *Trichoderma* mitogen-activated protein kinase signaling is involved in induction of plant systemic resistance. Appl Environ Microbiol 71:6241–6246

Xu Z, Hansen MA, Hansen LH, Jacquiod S, Sorensen SJ (2014) Bioinformatic approaches reveal metagenomic characterization of soil microbial community. PLoS One 9:e93445. https://doi.org/10.1371/journal.pone.0093445

Yedidia I, Benhamou N, Chet I (1999) Induction of defense responses in cucumber plants (*Cucumis sativus* L.) by the biocontrol agent *Trichoderma harzianum*. Appl Environ Microbiol 65:1061–1070

Yedidia I, Shoresh M, Kerem Z, Ben-hamou N, Kapulnik Y, Chet I (2003) Concomitant induction of systemic resistance to *Pseudomonas syringae* pv. *Lachrymans* in cucumber by *Trichoderma asperellum* (T-203) and accumulation of phytoalexins. Appl Environ Microbiol 69:7343–7353

Chapter 14
Trichoderma spp. in Consortium and Their Rhizospheric Interactions

Pratibha Sharma, P. P. Jambhulkar, M. Raja, S. K. Sain, and S. Javeria

Abstract Microbial consortium of efficient strains for biological control helps in improving microbial efficacy, reliability, and consistency under diverse soil and environmental conditions. Different genera in consortial formulation occupy varied niches in the root zone, thereby restricting competition among them. In this review we have discussed critically the need of consortial application of bioagents, their role in rhizosphere colonization, and disease suppression. Potential of consortial application of bioagents against fungal, bacterial, and nematodes can be utilized to fullest extent by selecting most potential strains and not by arbitrary use of consortia. It was also discussed here that out of all the consortial formulation applied globally, very few show synergistic effect in consortia. A numerical analysis model to assess synergism is discussed in this chapter. Thus, it is essential to select strains of microbes very carefully to develop efficient consortia to suppress crop diseases.

Keywords Biocontrol agents · Microbial consortia · Rhizosphere · *Trichoderma* spp.

P. Sharma (✉)
Division of Plant Pathology, ICAR-Indian Agricultural Research Institute, New Delhi, India

P. P. Jambhulkar
Department of Plant Pathology, Rani Lakshmi Bai Central Agricultural University, Jhansi, Uttar Pradesh, India

M. Raja
Department of Plant Pathology, SKN Agricultural University, Jobner, Rajasthan, India

S. K. Sain
ICAR-Central Institute for Cotton-Research Regional Station, Sirsa, Haryana, India

S. Javeria
Division of Seed Science and Technology, ICAR-Indian Agricultural Research Institute, New Delhi, India

A. K. Sharma, P. Sharma (eds.), *Trichoderma*, Rhizosphere Biology,
https://doi.org/10.1007/978-981-15-3321-1_14

14.1 Introduction

Microorganisms live in community and have both beneficial and non-beneficial effects in nature. Beneficial microorganisms play dynamic role in promoting plant and soil health in plant protection. Therefore, microbial consortia are replacing the use of single microbe in crops. The additive and synergistic effect of microorganisms in consortium for growth promotion and triggering of defense systems and metabolic pathways against pathogen have been reported. The study on host response to such microbial effects is in continuation. The *Trichoderma* spp. are the most common soil fungi, which have been isolated and used as biopesticide, biofertilizer, and soil amendments. This review is the compilation of different case studies on the use of *Trichoderma* spp. in microbial consortium against the major pathogen.

In this review, the outcomes of *Trichoderma*-based consortia which have shown their effective results in protecting the crop plants from various diseases have been presented. We have enlisted the combined use of *Trichoderma* spp. against various diseases along with the mode of application in a tabular form. The rationale for effective microbial consortia development and their capacity of bringing changes in host plants inducing improved systemic resistance and related biochemical pathways has also been discussed in this chapter. An explanation on the evaluation or assessment of microbial consortium on host and plant pathogens through mathematical models has been added.

14.2 Status of *Trichoderma* Research in India

To meet the ever-increasing demands for food and fiber, the green revolution has intensified agriculture worldwide; however, at the same time, this has damaged the natural ecosystems through indiscriminate uses of chemicals and fertilizers which are degrading the environment. Worldwide plant diseases act as major constraints in crop production and contributing in the economic crop losses before and after crop harvest. There is an urgent desire to control diseases for ensuring a steady and continuous supply of blameless produce for the continually increasing population in the world. The injudicious use of chemicals has caused negative impact on environment and resulted in increased chemical pesticides resistance in many harmful insect pests. Several ecofriendly alternative approaches have also been considered to minimize the negative effect on environment, and thus the use of biocontrol agents (BCAs) becomes one of the most important ones. *Trichoderma* spp. have been used widely as BCAs for management of several fungal plant pathogens causing diseases in crop plants through mycoparasitism.

In India, an overview of literature clearly indicated a spurt in their exploration and exploitation for the management of diseases of crops as indicated by a significant increase in the number of publications (Sharma et al. 2014). A critical appraisal further revealed that the use of *Trichoderma* spp. has been remarkably successful against several soilborne diseases of crop plants for which the sources of resistant have not yet been identified. Additionally, the biocontrol potential of *Trichoderma* spp. has been disclosed against numerous foliar diseases through seed or soil application indicating their effectiveness in inducing systemic disease resistance in the crop plants. Fascinatingly, *Trichoderma* spp. have been exploited as a plant growth-promoting agent similar to rhizobacteria. Therefore, *Trichoderma* spp. have been preferred as a potential integrated disease management (IDM) component. Integration of fertilizers or herbicides with BCAs to control plant diseases (Mukherjee et al. 1997) and in vitro and in vivo interactions of *Rhizobium-Trichoderma* were studied (Jayaraj and Ramabadran 1999). However, a variation in the biocontrol, pesticide compatibility, field performance, and rhizospheric ability of the isolates of *Trichoderma* species was also recorded. In screening different pesticides against *Trichoderma* species, low-toxic pesticides captan and mancozeb; medium-risk benomyl, iprodione, thiram, and pendimethalin; and highly toxic carbendazim and chlorpyrifos were identified (Sharma and Dureja 2004; Sharma et al. 2016). This review illustrates the usage of *Trichoderma* on different hosts against pathogens of several ranges in the country for the past two decades highlighting the significant achievements made by several groups.

14.3 Microbial Consortium: What Is It?

A microbial consortium can be defined as group of two or more microbial agents living together symbiotically. Consortiums can be endosymbiotic or ectosymbiotic. This microbial consortium concept was first introduced by Johannes Reinke in 1872. Later, the term proceeded to the term symbiosis that was introduced into biology. In natural environments, several microorganisms live in communities, and some of them have mutual beneficial interactions and provide beneficial effects to plants. When these microbes in a consortium are introduced to soil, they interact with a host plant, and as a result the natural soil conditions are partially mimicked. Therefore, the current investigation trend has focused toward investigating the part of microbial consortia in promoting plant health against several pathogens as well as growth. This paradigm shifts from the earlier investigations which involved single-microbe uses. Recently, research has indicated that the plant growth promotion and triggering defense responses in host plants during pathogen ingress mechanisms through microbial consortia. This is revealed that enhancement of the defense signaling cascades by microbial consortia enhances transcriptional activation of several metabolic pathways. This situation does not occur when a microbial consortium is used. However, along with time progress, an understanding had been developed about microbial consortium-induced plant defense responses. Further creation of evidence on host's responses against pathogen attacking the presence of microbial consortia at purposeful level is ongoing.

The outcomes of microbial consortia are used till now to protect crop plants against numerous plant pathogens. The research findings are explained for disease reduction when a microbial consortium was used, comparison of use of single microbe and in small microbial consortium, possible limitations in achieving desired product from the introduced consortia, and the justification for development of an operative microbial consortia which is able to induce enhanced systemic resistance. Lastly, some potential biotechnological applications to withstand the effect of microbe-induced defense responses in host plants are suggested (Sarma et al. 2015).

14.3.1 Trichoderma *and Rhizosphere Interactions*

Biological control in nature is through suppressive soils, monoculture decline, organic amendments and compost, physical and chemical treatments, and role of fauna in natural biological control. The genus *Trichoderma* spp., the most popular opportunistic and saprophytic soilborne fungi, is a multifunctional bioagent being utilized as biopesticide, biofertilizer, and soil amendments. *Trichoderma* spp. work through introduction/application at host and pathogen level through its direct and indirect modes of action, which have been observed between pathogen and biological control agent in the host (crop) rhizosphere. The competition for nutrients or space; antibiosis, the secretion of metabolites harmful to pathogens; and parasitism are the direct modes of action. The indirect modes of actions are induced resistance or perhaps plant growth promotion. Though, simultaneously or sequentially one antagonist may exhibit several modes of action.

The most commonly isolated *Trichoderma* species are essential as they can be applied in the fields in the form of biopesticide and biofertilizer and thus help in soil amendments. These species become opportunistic endophytes upon inoculation in seed or soil treatment (Leon et al. 2018). This improves plant nutrition and plant growth, and induced systemic resistance is achieved by various strains of *Trichoderma* species, i.e., *Trichoderma asperellum* strain T34 against *Pseudomonas syringae* pv. *lachrymans* a foliar pathogen of cucumber and (Pieterse et al. 2014; Segarra et al. 2007) tomato plant and against *P. syringae* pv. tomato and *Plectosphaerella cucumerina* in *Arabidopsis* (Segarra et al. 2009). The T34 strain showed improvement in tomato plant growth, nutrient uptake with induction of system resistance against *Botrytis cinerea* (Fernández et al. 2014). Induction of defense enzyme by *T. harzianum* strain Th3 against *F. oxysporum* f. sp. *lycopersici* causing wilt of tomato (Raja et al. 2016).

There are some information regarding the interaction of *Trichoderma* spp. and root exudates. Cucumber modified by *T. harzianum* T-E5 has root exudates, and the germination of soil pathogen *Fusarium oxysporum* f. sp. *cucumerium* gets reduced (Zhang et al. 2013). Also, the promotion of growth of numerous isolates of *T. harzianum*, *T. viride*, and *T. roseum* is reported due to root exudates of Bengal gram. There is very little knowledge related to the response of beneficial microorganisms against pathogen and how plants encourage this microorganism, importance of root exudates in defensive action of plants against foliar pathogens, and basically the reactions that are occurring in beneficial microbes.

14.3.2 Microbial Consortium and Disease Suppression

Plant growth-promoting microorganism (PGPM) or plant growth-promoting rhizobacteria (PGPR) live in association with strains of non-PGPR in soil or within the rhizosphere in different combinations and ecosystems (Vacheron et al. 2013). In order to get high-quality agricultural outcomes, the BCAs of various microorganisms species showing plant growth-promoting activities are mixed due to their community- or group-based living strategies.

The use of individual microorganisms for management of plant diseases related to plants is not more efficient in comparison to use of consortium of microorganisms that include different efficient strains. Further there is great importance of microorganism in a consortium as they help in increasing efficiency and dependability and magnify the consistent character of microorganism under soil condition and surroundings (Stockwell et al. 2011). Microorganism strains which are compatible with other microorganism and do not have any suppressive or inhibitory effect on strains of other microorganism when they are co-cultured in a common media in consortium can then promote growth of plants or may suppress diseases (Jain et al. 2012; Singh et al. 2013a, b). There are various positive factors of using variety of microorganism in a consortium as they may increase the efficiency of BCAs due to the nature of occupying different surroundings in the rhizosphere and thus are able to decrease competition among them. Further, enhancement of suppressing types of diseases is also regulated by a variety of biological controlled mechanisms that are offered by an individual microorganism component. Earlier various microbes such as *Trichoderma*, *Glomus*, *Rhizobium*, *Bacillus*, *Pseudomonas*, etc. were major microbes those were used in order to develop a consortium of microorganism (Duffy et al. 1996; Jetiyanon 2007; Srivastava et al. 2010; Kannan and Sureendar 2009).

The mixed usage of *Glomus*, *Trichoderma*, and fluorescent *Pseudomonas* helps in separation of *Fusarium* wilt which occurs in fields of tomato; basically combined usage is better as it is 50% more beneficial over single usage of *Glomus* (Srivastava et al. 2010). According to Duffy et al. (1996), *T. koningii* have the greater ability to suppress all pathogens of wheat when used with some *P. fluorescens* strains instead of using *T. koningii* alone. The same scenario can be seen with the strains of *Bacillus* strains IN937a and IN937b which enhanced the peroxidases and superoxide dismutase activities by 25–50% when used in combination in pepper and tomato against seed and soilborne pathogens (Jetiyanon 2007). The ability of microorganism to form phytohormones, their antagonistic activity against pathogens, and the ability to fix atmospheric nitrogen and phosphorus solubilizing action in soil all play a role in rational selection of these microorganisms to be used in consortia. After this selection, microorganisms which are efficient enough to enhance nutrient content (Sahni et al. 2008; Harman 2011) and are able to induce systemic resistance reselected and included in consortia of microbes (Bakker et al. 2007). The microorganism can perform their roles separately. Moreover, crop plants get benefit of availabilities of higher amount of P and N concentrations for better uptake that enhances plant yield and health only when a consortium of microorganism having compatibility are used together. Combinations with antagonist bioagents have the ability to produce and thus provide a plant that is disease-free. Nowadays, worldwide there is a need to minimize the usage of chemicals and fertilizers in agriculture; as this is of great concern, usage of microorganism in a consortium will play a great role in overcoming this problem. Usage of phosphorus solubilizing bacteria (PSB), *Trichoderma* and *Rhizobium*, together as consortia has showed increased uptake of N and P as well as biological control ability for soilborne pathogens in chickpea (Rudresh et al. 2005). However, the destiny of consortia microorganisms was not assessed in the soil rhizosphere according to most of the studies conducted earlier and so maximum emphasis must be provided to this aspect of research for improved utilization of microorganisms consortium in order to increase their efficiency.

14.4 Microbial Consortium and Plant Defense

Bacteria and fungi are microorganisms which are present in a vast variety and vary from pathogenic and non-beneficial to beneficial type for plants as they are in contact with them. Hence, in order to get benefit from microorganism, plants must identify between different microorganisms and start their respective mechanism to keep the pathogenic or unwanted species of microbes away and attract the beneficial or wanted species of microorganism. In order to defend themselves against pathogenic microbes, plants quickly activate their defensive method effectively by arresting the pathogen's growth. The efficient and fast defense mechanism of plants plays a domineering part in recognition and perception of pathogens (de Leon and Montesano 2013).

Various techniques have been established by plants in order to manage several abiotic and biotic stresses during evaluation. The defense mechanism of plants is triggered when pathogens affecting plants come in contact with them and the mechanism of defense occurs in a same fashion as it is induced by the beneficial microbes (Shoresh et al. 2010). Moreover, activation level and duration of response during defense mechanism are high when beneficial microbes present in rhizosphere which are used for biopriming. In recent time, our group has developed various efficient and successful microbial consortia against various pathogens which are soilborne in nature (*Sclerotium rolfsii* and *Sclerotinia sclerotiorum*). Microorganism consortium including effective potential BCAs such as *P. aeruginosa*, *B. subtilis*, and *T. harzianum* against *S. sclerotiorum* in pea (Jain et al. 2012) and *P. aeruginosa*, *Mesorhizobium* spp., and *T. harzianum* against *S. rolfsii* in chickpea (Singh et al. 2013a) has demonstrated host-mediated defense mechanism that includes reduction of plant disease. The pea plant treated with three microorganism consortia, including strains of *T. harzianum* TNHU27, *P. aeruginosa* PJHU15, and *B. subtilis* BHHU100, has found to increase the parameters of defense in treated plants by 1.4–2.3-fold when collected with *S. sclerotiorum*, while individual application of microorganism on pea plants increased only 1.1–1.7-fold having some parameter when compared to untreated counter parts. After the pathogen exposure, there is an accumulation of total phenolics, pathogenesis-related (PR) protein, and proline due to enzymatic activity that is of antioxidant enzyme and phenylpropanoid pathway which was activated by consortia of microorganism. Similarly, plants which were treated with consortia of these three microorganisms showed increased deposits of phenols by 1.4–4.6-fold compared to the only one or dual microorganism consortia (Jain et al. 2012).

According to Marimuthu et al. (2013), co-inoculation of *P. fluorescens* and *Azospirillum* had synergistic effect in nature on the cotton yields and provided protection against the root rot pathogen *Rhizoctonia bataticola*. However, with the plants maturity, the pollution of microorganism in rhizosphere decreases nearly by 50%. The application of consortia of two plant growth-promoting regulators, that is, PGPR strains, *Chryseobacterium balustinum* Aur 9 and *P. fluorescens* Aur 6, helped in blast disease management in rice (Lucas et al. 2009). In comparison to the single usage of microorganism, the severity of disease decreased to 50% when combination of strains was used. Similarly, different pathways of signaling, activated with increased effect of high ISR by MYB72 (that is independent of SAR) due to co-inoculation of *Pseudomonas* species Ps14 and *T. harzianum* Tr6 in cucumber against *Botrytis cinerea*. The disease index is decreased more than 40% by co-inoculation treatment with Ps14 and Tr6, in comparison with single-microbe inoculation (Alizadeh et al. 2013).

For decreasing the multiplication of root-knot nematode *Meloidogyne javanica*, usage of consortia of *Rhizobium* species and *Paecilomyces lilacinus* KIA was very efficient in the case of chickpea roots. As rhizobium has the capability to produce phytoalexin and antibiotics, *P. lilacinus* KIA has the ability to parasitize female nematodes and their eggs. The mixed effect of microorganism combination therefore contributes to a better plant growth and enhanced growth inhibition of nematodes (Siddiqui and Akhtar 2009). This consortium reduced the number of galls in root system by more than 20% with its single usage. Moreover, bad effects can be observed in management of diseases when BCAs were randomly combined (Xu et al. 2010a). In comparison with usage of single BCAs, results obtained by evaluation of three biocontrol products which are commercialized, taken two at a time for mixing, showed that incident of disease in strawberry due to *B. cinerea* was greater.

According to Bora et al. (2004) and Elliott et al. (2009), adverse effects of microorganism combination were also reported. Decreased ability of BCAs may be due to their antagonistic activities and incompatibility against each other (Xu et al. 2010b). Thus, in order to enhance the biocontrol efficiency, it is of high priority to achieve combination of microorganisms which are compatible, and hence it is important to address this issue through future research while developing a consortium of microorganism. We have clearly indicated that most of microorganisms were arbitrarily selected to perform consortia-related experiments in order to get results (Table 14.1). The method to select microorganisms for making an effective consortium is based on the ability to promote growth of the plants by microbe individually, with the thought of probability that consortia will not act against each other and thus will be effective against the pathogens which are tested. There were only few cases where consortia of microbes were developed according to their compatibility with other microbes in a consortia or consortium that was made on the basis of prior information that include effectiveness against the targeted pathogen.

Therefore, to reduce failures of experiments, it is of great importance to include compatibility testing, so that experiments related to consortium and only compatible microorganisms must be considered to reduce the chances of failure.

14.5 *Trichoderma* spp. Consortium in Disease Management

Various application methods of different species of *Trichoderma* have been described by various workers against many soilborne and foliar pathogens. A change came when the biocontrol scientists started application of *Trichoderma* spp. in combination of various microbes.

14.5.1 Nematode Management

Reddy et al. (1996) observed that application of *T. harzianum* at 2 or 4 g in combination with neem (*Azadirachta indica)*, karanj (*Pongamia pinnata*), and castor oil (*Ricinus communis)* cakes (at 20/40 g) can manage citrus nematode, *Tylenchulus semipenetrans*. The combination increased the growth of the crop and also reduced the population of *T. semipenetrans*. Meyer et al. (2001) reported that *Burkholderia cepacia* and *T. virens*, alone and in combinations, decreased biocontrol effectiveness compared to applications of individual microbes against *M. incognita* on bell pepper. Oyekanmi et al. (2007) observed the combination of different plant beneficial microbes for *M. incognita* management in soybean. Arbuscular mycorrhiza fungus (AMF) *G. mosseae* (200 spores/plant), the modulating bacterium *Bradyrhizobium japonicum* (10^6 cells/plant), and the nematode antagonistic fungus *T. pseudokoningii* (6.8×10^7 spores/plant) were tested in pot and field conditions. The consortium was found to be beneficial in reducing or *M. incognita* in soybean.

14.5.2 Fungal and Bacterial Disease Management

Elad et al. (1998) tested the combination of *T. harzianum* T39 and *Ampelomyces quisqualis* AQ10 on cucumber. Powdery mildew severity was reduced by *T. harzianum* T39 (TRICHODEX) up to 97%, but there was a decline in efficiency with the increase in epidemic. The performance of the bioformulation was not good on old leaves. *A. quisqualis* (AQ10) also controlled powdery mildew up to 98% and also effective on older leaves. The co-application of *T. harzianum* T39 and

Table 14.1 Consortial application *Trichoderma* spp. with other bacterial and fungal biocontrol agents

S. No.	Microbes used in consortium	Pathogens/diseases	Host	Mode of application	References
01	*Trichoderma harzianum*, neem, karanj, and castor oil cakes	*Tylenchulus semipenetrans*/citrus nematode	Citrus	Soil application	Reddy et al. (1996)
02	*Trichoderma harzianum* T39 and *Ampelomyces quisqualis* AQ10	*Sphaerotheca fusca* and *Botrytis cinerea*/powdery mildew and gray mold	Cucumber	Foliar application	Elad et al. (1998)
03	*Trichoderma virens* and *Burkholderia cepacia*	*Meloidogyne incognita*/root-knot nematode	Bell pepper	Soil application	Meyer et al. (2001)
04	*T. viride* pq 1 and *Pseudomonas fluorescens*, GB 4, GB 8, GB 10, and GB 27	*Sclerotium rolfsii*/stem rot	Groundnut	Seed coating	Manjula et al. (2004)
05	*Trichoderma viride* Pers. ex SF Gray and *Pseudomonas fluorescens* Migula	*Rhizoctonia solani* Kühn/sheath blight	Rice	Seed and soil application	Mathivanan et al. (2005)
06	*Trichoderma harzianum* and *T. viride*	*Sclerotinia sclerotiorum* and *Fusarium culmorum*/white rot and *Fusarium* rot	Cucumber	Soil application	Wojtkowisk-Gebarowska and Pietr (2006)
07	*Trichoderma harzianum* and *Streptomyces rochei*	*Phytophthora capsici*/root rot	Pepper	Soil application	Ezziyyani et al. (2007)
08	*Trichoderma harzianum* (ITCC-4572) and *Rhizobium*	*Sclerotium rolfsii*/stem rot	Groundnut	Soil application	Ganesan et al. (2007)
09	*Trichoderma pseudokoningii*, *Glomus mosseae*, and *Bradyrhizobium japonicum*	*Meloidogyne* spp./root-knot nematode	Soybean	Soil application	Oyekanmi et al. (2007)
10	*Trichoderma viride* (strains Tv1 and Tv13), *Pseudomonas fluorescens* (Pf1 and Py15), and *Bacillus subtilis* (Bs16)	*Macrophomina phaseolina*/dry root rot	Green gram	Seed treatment	Thilagavathi et al. (2007)
11	*Trichoderma harzianum*, *Acremonium strictum*, *Aspergillus niger*, *A. terreus*, and *Paecilomyces lilacinus*	*M. incognita*/root-knot nematode	Tomato		Goswami et al. (2008)

(continued)

Table 14.1 (continued)

S. No.	Microbes used in consortium	Pathogens/diseases	Host	Mode of application	References
12	*Bacillus subtilis* CA32 and *Trichoderma harzianum* RU01	Damping-off/*Rhizoctonia solani*	Brinjal and chili	Seed and soil application	Abeysinghe (2009)
13	*Trichoderma harzianum* and chitosan	*Leptographium procerum* and *Sphaeropsis sapinea*	Radiata pine		Chittenden and Singh (2009)
14	*T. viride* T1 and T17, *Trichoderma* spp., and *P. fluorescens*	*Fusarium oxysporum* f. sp. *vanillae*, *P. meadii*, and *Colletotrichum vanillae*	Vanilla	Soil and foliar application	Radjacommare et al. (2010)
15	Fluorescent *Pseudomonas*, *Trichoderma harzianum*, and *Glomus intraradices*	*Fusarium oxysporum* f. sp. *lycopersici*/tomato wilt	Tomato	Seed biopriming	Srivastava et al. (2010)
16	*Trichoderma viride*, *Bacillus subtilis*, and *Bacillus cereus*	*Rhizoctonia solani*/black scurf disease of potato	Potato	Seed treatment	Somani and Arora (2010)
17	*Trichoderma atroviride* strain 3A, *T. atroviride* strain 6, *Trichoderma harzianum*, strain SY, *Trichoderma pseudokoningii*, an unidentified strain *Trichoderma* sp. strain 2F, *T. harzianum* strain kmd, *B. subtilis* B69, *B. subtilis* B77, and *B. subtilis* B81	*Rhizoctonia solani*/damping-off	Dry bean	Seed treatment	Yobo et al. (2011)
18	*T. viride* (Tv1), *P. fluorescens* (Pf1), and *B. subtilis* (BS16)	*Lasiodiplodia theobromae*/collar and root rot disease	Physic nut	Soil application	Latha et al. (2011)
19	*T. harzianum* and *Glomus* (*G. constrictum*, *G. mosseae*, *G. claroideum, and G. intraradices*)	Fusarium wilt	Melon	Soil application	Martínez-Medina et al. (2011)
20	*Trichoderma harzianum* BHU-51, *Trichoderma harzianum* BHU-105	*Sclerotinia sclerotiorum*/rot of brinjal	Brinjal	Seed treatment	Singh and Singh (2012)

21	*Trichoderma harzianum* Tr6, and *Pseudomonas* sp. Ps14	*Fusarium oxysporum* f. sp. *radiciscucumerinum* and *Botrytis cinerea*	Cucumber and *Arabidopsis thaliana*	Soil inoculation	Alizadeh et al. (2013)
22	*Trichoderma viride* Pers. ex Gray (IARIP-1: MTCC No. 369), *T. virens* (Miller, Giddens, and Foster) Arx (IARI P-3: MTCC No. 5370), and *T. harzianum* Rifai (IARI P-4 MTCC No. 5371)	Fusarium wilt/*Fusarium oxysporum* f. sp. *ciceris*	Chickpea	Soil application and seed dressing	Dubey et al. (2013)
23	*Pseudomonas aeruginosa* PJHU15, *Trichoderma harzianum* TNHU27, and *Bacillus subtilis* BHHU100	Stem rot or white mold/*Sclerotinia sclerotiorum*	Pea	Seed coating	Jain et al. (2013)
24	*Bacillus firmus* E65, *Serratia marcescens* E31, *Pseudomonas aeruginosa* C32b, *Bacillus cereus* II.14	Rice blast disease/*Pyricularia oryzae*	Rice	Seed treatment	Suryadi et al. (2013)
25	*Trichoderma harzianum* (THU0816), *Pseudomonas aeruginosa* (PHU094), and *Mesorhizobium* sp. (RL091)	*Sclerotium rolfsii*	Chickpea	Seed and soil application	Singh et al. (2013a)
26	*Trichoderma harzianum* and *Pseudomonas* sp.	Collar rot/*Sclerotium rolfsii*	Tomato	Seed treatment	Singh et al. (2013b)
27	*T. atroviride* CH 101 (Ta CH 101) and CH 304.1 (Ta CH 304.1) and *T. virens* CH 303 (Tv CH 303), *B. subtilis* PCL1608 (Bs PCL1608), *P. chlororaphis* PCL1601 (Pc CL1601) and PCL1606 (Pc CL1606), and *P. pseudoalcaligenes* AVO110 (Pp VO110)	White root rot/*Rosellinia necatrix*	Avocado	Soil application	Ruano-Rosa et al. (2014)
28	*Trichoderma harzianum* (BHU-51), *Trichoderma harzianum* (BHU-105)	*Rhizoctonia solani*/damping-off	Tomato	Seedlings treatment	Singh et al. (2014a)
29	*Trichoderma harzianum* OTPB3 and *Bacillus subtilis* OTPB1 and *Pseudomonas putida* OPf1	Late blight/*Phytophthora infestans*	Tomato	Seed application	Kumar et al. (2015)

(continued)

Table 14.1 (continued)

S. No.	Microbes used in consortium	Pathogens/diseases	Host	Mode of application	References
30	*T. harzianum* T22, *T. atroviride* P1, *T. reesei* T34, *P. syringae*, *Pseudomonas fluorescens*, *P. corrugate*	Fruit decay/*Penicillium digitatum*	Citrus	Fruit dipping	Panebianco et al. (2015)
31	*Pseudomonas aeruginosa* (MBAA1), *Bacillus cereus* (MBAA2), and *Bacillus amyloliquefaciens* (MBAA3) and one fungi *Trichoderma citrinoviride* (MBAAT)	*Macrophomina phaseolina* and *Sclerotinia sclerotiorum*	Soybean	Seed coating	Thakkar and Saraf (2015)
32	*T. asperellum* (Prr2) + *Trichoderma* sp. (NRCB3)	Fusarium wilt/*Fusarium oxysporum* f. sp. cubense	Banana cv. Grand Naine	Soil application	Thangavelu and Gopi (2015)
33	*Trichoderma viride* (ITCC 2211), *Azotobacter chroococcum* (W5), *Pseudomonas fluorescens* (PF1), and *Bacillus subtilis* (RP24)	*Macrophomina phaseolina*/ashy stem (charcoal rot)	Cotton	Soil application	Triveni et al. (2015)
34	*Trichoderma viride* and *Pseudomonas fluorescens* CHA0	*Meloidogyne javanica*/root-knot nematodes	Tomato	Soil application	Saeedizadeh (2016)
35	*Trichoderma* spp.(TCMS 36 and SBIT 32 strain) copper and chitosan	*Phytophthora infestans*/late blight	Potato	Tuber treatment	Sajeesh et al. (2016)
36	*Trichoderma asperellum* T42 and *Pseudomonas fluorescens* OKC	*Fusarium oxysporum* f. sp. *ciceris*/wilt	Chickpea	Seed biopriming	Kumar et al. (2017)
37	*Trichoderma harzianum* (Th), *T. virens* (Tr), and *T. viride* (Tv) and consortium of AMF (Myc, mixture of *Funneliformis mosseae*, *Glomus cerebriforme*, *Rhizophagus irregularis*)	*Fusarium udum* Butler/*Fusarium* wilt	Pigeon pea		Dehariya et al. (2015)
38	*Trichoderma harzianum* strain Th3 and *Pseudomonas fluorescens* strain RRb11	*Magnaporthe oryzae* and *Xanthomonas oryzae* pv. *Oryzae*/blast and bacterial leaf blight	Rice	Seed and soil application	Jambhulkar et al. (2018)

39	*Trichoderma asperellum* (Tri), chitosan (Cs661), copper sulfate pentahydrate ($CuSO_4 \cdot 5H_2O$)	*Phytophthora infestans* (Mont.) de Bary/ late blight	Tomato	Seed treatment and foliar spray	Rautela et al. (2018)
40	*Trichoderma harzianum* KX523899 (T5) and *B. subtilis* MF352017 (Bs1)	*F. oxysporum* Schlecht end. Fr. f. sp. *ciceris*/*Fusarium* wilt	Chickpea	Seed treatment	Zaim et al. (2018)
41	*Trichoderma harzianum* (Th), *T. asperellum* (Ta), *T. virens* (Tvs1), *T. virens* (Tvs2), and *T. virens* (Tvs3)	*Pythium aphanidermatum* (Edson) Fitzpatrick/damping-off and root rot	Tomato	Seed and soil application	Elshahawy and El-Mohamedy (2019
Effect of *Trichoderma* consortium on host					
42	*Trichoderma harzianum*, *Glomus mosseae*, and *Bacillus coagulans*	Enhanced growth and nutrition	Weeping fig	Soil application	Srinath et al. (2003)
43	*T. harzianum*, PDBCTH 10; *T. viride*, PDBCTV 32, *T. virens*, PDBCTVs 12 and *Bacillus megaterium* subsp. *phospaticum* strain PB	Growth, nutrient uptake and yield of chickpea	Chickpea	Seed coating	Rudresh et al. (2005)
44	*Trichoderma harzianum* and *Piriformospora indica*	Increased root colonization and growth	Black pepper	Soil application	Anith et al. (2011)
45	*Trichoderma viride*, *Glomus intraradices*, *Acaulospora laevis*, and *P. fluorescens* (MTCC No. 103)	Growth and yield	Broccoli	Soil application	Tanwar et al. (2013)
46	*Trichoderma asperellum* T42, *Pseudomonas fluorescens* OKC and *Rhizobium* sp. RH4	Improves crop growth	Chickpea and common beans	Seed biopriming	Yadav et al. (2013)
47	*Trichoderma harzianum* TNHU27, *Bacillus subtilis* BHHU100, and *Pseudomonas aeruginosa* PJHU15	Modulation of nutritional and antioxidant potential	Pea	Seed coating	Jain et al. (2014)
48	*Trichoderma harzianum* (THU 0816), fluorescent *Pseudomonas* (PHU 094), and *Mesorhizobium* sp. (RL 091)	Enhance antioxidants	Chickpea	Seed dipping	Singh et al. (2014b)

(continued)

Table 14.1 (continued)

S. No.	Microbes used in consortium	Pathogens/diseases	Host	Mode of application	References
49	*Trichoderma atroviride* strain MUCL45632 and *Glomus intraradices* BEG72	Promote growth, yield and nutrient uptake of vegetable crops	Lettuce, tomato, melon, pepper, zucchini squash	Soil application	Colla et al. (2015)
50	*Trichoderma viride* and *Azotobacter chroococcum*	Modulating rhizosphere colonization, plant growth, soil nutrient availability, and plant defense enzyme activity	Chickpea	Seed and soil treatment	Velmourougane et al. (2017)
51	*Trichoderma virens* (GV41) or *Trichoderma harzianum* (T22)	Modulate rhizosphere microbial populations and improve N uptake efficiency, yield, and nutritional quality	Lettuce	Seedlings root dip	Fiorentino et al. (2018)

A. quisqualis AQ10 used against powdery mildew and fruit gray mold infection revealed that there was no powdery mildew reduction, but gray mold was observed to be controlled. The improvement in combined formulations was due to the use of additive oil (ADDQ). Spray of *T. harzianum* T39 was not effective as compared to soil treatment which reveals that there was no mycoparasitic activity, but induced resistance mechanism worked very well in cucumber against powdery mildew. Manjula et al. (2004) have reported that application of *P. fluorescens* GB 10 and *T. viride* pq 1 was comparatively better than Thiram® for protecting groundnut seedlings from stem rot disease. Hence, talc formulations of BCAs either alone or in combination can be recommended for the management of sheath blight of rice (Mathivanan et al. 2005). Wojtkowisk-Gebarowska and Pietr (2006) evaluated iprodione-resistant isolates of *Trichoderma* spp. for root colonization and growth stimulation of cucumber when applied alone and combined with fungicides. *T. harzianum* and *T. viride* iprodione-resistant mutants were observed to be competent, which can colonize and reduce the *S. sclerotiorum* and *F. culmorum* on colonies. Growth of cucumber crop enhanced to 41–100%. Compatible *T. harzianum* and *Streptomyces rochei* were tested against *Phytophthora capsici* causing root rot in pepper. Application of these combined formulation at 3.5×10^8 spores/ml of *T. harzianum* and 1.0×10^9 CFU/ml of *S. rochei* reduced 75% of the pathogen population in soil (Ezziyyani et al. 2007). Ganesan et al. (2007) reported combined application of *Rhizobium* and *T. harzianum* (ITCC—4572) significantly controls the stem rot (collar rot) caused by *Sclerotium rolfsi* and increases the plant growth by increasing higher shoot and root length and dry weight of the plants. Increase in peroxidase, polyphenol oxidase, and phenyl alanine ammonia lyase (defense-related enzymes) was significantly hgher in green gram due to application of *P. fluorescens* Pf1 + *T. viride* Tv1 stalk formulation. These combinations were effective in reducing dry root rot pathogen *M. phaseolina* incidence under both glasshouse and filed conditions followed by *P. fluorescens* Pf1+ *T. viride* Tv13 and *P. fluorescens* Py15+ *T. viride* Tv1 (Thilagavathi et al. 2007). Goswami et al. (2008) performed combined application of two fungal bioagents *T. harzianum* and *Acremonium strictum* for the management of *M. incognita* in tomato (vv. Pusa Ruby). The both bioagents were found to effective against both egg parasitic or opportunistic and toxic properties. The talc-based formulation (2×10^8 cfu/g) is found effective in diseases management under field conditions. Abeysinghe (2009) used two compatible biological control agents, *B. subtilis* CA32 and *T. harzianum* RU01, against *R. solani*, to control damping-off in brinjal and chili. Chittenden and Singh (2009) evaluated the control of two sap stain fungi of *Pinus radiata*, *Leptographium procerum*, and *Sphaeropsis sapinea* by using chitosan and *T. harzianum* in vitro. Chitosan is reported to restrict or delay the spore germination, but the combined use of chitosan and *T. harzianum* inhibited spore germination and colony development of these sap stain fungi. Radjacommare et al. (2010) evaluated the consortia of *P. fluorescens* strains and *Trichoderma* isolates against phytopathogenic fungi: *F. oxysporum* f. sp. *vanillae*, *P. meadii*, and *C. vanillae* in vanilla under field conditions. The combination of *P. fluorescens* strain Pf1 and *Trichoderma* isolates tricho1, and application of chitin (soil and foliar) shows significant disease suppression. The combination of *P. fluorescens* (P16), *T. harzianum* (T35), and arbuscular mycorrhizal fungus (AMF) was against *F. oxysporum* f. sp. *lycopersici* causing wilt in tomato (var. Pant Tomato-3). Seed biopriming with *P. fluorescens* (P16) and *T. harzianum* (T35) increased seed germination by 22–48% and also reduced days required for seed germination by 2.0–2.5 days. The combined use of *P. fluorescens*, *T. harzianum*, and AMF provided reduction in incidence and severity of disease by 74% and 67% in pots and field, respectively, and increased 20% yield. Adding cow dung compost with these combinations further reduced 81% and 74% disease incidence in pots and field, respectively, and also increased 33% yield (Srivastava et al. 2010). Bioformulation consortium of *T. viride* at 1×10^7 CFU/mg and *B. subtilis* strain B5 and *B. cereus* strain B4 at 1×10^{10} CFU/mg was used to manage *R. solani* causing black scurf disease in potato. The combination of these microbes shows significant reduction in incidence and severity index of black scurf of potato from 72% to 42% and severity index from 1.6 units to 0.6 units (Somani and Arora 2010). The rhizome rot of ginger caused by *Fusarium solani* and *Pythium myriotylum* was reported to be managed by the combined use of *T. harzianum*, *T. aureoviride* and *G. virens* (Gupta et al. 2010). The combination of BCAs, *B. subtilis* B69 and *B. subtilis* B81, *T. atroviride* strain 3A, *T. atroviride* strain 6, *T. harzianum* strain SY, and *T. harzianum* strain kmd enhanced root area and growth of the cucumber against *R. solani* damping-off in Yobo et al. (2011). The application of *Trichoderma* and *Bacillus* combinations was better than alone, which shows that there is a potential in using mixtures of these BCAs for improving plant growth and disease control.

The Physic nut (*Jatropha curcas*) tree suffers from collar rot and root rot pathogen (*Lasiodiplodia theobromae*). Different strains of *T. viride* (Tv1), *P. fluorescens* (Pf1), and *B. subtilis* (Bs16) were reported to be the best in exhibiting maximum inhibition against the physic nut collar and root rot pathogen *L. theobromae*. In another in vitro study, application of neem cake, farmyard manure (FYM), and zinc sulfate was most effective in reducing the pathogen growth. The microbial strains were compatible to neem cake and zinc sulfate. Compared to other treatments, the combined use of Pf1 + Tv1 + Bs16 + neem cake + FYM + zinc sulfate was superior in reducing the collar and root rot disease incidence in glasshouse and field experiments also (Latha et al. 2011). The maximum reduction in *S. sclerotiorum* causing rot of brinjal was also recorded in the seedlings treated with consortium of *T. harzianum* isolates BHU-51 and BHU-105 (Singh and Singh 2012). Singh et al. (2013a) reported that the seed and seedlings of tomato crop treated with a combination of *T. harzianum* and *Pseudomonas* sp. are very effective against *S. rolfsii* rot. This reduced disease by 53.23% and significantly increased the yield from 254.07 to 45.29 tons/ha. Shanmugam et al. (2013) considered effective strain mixtures of *T. harzianum* isolate S17TH and a *Burkholderia cepacia* isolate TEPFSungal for rhizome rot control in polyhouse and field conditions. The combination of *T. harzianum* (BHU-51) and *T. harzianum* (BHU-105) shows maximum disease reduction and increased mineral content in treated plants. The consortium showed significant increase in shoot length, chlorophyll content, and yield (Singh et al. 2014a). Combined application of *T. asperellum* (Prr2) and Trichoderma sp. (NRCB3) was effective for the management *F. oxysporum f. sp. cubense* in banana.

Multiple biocontrol potentials of these *Trichoderma* isolates were tested against Foc and also promote the growth of the host. In greenhouse conditions, the *Trichoderma* sp. NRCB3 + endophytic *T. asperellum* Prr2 reduced *Fusarium* wilt disease by 100% and increased plant growth up to 250% (Thangavelu and Gopi 2015).

Thakkar and Saraf (2015) studied the microbial consortium of *P. aeruginosa* (MBAA1), *B. cereus* (MBAA2), *B. amyloliquefaciens* (MBAA3), and *T. citrinoviride* (MBAAT) for the management of *M. phaseolina* and *S. sclerotiorum* causing root rot and stem rot in *G. max*. Seed treatment with *P. aeruginosa* (MBAA1) and *T. citrinoviride* (MBAAT) showed the considerable increase in seed germination, vigor index (1863.2), and chlorophyll contents in host. Panebianco et al. (2015) demonstrated the efficacy of compatible BCAs *P. syringae* 48SR2 and *T. atroviride* P1 for integrated management of green mold decay of lemon and orange fruits. The conidia dip treatments alone significantly reduced disease incidence and severity. Kumar et al. (2015) developed consortium of *B. subtilis* OTPB1 and *T. harzianum* OTPB3 by using seed coating formulation for the management of late blight in tomato. The microbial consortium formulation-treated seedlings significantly reduced lesion size caused by *Alternaria solani* and *P. infestans* on intact tomato leaves by 78.7% and 82.1%. Microbial consortium formulation in vegetable crops (brinjal, beans, bitter gourd, bottle gourd, cabbage, chili, carrot, cauliflower, pumpkin, ridged gourd), fruit crop (papaya in plastic trays in glasshouse), tuber crops (potato), ginger and turmeric increased plant growth under greenhouse conditions and plant growth and yield under field conditions. Triveni et al. (2015) developed biofilms by using *T. viride* with *Azotobacter chroococcum*, *P. fluorescens*, and *B. subtilis* for the management of *M. phaseolina* in cotton. Sajeesh et al. (2016) reported copper, chitosan, and *Trichoderma* for management of potato late blight disease in field under hill condition. These combinations revealed that triple combination of chitosan (500 ppm), copper (500 ppm.), and copper-tolerant *Trichoderma* (1%) was highly effective in managing late blight disease of potato. Saeedizadeh (2016) applied the combination of *T. viride* and *P. fluorescens* CHA0 against *M. javanica* in tomato and found enhanced biocontrol efficacy of these BCAs as compared to the individual treatments of fungus and bacterium. In cases of the cultivars Mobile and Bonny, treatment with *T. viride* exhibits potential control in root-knot nematode.

Kumar et al. (2017) used the combination of *P. fluorescens* OKC and *T. asperellum* T42 for the management of *F. oxysporum* f. sp. *ciceris* (Foc) in chickpea. Seed biopriming with these bioagents had beneficial roles in enhancing plant secondary metabolites such as phenolics, and some of which are also secreted through root exudation. Seed treatment with *B. subtilis* MF352017 (Bs1) and *T. harzianum* (T5) shows significant protection of chickpea seedlings from *F. oxysporum* f. sp. *ciceris* as compared to untreated plants (Zaim et al. 2018). The application of *B. subtilis* MF352017 (Bs1) and *T. harzianum* (T5) suppressed 93.67% disease and also enhanced plant growth and dry weights of shoot and root effectively. Rautela et al. (2018) evaluated copper compounds, viz., CuOH (technical grade), CuOH (Kocide), and copper sulfate pentahydrate ($CuSO_4 \cdot 5H_2O$) in combination with chitosan (Cs661) and *T. asperellum* (Tri) against late blight disease of tomato. Minimum disease severity (2.33%) was recorded in plots treated with CuOH (TG) + Cs661 (Ac) + Tri, whereas maximum disease severity was recorded in control plot (40%) after 95 DAT. Combined application of *T. harzianum* strain Th3 and *P. fluorescens* strain RRb11 and carbendazim against *Magnaporthe oryzae* and *Xanthomonas oryzae* pv. *oryzae* was evaluated by Jambhulkar et al. (2018). Combined application of *T. harzianum* Th3 and *P. fluorescens* RRb11 reduced severity of *M. oryzae* 69.5% displaying a synergy factor (SF) of 1.29. Dehariya et al. (2015) studied the combination of *T. harzianum* (Th), *T. virens* (Tr), and *T. viride* (Tv) and consortium of AMF (Myc, mixture of *Funneliformis mosseae*, *Glomus cerebriforme*, *Rhizophagus irregularis*) against *Fusarium* wilt of pigeon pea caused by *F. udum*. Elshahawy and El-Mohamedy (2019) evaluated five *Trichoderma* isolates and found that combined application decreased *Pythium* damping-off and root rot by 57.2% and increased the survival of tomato plants by 87.5%. These isolates stimulated systemic defense responses in tomato plants even under field by activating peroxidase, polyphenol oxidase, and chitinase activity.

14.5.3 Consortia Application of Trichoderma *spp. for Betterment of Host*

Trichoderma spp.-based consortium played crucial role in the growth of host. The extensive studies were conducted both in basic and applied aspect. The introduction of this consortium to host plants is done through seed, soil, and foliar treatment. Srinath et al. (2003) investigated the combined effect of PDPR *B. coagulans* and *T. harzianum* micropropagated *Ficus benjamina* on growth and nutrition. The height of AM-inoculated plantlets was further increased when they were co-inoculated with either *B. coagulans* or *T. harzianum*. Maximum height was detected in plantlets treated with *G. mosseae* + *B. coagulans* + *T. harzianum*. The combined efficacy of *T. harzianum* PDBCTH 10, *T. viride* PDBCTV 32, *T. virens* PDBCTVs 12, and *Bacillus megaterium* subsp. *phospaticum* strain PB has been studied in chickpea. The significant increase in N uptake in shoots (120 mg/plant) was recorded in PSB, *T. harzianum*, rock phosphate, and *Rhizobium*-treated plant under glasshouse conditions. Highest nitrogen uptake to the shoot (1079 mg/plant) and the peas (1600 mg/plant) was recorded in PSB + *T. harzianum* + rock phosphate + *Rhizobium* application, while N uptake into the root was highest (13.7 mg/plant) in *T. virens* + rock phosphate + *Rhizobium* application in field conditions (Rudresh et al. 2005). In black pepper, the application of *T. harzianum* and *Piriformospora indica* has increased root colonization and promoted growth of the plants (Anith et al. 2011).

The interactive effect of *T. viride*, *G. intraradices*, *Acaulospora laevis*, and *P. fluorescens* (MTCC No. 103) was evaluated on growth and yield of broccoli on growth, mycorrhization, chlorophyll content, nutrient uptakes, and yield of broccoli plants (Tanwar et al. 2013). Seed biopriming with potential BCAs *T. asperellum* T42, *P. fluorescens* OKC, and *Rhizobium* sp. RH4 improved seed germination and seedling vigor in chickpea and rajma (Yadav et al. 2013). Microbial consortium of

T. harzianum TNHU27, *B. subtilis* BHHU100, and *P. aeruginosa* PJHU15 showed increased total phenolic, flavonoid, ascorbic acid and protein contents, free radical scavenging, hydroxyl radical scavenging, iron chelation activities, and reduced power in the seeds and pericarp of pods. Accumulation of phenolic compounds, viz., shikimic acid, gallic acid, tannic acid, syringic acid, *p*-coumaric acid, quercetin, and kaempferol from seed extracts of different treatments was also observed (Jain et al. 2014). Singh et al. (2014b) observed enhancement in antioxidants synthesis by using beneficial compatible microbes, viz., *T. harzianum* (THU 0816), fluorescent *Pseudomonas* (PHU 094), and *Mesorhizobium* sp. (RL 091). PHU 094 + THU 0816 also enhanced antioxidant and phenolic content in pericarp.

Colla et al. (2015) evaluated combined effect of *T. atroviride* strain MUCL45632 and *G. intraradices* BEG72 in vegetable crops, namely, lettuce, tomato, melon, pepper, and zucchini squash, and found that the shoot dry weight was rapidly increased by 167%, 56%, 115%, 68%, and 58%, respectively, and also observed increase in level of nutrient uptake (e.g., P, Mg, Fe, Zn, and B). Velmourougane et al. (2017) developed biofilms from *A. chroococcum* and *T. viride* and evaluated its growth promotion traits in chickpea. Tv-Az biofilm exhibited improved colonization in the rhizosphere and rhizoplane of the host compared to individual application. The 30–45% enhancement was observed in root volume, root-shoot ratio, and root protein exudation of fumaric acid and succinic acid and a twofold increase in plant enzyme activity. Fiorentino et al. (2018) conducted greenhouse experiments in iceberg lettuce (*Lactuca sativa* L.) and rocket (*Eruca sativa* Mill.) for the assessment of modulation in rhizosphere microbial populations and improve N uptake efficiency, yield, and nutritional quality by applying biostimulant fungi *T. virens* (GV41) or *T. harzianum* (T22). *T. virens* GV41 was found to be the best microbial-biostimulant and enhanced crop growth and nutrient uptake on both lettuce and rocket. This also increased N fertilizer use efficiency of lettuce through favoring the uptake of native soil N by both lettuce and rocket and improved N uptake under low N availability conditions.

14.6 Mathematical Models in Microbial Consortia Studies

The efficacy of microbial consortium depends on the synergistic effects of two or more biocontrol agents. Based on theoretical definitions of Bliss independence or Loewe additivity, the combination of biocontrol agents will be formulated. The case studies on the statistical analysis for the independence-synergistic-antagonistic interactions were found very rarely among the biocontrol agents. Xu et al. (2011a) reviewed total 465 publications and found only 2% publications showed evidence of synergistic effects among biocontrol agents. Most of the research findings indicated the antagonistic interactions of biocontrol agents, and the synergistic interactions were rare events. Since decades several research hypotheses based on the biocontrol mechanisms of microbial consortium were formulated for the biological control of plant diseases.

According to model assumptions, Xu et al. (2011b) suggested that consortial application of two or more biocontrol agents having different biocontrol mechanisms resulted in control efficacies which are similar to individual. These findings suggested that there is urgent need to assess the results of experiments related to compatibility among the BCAs. Xu and Jeger (2013) proposed numerical simulations model to analyze the effects of varied conditions in biocontrol efficacy during single or consortial application of biocontrol agents and also emphasized the effect of edaphic factors like temperature in biocontrol mechanisms of BCAs. The results showed the variation in temperature affect the antagonistic mechanisms rather synergistic interactions among the biocontrol microbes. In addition to the numerical evaluation of biocontrol mechanisms in combined application, Jambhulkar et al. (2018) assessed the synergistic and antagonistic efficacy of *Trichoderma harzianum* Th3, *Pseudomonas fluorescens* RRb11, and the fungicide carbendazim against rice blast (RB) pathogen, *Magnaporthe oryzae*, and bacterial leaf blight (BLB) pathogen *Xanthomonas oryzae* pv. *oryzae*. The synergistic, additive, or antagonistic effects were evaluated by using numerical hypotheses of "Bliss independence" among the BCAs. The combined application of *Trichoderma harzianum* Th3 and *Pseudomonas fluorescens* strain RRb11 was found to have synergistically reduced severity of rice blast by 69.5% and the synergy factor (SF) 1.29 but antagonistic with respect to bacterial leaf blight (BLB) (Table 14.2). The synergism in consortial application of biocontrol agents needs to be understood in IDM and IPM very closely.

14.7 Conclusion

Recently, the number of consortial bioformulations has been evaluated against various crop diseases. It is observed that simultaneous use of biocontrol agents having different mechanisms satisfy the norms of integrated pest management for pest suppression. The benefit of such concept is that if one or other mechanism is not effective, the other mechanism can compensate the absence of former. Although it is equally important to have efficacy of consortia of biocontrol agents for disease suppression, at the same time, it is necessary to learn how the consortia influence plant metabolism and molecular mechanism of such interaction. This study is relatively unexplored and requires investigations on different aspects of the application. Efforts are needed to identify compatible strains of biocontrol agents which can positively influence the host physiological and transcriptional regulations for development of potent and cost-effective commercial consortial bioformulation. Also there is need to develop methodologies to assess probable impacts of the consortia on host plants through synergistic acts of the microbes.

Table 14.2 Effect of combination of BCAs and carbendazim on synergistic and antagonistic efficiency against blast and BLB of rice (source: Jambhulkar et al. 2018)

Treatments	Observed efficiency (%)	Expected efficiency (%)[a]	Synergy factor[b]	Interaction
Blast				
T. harzianum Th3 + *P. fluorescens* RRb11 + Carbendazim ($E_{Th3,RRb11,C'zim}$)	62.4	81.2	0.77	Antagonistic
*T. harzianum*Th3 + *P. fluorescens* RRb11($E_{Th3,RRb11}$)	69.5	54.0	1.29	Synergistic
T. harzianum Th3 + Carbendazim ($E_{Th3{\cdot}C'zim}$)	51.8	76.0	0.68	Antagonistic
Bacterial leaf blight				
T. harzianum Th3 + *P. fluorescens* RRb11 + Carbendazim ($E_{Th3,RRb11,C'zim}$)	53.9	76.2	0.71	Antagonistic
T. harzianum Th3 + *P. fluorescens* RRb11 ($E_{Th3,RRb11}$)	50.2	70.4	0.71	Antagonistic
T. harzianum Th3 + Carbendazim ($E_{Th3{\cdot}C'zim}$)	19.5	43.7	0.45	Antagonistic

[a]Expected efficiency calculated by formula $E_{Th3,RRb11,C'zim} = E_{Th3} + E_{RRb11} + E_{C'zim} - (E_{Th3} \times E_{RRb11}) - (E_{Th3} \times E_{C'zim}) - (E_{RRb11} \times E_{C'zim}) + (E_{Th3} \times E_{RRb11} \times E_{C'zim})$
[b]Synergy factor (SF) = E(obs)/E(exp), where E(obs) = observed control efficacy by mixture and E(exp) = expected control efficacy by the mixture

Acknowledgments The corresponding author is highly thankful to DG, Indian Council of Agriculture Research (ICAR) and Agriculture Education Division, ICAR, New Delhi.

References

Abeysinghe S (2009) Effect of combined use of *Bacillus subtilis* CA32 and *Trichoderma harzianum* RUOI on biological control of *Rhizoctonia solani* on *Solanum melongena* and *Capsicum annuum*. Plant Pathol J 8:9–16

Alizadeh H, Behboudi K, Ahmadzadeh M, Javan-Nikkhah M, Zamioudis C, Pieterse CMJ, Bakker PAHM (2013) Induced systemic resistance in cucumber and *Arabidopsis thaliana* by the combination of *Trichoderma harzianum* Tr6 and *Pseudomonas* sp. Ps14. Biol Control 65:14–23

Anith KN, Faseela KM, Archana PA, Prathapan KD (2011) Compatibility of *Piriformospora indica* and *Trichoderma harzianum* as dual inoculants in black pepper (*Piper nigrum* L.). Symbiosis 55 (1):11–17

Bakker PAHM, Pieterse CMJ, Van Loon LC (2007) Induced systemic resistance by *fluorescent Pseudomonas* spp. Phytopathology 97:239–243

Bora T, Ozaktan H, Gore E, Aslan E (2004) Biological control of *Fusarium oxysporum* f. sp. *melonis* by wettable powder formulations of the two strains of *Pseudomonas putida*. J Phytopathol 152:471–475

Chittenden C, Singh T (2009) In vitro evaluation of combination of *Trichoderma harzianum* and chitosan for the control of sap stain fungi. Biol Control 50(3):262–266

Colla G, Rouphael Y, Di Mattia E, El-Nakhel C, Cardarelli M (2015) Co-inoculation of *Glomus intraradices* and *Trichoderma atroviride* acts as a biostimulant to promote growth, yield and nutrient uptake of vegetable crops. J Sci Food Agr 95(8):1706–1715

de Leon IP, Montesano M (2013) Activation of defense mechanisms against pathogens in mosses and flowering plants. Int J Mol Sci 14:3178–3200

Dehariya K, Sheikh IA, Vyas D, Shukla A (2015) Trichoderma and arbuscular mycorrhizal fungi based biocontrol of *Fusarium udum* butler and their growth promotion effects on pigeon pea. J Agric Sci Technol 17:505–517

Dubey SC, Tripathi A, Singh B (2013) Integrated management of Fusarium wilt by combined soil application and seed dressing formulations of Trichoderma species to increase grain yield of chickpea. Int J Pest Manage 59(1):47–54

Duffy BK, Simon A, Weller DM (1996) Combination of *Trichoderma koningii* with fluorescent pseudomonads for control of take-all on wheat. Phytopathology 86:188–194

Elad Y, Kirshner B, Yehuda N, Sztejnberg A (1998) Management of powdery mildew and gray mold of cucumber by *Trichoderma harzianum* T39 and *Ampelomyces quisqualis* AQ10. Biol Control 43(2):241–251

Elliott M, Shamoun SF, Sumampong G, James D, Masri S, Varga A (2009) Evaluation of several commercial biocontrol products on European and North American populations of *Phytophthora ramorum*. Biocontrol Sci Tech 19:1007–1021

Elshahawy IE, El-Mohamedy RS (2019) Biological control of Pythium damping-off and root-rot diseases of tomato using *Trichoderma* isolates employed alone or in combination. J Plant Pathol 101(3):597–608

Ezziyyani M, Requena ME, Egea-Gilabert C, Candela ME (2007) Biological control of Phytophthora root rot of pepper using *Trichoderma harzianum* and *Streptomyces rochei* in combination. J Phytopathol 155(6):342–349

Fernández E, Segarra G, Trillas MI (2014) Physiological effects of the induction of resistance by compost or *Trichoderma asperellum* strain T34 against *Botrytis cinerea* in tomato. Biol Control 78:77–85

Fiorentino N, Ventorino V, Woo SL, Pepe O, De Rosa A, Gioia L, Romano I, Lombardi N, Napolitano M, Colla G, Rouphael Y (2018) *Trichoderma*-based biostimulants modulate rhizosphere microbial populations and improve N uptake efficiency, yield, and nutritional quality of leafy vegetables. Front Plant Sci 9:743. https://doi.org/10.3389/fpls.2018.00743

Ganesan S, Kuppusamy RG, Sekar R (2007) Integrated management of stem rot disease (*Sclerotium rolfsii*) of groundnut (*Arachis hypogaea* L.) using Rhizobium and *Trichoderma harzianum* (ITCC-4572). Turk J Agric For 31(2):103–108

Goswami J, Pandey RK, Tewari JP, Goswami BK (2008) Management of root knot nematode on tomato through application of fungal antagonists, *Acremonium strictum* and *Trichoderma harzianum*. J Environ Sci Health B 43(3):237–240

Gupta M, Dohroo NP, Gangta V, Shanmugam V (2010) Effect of microbial inoculants on rhizome disease and growth parameters of ginger. Indian Phytopathol 63(4):438–441

Harman GE (2011) Multifunctional fungal plant symbionts: new tools to enhance plant growth and productivity. New Phytol 189:647–649

Jain A, Singh S, Sarma BK, Singh HB (2012) Microbial consortium-mediated reprogramming of defence network in pea to enhance tolerance against *Sclerotinia sclerotiorum*. J Appl Microbiol 112:537–550

Jain A, Singh A, Singh S, Singh HB (2013) Microbial consortium-induced changes in oxidative stress markers in pea plants challenged with *Sclerotinia sclerotiorum*. J Plant Growth Regul 32 (2):388–398

Jain A, Singh A, Chaudhary A, Singh S, Singh HB (2014) Modulation of nutritional and antioxidant potential of seeds and pericarp of pea pods treated with microbial consortium. Food Res Int 64:275–282

Jambhulkar PP, Sharma P, Raja M, Lakshman DK, Rokadia P, Jambhulkar N (2018) Assessing synergism of combined applications of *Trichoderma harzianum* and *Pseudomonas fluorescens* to control blast and bacterial leaf blight of rice. Eur J Plant Pathol 152(3):747–757

Jayaraj J, Ramabadran R (1999) Rhizobium-Trichoderma interaction *in vitro* and *in vivo*. Indian Phytopathol 52(2):190–192

Jetiyanon K (2007) Defensive-related enzyme response in plants treated with a mixture of *Bacillus strains* (IN937a and IN937b) against different pathogens. Biol Control 42:178–185

Kannan V, Sureendar R (2009) Synergistic effect of beneficial rhizosphere microflora in biocontrol and plant growth promotion. J Basic Microbiol 49:158–164

Kumar SM, Chowdappa P, Krishna V (2015) Development of seed coating formulation using consortium of *Bacillus subtilis* OTPB1 and *Trichoderma harzianum* OTPB3 for plant growth promotion and induction of systemic resistance in field and horticultural crops. Indian Phytopathol 68(1):25–31

Kumar M, Patel JS, Kumar G, Sarka A, Singh HB, Sarma BK (2017) Studies on Pseudomonas and Trichoderma-mediated root exudation pattern in chickpea against *Fusarium oxysporum* f. sp. *ciceris*. J Agric Sci Technol 19:969–978

Latha P, Anand T, Prakasam V, Jonathan EI, Paramathma M, Samiyappan R (2011) Combining Pseudomonas, Bacillus and Trichoderma strains with organic amendments and micronutrient to enhance suppression of collar and root rot disease in physic nut. Appl Soil Ecol 49:215–223

Leon VC, Raja M, Pandian RTP, Kumar A, Sharma P (2018) Studies on opportunistic endophytism of *Trichoderma* species in rice. Indian J Exp Biol 56:121–128

Lucas JA, Solano BR, Montes F, Ojeda J, Megias M, Gutierrez Manero FJ (2009) Use of two PGPR strains in the integrated management of blast disease in rice (*Oryza sativa*) in southern Spain. Field Crops Res 114:404–410

Manjula K, Kishore GK, Girish AG, Singh SD (2004) Combined application of *Pseudomonas fluorescens* and *Trichoderma viride* has an improved biocontrol activity against stem rot in groundnut. Plant Pathol J 20(1):75–80

Marimuthu S, Ramamoorthy V, Samiyappan R, Subbian P (2013) Intercropping system with combined application of Azospirillum and *Pseudomonas fluorescens* reduces root rot incidence caused by *Rhizoctonia bataticola* and increases seed cotton yield. J Phytopathol 161:405–411

Martínez-Medina A, Roldán A, Pascual JA (2011) Interaction between arbuscular mycorrhizal fungi and *Trichoderma harzianum* under conventional and low input fertilization field condition in melon crops: growth response and Fusarium wilt biocontrol. Appl Soil Ecol 47(2):98–105

Mathivanan N, Prabavathy VR, Vijayanandraj VR (2005) Application of talc formulations of *Pseudomonas fluorescens* Migula and *Trichoderma viride* Pers. ex SF Gray decrease the sheath blight disease and enhance the plant growth and yield in rice. J Phytopathol 153 (11–12):697–701

Meyer SL, Roberts DP, Chitwood DJ, Carta LK, Lumsden R (2001) Application of *Burkholderia cepacia* and *Trichoderma virens*, alone and in combinations, against *Meloidogyne incognita* on bell pepper. Nematropica 31(1):75–86

Mukherjee PK, Haware MP, Raghu K (1997) Induction and evaluation of benomyl-tolerant mutants of *Trichoderma viride* for biological control of Botrytis grey mould of chickpea. Indian Phytopathol 50(4):485–489

Oyekanmi EO, Coyne DL, Fagade OE, Osonubi O (2007) Improving root-knot nematode management on two soybean genotypes through the application of *Bradyrhizobium japonicum*, *Trichoderma pseudokoningii* and *Glomus mosseae* in full factorial combinations. Crop Prot 26(7):1006–1012

Panebianco S, Vitale A, Polizzi G, Scala F, Cirvilleri G (2015) Enhanced control of postharvest citrus fruit decay by means of the combined use of compatible biocontrol agents. Biol Control 84:19–27

Pieterse CMJ, Zamioudis C, Berendsen RL, Weller DM, Van Wees SCM, Bakker PAHM (2014) Induced systemic resistance by beneficial microbes. Annu Rev Phytopathol 52:347–375

Radjacommare R, Venkatesan S, Samiyappan R (2010) Biological control of phytopathogenic fungi of vanilla through lytic action of *Trichoderma* species and *Pseudomonas fluorescens*. Arch Phytopathol Plant Protect 43(1):1–17

Raja M, Thava Prakasa Pandian R, Sharma M, Jambhulkar PP, Sharma P (2016) Study of induced systemic resistance in tomato against *Fusarium oxysporum* f. sp. *Lycopersici* causing wilt of tomato. Indian Phytopathol 69(4):539–542

Rautela A, Shukla N, Ghatak A, Tewari AK, Kumar J (2018) Field evaluation of different copper sources in a consortium of "Copper-Chitosan-Trichoderma" for management of late blight disease of tomato. J Pharmacogn Phytochem 7(4):1260–1266

Reddy PP, Rao MS, Nagesh M (1996) Management of citrus nematode, *Tylenchulus semipenetrans*, by integration of *Trichoderma harzianum* with oil cakes. Nematol Mediterr 24 (2):265–267

Reinke J (1872) Ueber die anatomischen Verhältnisseeiniger Arten von *Gunnera* L. Nachrichten von der Königl. Gesellschaft der Wissenschaften und der Georg-Augusts-Universitätzu Göttingen 9:100–108

Ruano-Rosa D, Cazorla FM, Bonilla N, Martin-Perez R, De Vicente A, Lopez-Herrera CJ (2014) Biological control of avocado white root rot with combined applications of *Trichoderma* spp. and rhizobacteria. Eur J Plant Pathol 138(4):751–762

Rudresh DL, Shivaprakash MK, Prasad RD (2005) Effect of combined application of Rhizobium, phosphate solubilizing bacterium and *Trichoderma* spp. on growth, nutrient uptake and yield of chickpea (*Cicer arietinum* L.). Appl Soil Ecol 28:139–146

Saeedizadeh A (2016) *Trichoderma viride* and *Pseudomonas fluorescens* CHA0 against *Meloidogyne javanica* in the rhizosphere of tomato plants. Hellenic Plant Protect J 9(1):28–34

Sahni S, Sarma BK, Singh DP, Singh HB, Singh KP (2008) Vermicompost enhances performance of plant growth-promoting rhizobacteria in *Cicer arietinum* rhizosphere against *Sclerotium rolfsii*. Crop Prot 27:369–376

Sajeesh PK, Bhardwaj NR, Balodi R, Kumar J (2016) Field evaluation of triple combination of copper, chitosan and Trichoderma for management of late blight disease of potato under hill condition. Adv Life Sci 5(7):2771–2778

Sarma BK, Yadav SK, Singh S, Singh HB (2015) Microbial consortium-mediated plant defense against phytopathogens: readdressing for enhancing efficacy. Soil Biol Biochem 87:25–33

Segarra G, Casanova E, Bellido D, Odena MA, Oliveira E, Trillas I (2007) Proteome, salicylic acid, and jasmonic acid changes in cucumber plants inoculated with *Trichoderma asperellum* strain T34. Proteomics 7:3943–3952

Segarra G, Van der Ent S, Trillas I, Pieterse CMJ (2009) MYB72, a node of convergence in induced systemic resistance triggered by a fungal and a bacterial beneficial microbe. Plant Biol 11:90–96

Shanmugam V, Gupta S, Dohroo NP (2013) Selection of compatible biocontrol strain mixture based on co cultivation for control of rhizome rot of ginger. Crop Prot 43:119–127

Sharma P, Dureja P (2004) Evaluation of *T. harzianum* and *T. viride* isolates at BCA pathogen crop interface. J Mycol Plant Pathol 34:47–55

Sharma P, Sharma M, Raja M, Shanmugam V (2014) Status of *Trichoderma* research in India: a review. Indian Phytopathol 67(1):1–19

Sharma P, Sharma M, Raja M, Singh DV, Srivastava M (2016) Use of *Trichoderma* spp. in biodegradation of Carbendazim. Indian J Agric Sci 86(7):59–62

Shoresh M, Harman GE, Mastouri F (2010) Induced systemic resistance and plant responses to fungal biocontrol agents. Annu Rev Phytopathol 48:1–23

Siddiqui ZA, Akhtar MS (2009) Effect of plant growth promoting rhizobacteria, nematode parasitic fungi and root-nodule bacterium on root-knot nematodes *Meloidogyne javanica* and growth of chickpea. Biocont Sci Technol 19:511–521

Singh SP, Singh HB (2012) Effect of consortium of *Trichoderma harzianum* isolates on growth attributes and *Sclerotinia sclerotiorum* rot of brinjal. Veg Sci 39(2):144–148

Singh A, Sarma BK, Upadhyay RS, Singh HB (2013a) Compatible rhizosphere microbes mediated alleviation of biotic stress in chickpea through enhanced antioxidant and phenylpropanoid activities. Microbiol Res 168:33–40

Singh A, Jain A, Sarma BK, Upadhyay RS, Singh HB (2013b) Rhizosphere microbes facilitate redox homeostasis in *Cicer arietinum* against biotic stress. Ann Appl Biol 163:33–46

Singh A, Jain A, Sarma BK, Upadhyay RS, Sing HB (2014a) Rhizosphere competent microbial consortium mediates rapid changes in phenolic profiles in chickpea during *Sclerotium rolfsii* infection. Microbiol Res 169(5–6):353–360

Singh A, Jain A, Sarma BK, Upadhyay RS, Singh HB (2014b) Beneficial compatible microbes enhance antioxidants in chickpea edible parts through synergistic interactions. LWT-Food Sci Technol 56(2):390–397

Somani AK, Arora RK (2010) Field efficacy of *Trichoderma viride*, *Bacillus subtilis* and *Bacillus cereus* in consortium for control of *Rhizoctonia solani* causing black scurf disease of potato. Indian Phytopathol 63(1):23–25

Srinath J, Bagyaraj DJ, Satyanarayana BN (2003) Enhanced growth and nutrition of micropropagated *Ficus benjamina* to *Glomus mosseae* co-inoculated with *Trichoderma harzianum* and *Bacillus coagulans*. World J Microbiol Biotechnol 19(1):69–72

Srivastava R, Khalid A, Singh US, Sharma AK (2010) Evaluation of arbuscular mycorrhizal fungus, fluorescent *Pseudomonas* and *Trichoderma harzianum* formulation against *Fusarium oxysporum* f. sp. *lycopersici* for the management of tomato wilt. Biol Control 53:24–31

Stockwell VO, Johnson KB, Sugar D, Loper JE (2011) Mechanistically compatible mixtures of bacterial antagonists improve biological control of fire blight of pear. Phytopathology 101:113–123

Suryadi Y, Susilowati D, Riana E, Mubarik NR (2013) Management of rice blast disease (*Pyricularia oryzae*) using formulated bacterial consortium. Emir J Food Agric 25(5):349–357

Tanwar A, Aggarwal A, Kaushish S, Chauhan S (2013) Interactive effect of AM fungi with *Trichoderma viride* and *Pseudomonas fluorescens* on growth and yield of broccoli. Plant Prot Sci 49(3):137–145

Thakkar A, Saraf M (2015) Development of microbial consortia as a biocontrol agent for effective management of fungal diseases in Glycine max L. Arch Phytopathol Plant Prot 48(6):459–474

Thangavelu R, Gopi M (2015) Combined application of native *Trichoderma* isolates possessing multiple functions for the control of Fusarium wilt disease in banana cv. Grand Naine. Biocont Sci Technol 25(10):1147–1164

Thilagavathi R, Saravanakumar D, Ragupathi N, Samiyappan R (2007) A combination of biocontrol agents improves the management of dry root rot (*Macrophomina phaseolina*) in green gram. Phytopathol Mediterr 46(2):157–167

Triveni S, Prasanna R, Kumar A, Bidyarani N, Singh R, Saxena AK (2015) Evaluating the promise of Trichoderma and Anabaena based biofilms as multifunctional agents in *Macrophomina phaseolina*-infected cotton crop. Biocont Sci Technol 25(6):656–670

Vacheron J, Desbrosses G, Bouffaud ML, Touraine B, Moenne-Loccoz Y, Muller D, Legendre L, Wisniewski-Dye F, Prigent-Combaret C (2013) Plant growth-promoting rhizobacteria and root system functioning. Front Plant Sci 4:356

Velmourougane K, Prasanna R, Singh S, Chawla G, Kumar A, Saxena AK (2017) Modulating rhizosphere colonisation, plant growth, soil nutrient availability and plant defense enzyme activity through *Trichoderma viride-Azotobacter chroococcum* biofilm inoculation in chickpea. Plant Soil 421(1–2):157–174

Wojtkowisk-Gebarowska E, Pietr SJ (2006) Colonization of roots and growth stimulation of cucumber by iprodione-resistant isolates of *Trichoderma* spp. applied alone and combined with fungicides. Phytopathol Pol 41:51–64

Xu XM, Jeger MJ (2013) Combined use of two biocontrol agents with different biocontrol mechanisms most likely results in less than expected efficacy in controlling foliar pathogens under fluctuating conditions: a modeling study. Phytopathology 103:108–116

Xu XM, Robinson JD, Jeger M, Jeffries P (2010a) Using combinations of biocontrol agents to control *Botrytis cinerea* on straw-berry leaves under fluctuating temperatures. Biocont Sci Technol 20:359–373

Xu XM, Salama N, Jeffries P, Jeger MJ (2010b) Numerical studies of biocontrol efficacies of foliar plant pathogens in relation to the characteristics of a biocontrol agent. Phytopathology 100:814–821

Xu XM, Jeffries P, Pautasso M, Jeger MJ (2011a) Combined use of biocontrol agents to manage plant diseases in theory and practice. Phytopathology 101:1024–1031

Xu XM, Jeffries P, Pautasso M, Jeger MJ (2011b) A numerical study of combined use of two biocontrol agents with different biocontrol mechanisms in controlling foliar pathogens. Phytopathology 101:1032–1044

Yadav SK, Dave A, Sarkar A, Singh HB, Sarma BK (2013) Co-inoculated biopriming with *Trichoderma*, *Pseudomonas* and Rhizobium improves crop growth in *Cicer arietinum* and *Phaseolus vulgaris*. Int J Agric Environ Biotechnol 6(2):255–259

Yobo KS, Laing MD, Hunter CH (2011) Effects of single and combined inoculations of selected *Trichoderma* and *Bacillus* isolates on growth of dry bean and biological control of *Rhizoctonia solani* damping-off. Afr J Biotechnol 10(44):8746–8756

Zaim S, Bekkar AA, Belabid L (2018) Efficacy of *Bacillus subtilis* and *Trichoderma harzianum* combination on chickpea Fusarium wilt caused by *F. oxysporum* f. sp. *ciceris*. Arch Phytopathol Plant Prot 51(3–4):217–226

Zhang F, Zhu Z, Yang X, Ran W, Shen Q (2013) *Trichoderma harzianum* T-E5 significantly affects cucumber root exudates and fungal community in the cucumber rhizosphere. Appl Soil Ecol 72:41–48

Chapter 15
Trichoderma Interactions in Vegetable Rhizosphere Under Tropical Weather Conditions

William Rivera-Méndez

Abstract The use of *Trichoderma* in world agriculture has established itself as an alternative for sustainable production. Not always bioproducts developed in other latitudes have managed to respond adequately when they are introduced in soils of tropical climate. It has been mentioned that poor adaptation to these conditions causes some isolates not to be incorporated into the soil microbiome; however, several factors regulate the structure of a microorganism's community associated with a crop. This chapter details aspect related to research with native isolates with the aim of improving the inoculation techniques of the fungus and its establishment in rhizosphere of onion, garlic, and sweet pepper crops, under tropical environmental conditions in Central America. Tests carried out over several years have shown that even native isolates may have an irregular behavior in terms of their capacity for mycoparasitism, growth regulation, and influence on yield. It is necessary to deepen into the way in which biological control mechanisms of each *Trichoderma* species and isolates are affected by the soil dynamics, the climatic factors, the systems and intensity of the crop, and the forms of inoculation.

Keywords Vegetable rhizosphere · Tropical weather · Soil microbiome · Growth regulation · Yield · Onion · Sweet pepper

15.1 Introduction

Trichoderma genus has been studied for a long time, and its role in the agriculture now is better understood. It has been used in various crops worldwide for crop protection reducing agrochemical dependence and associated risk to the environment and human beings. The *Trichoderma* species show different biocontrol mechanisms to help plants to regulate the growth, inducing the defense routes and

W. Rivera-Méndez (✉)
Instituto Tecnológico de Costa Rica, Centro de Investigación en Biotecnología, Cartago, Costa Rica
e-mail: wirivera@itcr.ac.cr

A. K. Sharma, P. Sharma (eds.), *Trichoderma*, Rhizosphere Biology,
https://doi.org/10.1007/978-981-15-3321-1_15

reducing the pathogen populations. In many cases, this fungus has allowed obtaining better crops through the plant-microorganism interaction. The farmers can take advantage of these relations, but it is necessary that the interaction be effectively established, and in field experiences a lot of times it does not occur.

The versatility shown by different species of *Trichoderma* genus to establish interactions with plants through the rhizosphere has been clearly established over the last few years (Mukherjee et al. 2012). These interactions have shown important effects in the exploration and colonization capacities of soils by plant roots, in resistance to diseases, and in the regulation of plant growth and tolerance to stress factors. Molecular and biochemical studies establish that the relationships occur through mechanisms of recognition and induction of metabolic pathways that cause the secretion of plant growth regulators, the activation of RSI mechanisms or SA-dependent resistance and that trigger the phenomenon called priming (Martínez-Medina et al. 2013).

The scientific evidence that has clarified the role of *Trichoderma* in the ecosystems, particularly in the rhizosphere of different plant species, obviously, comes from work done in vitro and greenhouse test mostly, which has gradually led to development of products that have been tested in field and that are offered to farmers and production companies around the world.

For countries that are in areas of tropical climate, many of biological control products developed by companies from other latitudes do not adapt effectively to native agroecosystems (Rivera-Méndez 2016), or studies conducted under in vitro conditions are difficult to adapt to field tests. In addition, in many places, bioproducts use semi-handcrafted production technologies and lack information about the different biological control mechanisms that microorganisms develop (Woo et al. 2014), which influences their effectiveness. In addition, environmental and edaphic factors make the rhizosphere's composition, and therefore its functioning can be completely different, even within the same life zone or climatic condition.

Undoubtedly, knowledge of how members of the *Trichoderma* genus can carry out their interactions with plants and with other species in rhizospheric communities of soils of tropical regions is fundamental to establish sustainable production systems that contribute to worldwide food production, even in conditions of climate change, salt stress, water stress, and various environmental disturbances.

This chapter shows experiences using different *Trichoderma* species in two crop models, in greenhouse or in outdoor, in a tropical weather zone, to quantify effects on the growth regulation and in yield and measuring root colonization percentage as indicator of an endophytic interaction. For improving the establishment of *Trichoderma* in soil, it is necessary to understand the soil microbiome composition, the rhizosphere microbiome associated with plants, and factors that are involved in the community structure and function.

15.2 Climatic Conditions in Intertropical Life Zone and Its Influence on Soil Microorganisms' Communities

Tropical climatic conditions shape the environmental conditions for the life development in a wide area in the planet. It is generally recognized that these conditions developed between the Tropics of Cancer and Capricorn, giving rise to the intertropical life zone. However, the geographical characteristics of this area are complex and influence environmental conditions, creating different weather patterns. Geerts and Linacre (2002) characterized tropic as a region with constant rains, stable temperature, and lack of aridity. Another peculiarity of intertropical zone is the establishment of microclimates in very small geographic areas and with a lot of atmospheric variability in zones of a few kilometers in length.

The quantity and availability of water in tropical soils vary according to soil type, the geographical features, vegetation cover, and quantity and intensity of rainfall received. Another important factor of tropical climate that affects the living conditions is that throughout the year there are only two climatic seasons: rainy and dry (Uzo and Currah 2018). These conditions have allowed appearance of microorganisms that have developed favorable adaptations for survival. Particularly, in *Trichoderma* have been found many genes involved in processes of tolerance to water stress and soil anaerobiosis, adherence to cell walls and root hairs, colonization of cortical tissues, competition, accelerated growth, its establishment as an endophyte, and the resistance of its structures to environmental factors such as temperature and ultraviolet radiation (Kashyap et al. 2017).

Theoretically, composition and functioning of rhizosphere communities will be greatly affected by climatic and soil conditions (Nuccio et al. 2016; Singh et al. 2018), as well as a series of interrelated biotic and abiotic factors, and it would be expected then that diversity of a microorganisms' community is affected by geography. However, some recent findings (Fierer and Jackson 2006) suggest that, in the case of the diversity of soil bacterial communities, this is highly influenced by the soil type and its pH rather than by conditions such as the site latitude. In the case of organisms as fungi, Green et al. (2004) showed indications that although the local diversity between sites may be high, between regions it is moderate because biodiversity turnover (coefficient of emergence/extinction) is low over long geographic distances. Therefore, it is possible that the differences found in values of diversity and richness in several types of ecosystems and soils in the world do not correspond with predictors of diversity such as distance and other geographic factors.

However, climatic and biotic factors are still an important part in the establishment of a microorganism in soil, especially if it has been inoculated artificially, as in the case of inoculative applications of *Trichoderma*. Elements such as water quantity and availability, its movement, soil temperature and its variation, the relationships with other organisms, the inoculation form, and biological capacities of the fungus are factors that will determine its successful inclusion in the structure of community associated with plant roots (Bailey et al. 2008; López-Bucio et al. 2015). It has also been proven that although the depth, at least in the first 20 cm from surface, does not affect the diversity and abundance of microorganisms, the soil tillage and the amount of organic matter in the plowed layer, effectively, modify the occurrence and distribution of *Trichoderma* and possibly other fungi (Okoth et al. 2009).

15.3 Characteristics of Microbiomes in Tropical Soils

In tropics, we can find a great variety of biomes. The main ones include jungles, cloud forest, dry, humid and very humid tropical forest, Paramo, deserts, and savanna. The wide variety of crops that have been established in the intertropical strip gives an idea of complications to define generalities about rhizosphere communities that are associated in plants.

The application of *Trichoderma* in crops has been increasing year after year (Singh et al. 2017). The tropics are not the exception, and it is necessary to point out the crops in which the applications of this fungus have had the greatest positive impact in recent years. In the tropics, it is applied to various vegetables, fruits with high sugar content, tubers and roots, cereals in flooded and dry crops, grasses and ground covers and legumes (Ghini et al. 2011; Sanchez 2019; Bettiol et al. 2014). Of course, variety of crops is very large, but Fig. 15.1 shows some examples of the highest production or area sown.

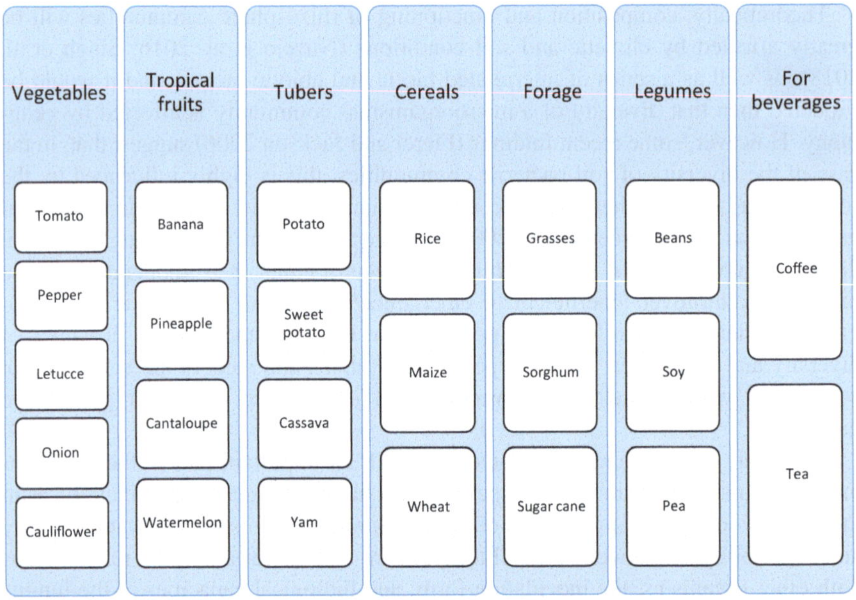

Fig. 15.1 Major tropical crops with *Trichoderma*'s positive experiences for biocontrol in tropical regions. (Data extracted from Ghini et al. 2011, Sanchez 2019, Bettiol et al. 2014)

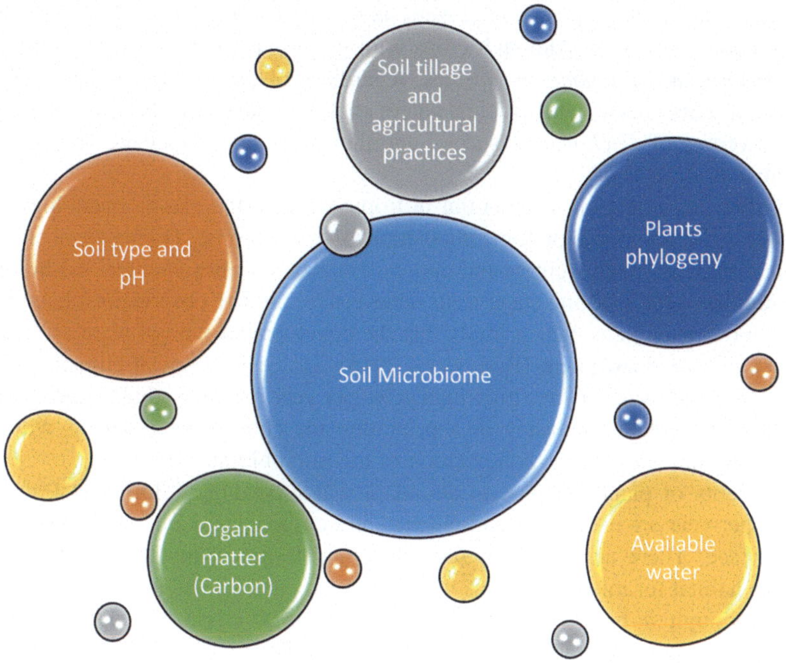

Fig. 15.2 Key factors identified that affect the microbiome composition in tropical soils

The complexity of soil microbiomes has been widely evaluated. It is currently accepted that there is no such thing as a "typical" microbiome. Each one can have a different composition, with relative abundances distinct from various taxa, with a predominance of different phyla and genera of microorganisms (Fierer 2017). However, many research efforts are made to better understand the soil microbiology; it is in this context that we seek to establish concepts such as core microbiome (Shade and Handelsman 2011).

Some factors that seem to have the greatest influence on composition of tropical soil microbiomes are the availability of water near roots, dominant plant species and ramification of their root systems, the type and pH of soil, the organic matter available, the tillage of soil, and plant phylogeny (Fig. 15.2).

The availability of moisture in soil or free water, for example, shows a direct relationship with the total biomass of soil, so in tropical zone (except deserts), where rainfall is abundant, there is the greatest microbial biomass accumulation per soil gram, with values that can reach up to 250 g/m^2 (Serna-Chavez et al. 2013).

Some authors have established that the most determining factor in conformation of a rhizospheric microbiome is the soil type, while host plant phylogeny (by effect of root exudates) has been considered a factor with much less influence (Bulgarelli et al. 2013). However, recent studies with non-angiosperm or taxonomically more distant plants show that effect of phylogeny is stronger than the observed in plants used as an experimental model as *Arabidopsis* (Yeoh et al. 2017). It would be expected that in native plants, phylogeny is highly determinant for microorganism communities' structure. Although crops with global distribution, such as corn and rice, share many *Trichoderma* species, regardless of the location where they are grown, in places such as the tropics, where endemic plants are abundant, there is a great diversity of *Trichoderma* species and their populations they are very specific (Zachow et al. 2016).

In other investigations carried out in tropical crops, the plant's phenology also showed to be a key factor for composition of microbiome. Gomes et al. (2003) pointed out that fungal root communities of maize varieties changed as the plant matured, from the seedling stage to its senescence. Similar observations have been made in other latitudes, which imply that the development state of plant affects the composition of rhizosphere (Berg et al. 2015). However, tropical climatic conditions, which allow up to three growing cycles per year for some edible plant species, should affect either decreasing or accelerating the change of organisms within a community and varying the composition of the microbiome.

The roots of plants adapted to abundant water conditions during at least one season per year are usually much branched and cover a large amount of area along soil surface. There are abundant secondary and tertiary hairs that create a very suitable habitat for microorganisms in terms of temperature and humidity conservation. Although it has been determined that oxygen levels are highly variable depending on soil structure and its depth, it is not clear whether at least in the first 20 cm, the abundance of microorganisms is affected (Okoth et al. 2009), but it does seem clear that the rhizosphere composition suffers alterations in its structure (Wang et al. 2017; Fierer 2017).

The use of soil for agricultural purposes and the intensity with which it is tilled directly affect the microbiome of rhizosphere. In soils from Europe with temperate conditions and Mediterranean climates, several studies have shown that general soil diversity is reduced by changing land use or by increasing tillage (de Vries et al. 2013). However, some researchers have pointed out that for the tropics this does not mean a reduction in alpha diversity, so cultivated or pasture sites may have higher rates of richness and functional diversity than tropical forests such as the Amazon (Mendes et al. 2015). Other authors even point out that the change in land use modifies the assembly process of the community, but not the species that make it up (Mayfield et al. 2010).

The other factor, sometimes forgotten, is soluble secretomes produced by microorganisms present in soil. Molecules exuded by every single species can affect the chances of survival of other new species that is trying to adapt to soil conditions. About it, the *Trichoderma*'s secretome and its properties are broadly studied (Mendoza-Mendoza et al. 2018; Lamdan et al. 2015).

Trichoderma has been a genus frequently prospected in metagenomic studies due to its universal beneficial fungal character, high diversity, and widely developed taxonomic classification. Their presence in tropical agroecosystems and their permanence in rhizosphere can be a key factor for environmental and economic sustainability of crops given the microbiological richness of ecosystems and ignorance of these.

15.4 Experiences Using *Trichoderma* in Tropical Agroecosystems in Costa Rica, Central America

The Central American region is an area located right in center of American continent and is located between the Equator and Tropic of Cancer. The biomes and ecosystems found are very diverse and are subject to intense pressure due to the expansion of agricultural frontier (Barlow et al. 2018). The biodiversity found is very big, but this is constantly threatened by the planting of monocultures for export, erosion, and conflicts of a social and cultural nature.

The main crops that predominate under an intensive production system are vegetables; fruits such as bananas, pineapples, and melons; tubers such as cassava and sweet potato; coffee; oil palm; sugarcane; wood; and large areas of pastures. Grains such as rice and corn and a series of medicinal and endemic plants used as food, condiments, and fiber sources are also produced to a lesser extent. It is not possible to describe the introduction of *Trichoderma* in all these crops, so two vegetables have been selected as models of fungus introduction in these crops.

Costa Rica is a small country, with intensive agriculture, focused in fruits and vegetables. Some experiments were carried out in a province named Cartago, testing *Trichoderma* isolates in onion, garlic, and sweet pepper, at in vitro and in vivo test.

15.5 *Trichoderma* in Onion and Garlic Rhizosphere

The communities established around onion and garlic's seedlings root were checked using microbiological and molecular approach, through isolation, sequencing, and metagenomic analysis. Soil managed under conventional conditions (only agrochemicals) showed a predominance of ascomycetes and basidiomycetes populations but particularly concentrated in phytopathogenic species. Among these was the detection of large amounts of DNA corresponding to high inocula of fungi such as *Setophoma terrestris*, *Stromatinia cepivora*, *Sclerotium rolfsii*, *Penicillium* spp., *Rhizoctonia solani*, and *Phoma* spp. As for bacteria, *Acidobacteria* and *Proteobacteria phyla* predominated, and *Actinobacteria* to a lesser extent.

Similar results were obtained by Gardner et al. (2011). They characterized a sample of onion rhizosphere in Alabama using markers for 17 bacterial families, and among the most abundant were *Flavobacteriaceae*, *Sphingobacteriaceae*, *Flexibacteraceae*, and *Xanthomonadaceae*. *Rhodocyclaceae*, *Streptomycetaceae*, and *Burkholderiaceae* were found in intermediate or low quantities. Many of the microorganisms isolated from onion cultivars are predominantly saprophytic or pathogenic, for example, bacteria of *Burkholderia cepacia* complex (Jacobs et al. 2008) and *Pseudomonas fluorescens* (Afsharmanesh et al. 2010). About the fungi, the mentioned ones are frequently pathogens typical of onion crops such as *Sclerotium cepivorum* (currently *Stromatinia*), *Sclerotium rolfsii*, and *Fusarium oxysporum* (Özer and Köycü 2004).

Most of fungi found are universal soil pathogens, but in the case of necrotrophic genera *Stromatinia cepivora* and *Sclerotium rolfsii*, they cause the two main diseases found worldwide in these crops. In addition, species of *Penicillium* sp. are occasional pathogens at field and post-harvest levels. The fungus *Setophoma terrestris* is a biotrophic pathogen that reduces productivity, and its incidence is difficult to quantify.

In this area, at least three different *Trichoderma* species were identified, from a total of 80 isolates obtained only from onion rhizosphere, so it is to be assumed that a more exhaustive examination to analyze diversity will find a significant species variety, with different agricultural potential according to biological control mechanisms that each species possesses. The species obtained were *T. asperellum*, *T. harzianum*, and *T. virens*, and some molecular soil analysis showed the presence of *T. koningii* (Alvarado-Marchena and Rivera-Méndez 2016). These species were mainly associated with a high presence of *Stromatinia* sclerotia and seedling roots less than a month after being transplanted. Although a common technique to achieve *Trichoderma* isolation in relatively humid environments is the use of traps with sclerotia or solid substrates, in the case of rhizospheric organisms, it is necessary to use soil or directly sections of root tissue.

The analysis of how each of these isolated species affect the plant and the other associated organisms in the rhizosphere should be the starting point before deciding introduction, re-inoculation, or massive multiplication to prove the effect in field. Description of biocontrol mechanisms that each isolate exhibits may or may not coincide with those that have been determined for a species in laboratory tests or by evaluation of their genes. In the case of microorganisms, it is always important to consider biochemical and molecular specificity that each one may have, the effects of environmental interaction, their genetics, and phenomena such as coevolution with other organisms.

The ability to promote the growth of onion plants was evaluated by measuring elongation of main root. In comparing with the control, *T. asperellum* caused growth inhibition, and *T. virens* did not show a positive effect on root growth. *T. harzianum* increases the final dry weight and the root length. This was verified both in trays with sterile substrate and in seeds grown in vitro.

Upon further analysis of the impact on growth, inocula of *T. asperellum* at a final dose of 2.0×10^6 conidia/ml and 2.0×10^5 conidia/ml, as well as filtrates obtained from liquid cultures, at 100 and 50% of the initial concentration, were evaluated on onion seedlings to see their effect on rooting, root length and weight, and the aerial part. Results obtained showed that filtrate at 50% the dose and the application of conidia to 2.0×10^6 conidia/ml have a similar effect in terms of root elongation obtained and fresh and dry weight of root and aerial part (Rivera-Méndez et al. 2018a).

Some species of *Trichoderma* have been shown to be effective as biocontrol agents against pathogens typical of Amaryllidaceae, for example, *T. koningii* and *T. harzianum* have been used for the control of *S. cepivora* (Metcalf et al. 2004; Miranda et al. 2006), *S. rolfsii* (Viterbo et al. 2002), and *T. viride* against *F. oxysporum* (Coşkuntuna and Özer 2008). In vitro tests using the dual culture methodology evaluated the effectiveness of three strains mentioned against *S. cepivorum* and *S. terrestris*, and the effect of secondary metabolites diffusing in culture medium fermented in solid or in liquid was determined (Table 15.1).

In the dual culture assays, *T. virens* had the highest control percentages against both pathogens, with very similar results for *T. harzianum*. *T. asperellum* was less effective. In the tests using cellophane membranes, *T. asperellum* obtained the best percentage against *S. terrestris*, which indicates that metabolites that it secretes into culture medium have a very selective inhibitory effect toward this pathogen. Although many of isolates of *T. asperellum* have shown good percentages of inhibition against *S. cepivorum*, only 1 of 80 tested had better control than *T. harzianum* isolates. Therefore, it is likely that although this single strain has a greater capacity for control, in general terms *T. harzianum* and *T. virens* seem to be better agents for mycoparasitism (Rivera-Méndez et al. 2018b).

In this case, the observations seem to be confirmed by comparing the genomes of various *Trichoderma* species. *T. harzianum*, *T. virens*, and *T. asperellum* are quite different in terms of their strategies of mycoparasitism, root infection, growth promotion, and stimulation of plant defenses (Mukherjee et al. 2013). Although all are considered suitable species for biocontrol, their effectiveness can be very different, depending on the pathogen they face and conditions that mediate in that interaction. The above is valid for other *Trichoderma* species, which through the comparison of their transcriptomes reveal the expression of different mycoparasitism strategies (Atanasova et al. 2013).

Evaluation of performance in field tests is an important parameter to analyze the interaction effect of organism inoculated in the plants. The evaluations carried out in four cultivation cycles in experimental farms located at 2000 m altitude in garlic and onion crops showed interesting results when using the isolated *Trichoderma* species, *Bacillus licheniformis*, and a treatment based on the mixture of all microorganisms evaluated.

Garlic field tests were done during 2 consecutive years. *T. asperellum*, *T. virens*, and the mixture showed the best yields with difference over other treatments, although *T. virens* also exhibited the worst performance in plant protection against pathogens. *T. asperellum* offered the best plant protection against *S. terrestris*. In this case, *T. harzianum* and *B. licheniformis* improved the yield of control but did not equal the effectiveness of other species, although they had a good protective effect against both pathogens (Fig. 15.3).

Onion trials showed that any of microorganisms used was effective in reducing *Setophoma* infection, but the "mixed" treatment provided the best protection against *Stromatinia* while significantly improving performance (Fig. 15.4). Analysis of data from four growth cycles has shown an increase in the effectiveness of treatments that include mixtures of microorganisms. Although it is also important to verify interspecific compatibility and its effect on productivity. Although *Trichoderma* is a very versatile genus, its functions in agroecosystems must be analyzed from the point of view of communities' structure and microorganisms associated with the soil and the crops rhizosphere. Recently, Rivera-Méndez et al. (2020) show the effect of

Table 15.1 In vitro antagonism assays and enzymatic activity in two fermentation ways

Species	Dual culture assays Growth inhibition at 10 days (%)		Cellophane membranes assays Growth inhibition at 5 days (%)		Enzymatic activity in liquid medium (μmol/ml/min)	Enzymatic activity in solid substrate (μmol/ml/min)
	S. terrestris	*S. cepivorum*	*S. terrestris*	*S. cepivorum*		
T. harzianum	95 ± 0.2a	90 ± 0.2b	75.5 ± 1.2b	98.1 ± 0.1a	12	30
T. virens	100 ± 0.0a	100 ± 0.0a	65.05 ± 1.5c	96.28 ± 0.2a	13	33
T. asperellum	84 ± 0.6b	82 ± 0.4c	88.03 ± 0.6a	88.29 ± 0.3b	11	27

Values followed by different letters are significantly different according to the Fisher test ($P < 0.05$)

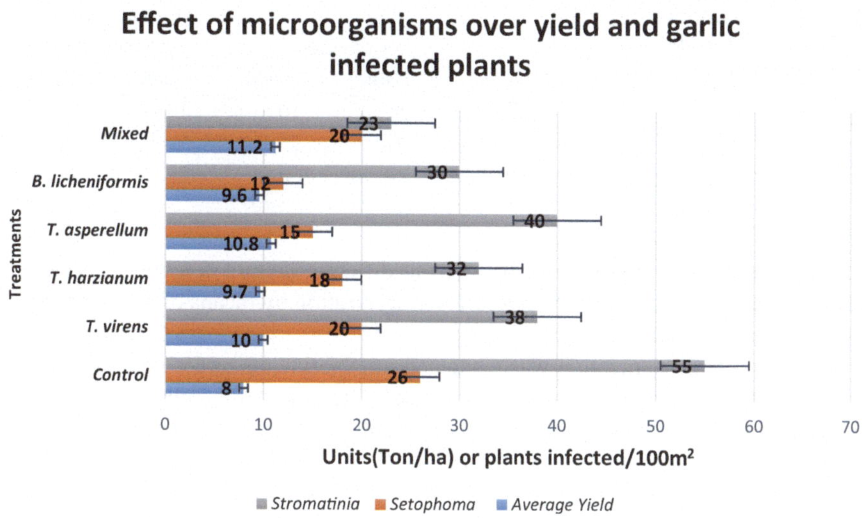

Fig. 15.3 Microorganisms' effect over yield and plant protection in field trials in garlic crop

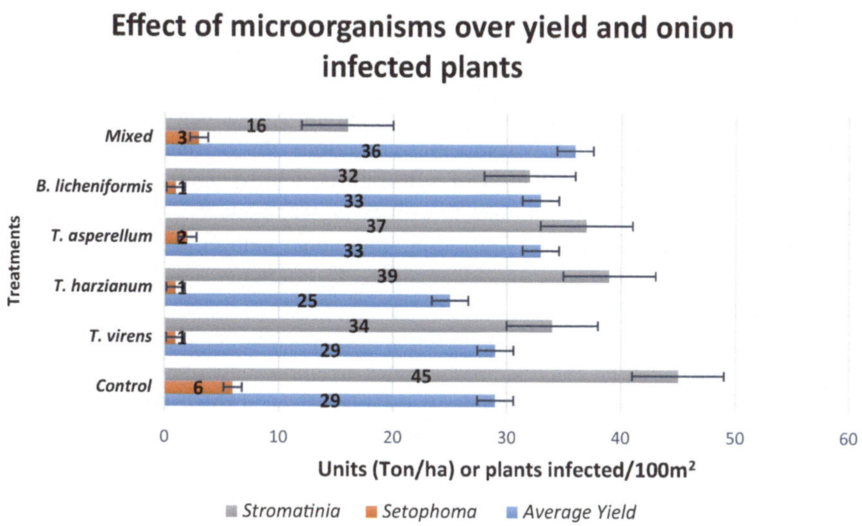

Fig. 15.4 Microorganism's effect over yield and plant protection in field trials in onion crop

T. asperellum as defense response inducer against *S. cepivorum* and the increase in the onion's final yield, in greenhouse and field assays.

An aspect that is often overlooked after inoculation processes of organisms at rhizospheric level is to verify if the organism really managed to adapt to soil conditions and, in the case of *Trichoderma*, if it also has a behavior as an endophyte. Quantification of root segments colonized by *Trichoderma* can be an indicator of

Table 15.2 Colonization percentages showed by *Trichoderma* in roots

	Field % Colonized roots		Greenhouse % Colonized roots	
Species	Garlic	Onion	Garlic	Onion
T. harzianum	20	10	45	50
T. virens	20	25	55	35
T. asperellum	45	50	75	95

inoculation success over time or by varying different culture conditions. This can also be achieved by quantifying the DNA concentration by means of real-time PCR, after obtaining an adjusted curve.

Percentage of colonization was evaluated in experiments under field and greenhouse conditions, after inoculation of onion seedlings and garlic bulbs by immersing them for 5 min in a conidia solution at a concentration of 1.0×10^7 conidia/ml and subsequent drying at room temperature for 15 h. At the end of cultivation cycle (4 months in tropical climate for onion variety F515® and a wild garlic cultivar), 100 root segments were randomly recovered at application sites of each treatment, during two growing cycles. Each segment was disinfected and cultivated to determine how many of these segments the inoculated microorganism could recover. Results obtained are shown in Table 15.2.

Colonization percentages are very different between field and greenhouse tests, as it is obvious that recovery in field trials is much lower, which is explained by the difficulty that *Trichoderma* experiences to settle in soil under natural farming conditions and exposure to environmental conditions. In greenhouse conditions that are controlled, the composition of substrate is defined, and microorganisms associated with it are part of a community with a simpler structure. *T. asperellum* had the highest percentages of root colonization, both under field or greenhouse conditions and regardless of the crop. Meanwhile, *T. harzianum* was the one that presented the lowest percentages.

After 7 years doing research on reproduction, inoculation, incorporation, and effectiveness of different *Trichoderma* isolates in garlic and onion crops, it has created a management plan to facilitate its establishment in rhizosphere. With this plan they can take advantage of capabilities of each species, either as a regulator of growth and productivity, as a stimulant of plant resistance, or as a mycoparasite. In this way, onion seeds (without any cover chemical treatment) are immersed for 5 min in a conidia solution and left to dry overnight prior to sowing in substrate. At the time of transplanting the seedlings, they are again immersed for 5 min in a conidia solution of *T. asperellum* before sowing in the soil. At 45 and 60 days after sowing, sprays of a combined conidia solution of *T. harzianum* and *T. virens* are made. The foliar fertilization and pathogen control programs have been the conventional ones (use of agrochemicals ad synthetic fertilizers). Under this model of biocontrol, a reduction of losses by white root and southern blight of 60% and a gain in fresh weight in harvest of 23% have been achieved.

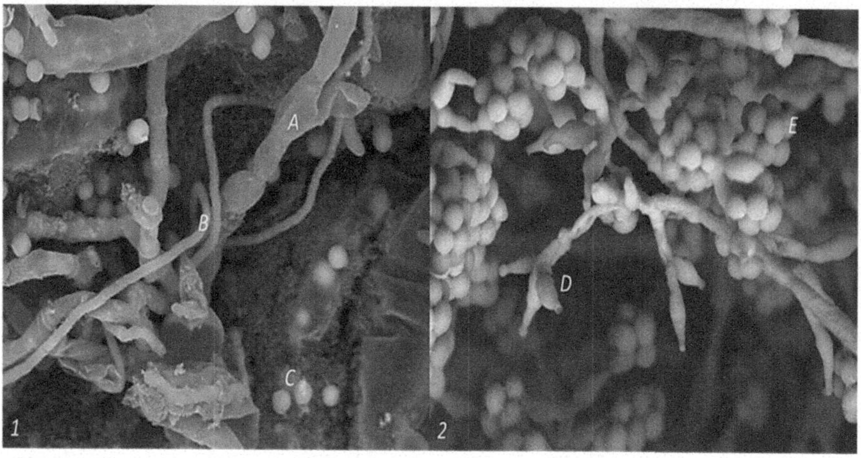

Fig. 15.5 *Trichoderma* observed from roots. (1) *T. harzianum* hyphae (A) around *S. cepivora* mycelium (B), near *T. harzianum* conidia (C) over root surface. (2) *T. asperellum* conidiophore (D) and conidia grouping (E) protruding from root tip

The ability of *Trichoderma* to survive and grow in agroecosystems can be observed with electronic microscopy images (Fig. 15.5). Soils, particularly, in onion and garlic rhizospheres frequently show *T. asperellum* hyphae over root tips and its mycelium. *T harzianum* and *T. virens* are more abundant over pathogen structures than roots.

Knowledge about *Trichoderma*'s biocontrol mechanism, its genome and secretome, and its characteristics as endophyte, saprophyte, and mycoparasite should be the base analysis for the development of strategies for application to crops that serve to reshape associated rhizospheric communities.

15.6 Use of *Trichoderma* in Sweet Pepper (*Capsicum annuum*) Rhizosphere

The sweet pepper (*Capsicum annuum*) is a vegetable very appreciated for its culinary characteristics, since it does not have spiciness and its sweet flavor comes in fruits with different shapes and colors. Its cultivation has spread widely throughout the world, and in Latin America it is a common component of local production agroecosystems and subsistence. Preferably, its cultivation is carried out under greenhouse or in small spaces in outdoor. It is part of the Solanaceae family, and because of its similarity to tomato, it is usually used in plant pathogen and biocontrol interaction research.

Many previous studies have established a beneficial effect of *Trichoderma* on several pathogens that affect this crop, especially the most representative, *Fusarium oxysporum*, *Phytophthora capsici*, and *Rhizoctonia solani*, so its application is common as part of integrated management programs of these diseases. However, its effects on regulation of plant growth and on final yield have not been clearly examined. In tropical systems, sweet pepper is a plant that can produce fruits for up to 2 years, so estimation of yields must be analyzed during a certain time interval.

A soil analysis was conducted to determine the basic composition of communities associated with sweet pepper crops. It showed a prevalence of ascomycetes of very different genera, with prevalence of the class *Sordariomycetes* (*Colletotrichum* spp. and *Fusarium* spp.). Attempts to isolate genera of beneficial fungi of same taxonomic class (such as *Trichoderma* spp. and *Purpureocillium* spp.) were unsuccessful, and in molecular tests, it was not possible to detect them in soil. Regarding bacteria, the phyla *Proteobacteria* and *Actinobacteria* predominated. Composition obtained has great similarities with those used as a reference and comes from soils of clayey-silty type, cultivated intensively and with conventional use of pesticides and fertigation by dripping. Plastic cover protected the soil, and no amendments were made with organic matter before its preparation. Kolton et al. (2011) have indicated the predominance of bacterial phyla *Proteobacteria*, *Bacteroidetes*, *Actinobacteria*, *Firmicutes*, and others, decreasing in this order, in the microbiome of sweet pepper plants. Jung et al. (2015) presented similar results, where the rhizosphere of red sweet pepper had a bacterial structure dominated by *Proteobacteria*, *Actinobacteria*, and *Firmicutes*. The descriptions of fungi associated with the rhizosphere of this crop focus their attention on pathogens, isolated either by culture or by identification with T-RFLPs, which include *Colletotrichum*, *Fusarium*, *Alternaria*, *Sclerotinia*, and *Rhizoctonia* (Kim et al. 2010).

For growth regulation assays, 12 isolates of *T. asperellum* and 3 of *T. harzianum* were used. The effect on root growth was preliminarily tested using in vitro systems, by culture in plates with use of microdroplets of a concentrated fungus solution at 1.0×10^7 conidia/ml. The results obtained were very different for all isolates, but there was also a great difference between the two species. In summary, *T. harzianum* did not exert a growth-promoting effect on root length, while *T. asperellum* varied its response with respect to the isolate. Other beneficial microorganism species such as *Bacillus subtilis*, *Penicillium pimiteouiense*, and *Bacillus licheniformis* were also tested in the trials, but the best results of root elongation and lateral root production always occurred in the interaction with *T. asperellum*.

In greenhouse experiments, effect of growth regulating of root and aerial part of plant was tested by inoculating *T. asperellum* in a substrate composed of organic fertilizer, soil, and rice husks in a 1:2:1 ratio (Brenes-Madriz et al. 2019). For this, the best of isolates obtained in the in vitro tests was used. The variety Natalie® was used for chili because it is one of the most used in Central America. The seeds were washed to eliminate chemical treatment in the cover, and they were planted in 288-well trays with a mechanical seed drill. Tests were carried out during two growing cycles per year, for 2 consecutive years. In two growing cycles of first year, four treatments were tested: a control (water), Vitavax (carboxin + captan, commercial mixture used in the sowing of seeds), a conidia solution of *T. asperellum* at 1.0×10^7 conidia/ml, and another solution at 1.0×10^6 conidia/ml. In the two cultivation cycles of the second year, two other treatments were added: a

Table 15.3 Growth regulation effect in greenhouse experiments using sweet pepper

Year	Treatments	Root length (mm)	Aerial part length (mm)	Dry root weight (mg)	Dry aerial part weight (mg)
1	Vitavax® (carboxin + captan)	106.9b	97.8[a]	153.9b	245.2c
	T. asperellum (1.0×10^6 conidia/ml)	*118.5a*	*89.0b*	*171.3a*	*311.4a*
	T. asperellum (1.0×10^7 conidia/ml)	*112.2ab*	*83.9c*	*162.9a*	*280.0b*
	Control (water)	*99.7c*	*90.8b*	*167.9a*	*280.7b*
2	Vitavax® (carboxin + captan)	*101.2b*	*90.3[a]*	*45.8d*	*114.7d*
	T. asperellum (1.0×10^6 condiai/ml)	*103.8b*	*92.6[a]*	*69.9a*	*143.0b*
	T. asperellum (1.0×10^7 conidia/ml)	*106.8ab*	*80.2b*	*47.0d*	*137.6b*
	T. asperellum filtrate 50%	*111.5a*	*88.9[a]*	*64.0a*	*163.3a*
	T. asperellum filtrate 100%	*88.7c*	*88.7[a]*	*58.3b*	*127.6c*
	Control (water)	*113.3a*	*74.8c*	*51.0c*	*109.2d*

Values followed by different letters are significantly different according to the Fisher test ($P < 0.05$)

T. asperellum filtrate without spores and a 50% v/v dilution of this filtrate. Table 15.3 represents the means obtained during trials for each treatment in each year. Each year represents two crop cycles, one in the dry season and another in rainy season, because when analyzed separately there was no difference between the values obtained.

With respect to the first-year experiments, it was observed that the greatest effect on root length was *T. asperellum* with the lowest concentration of conidia (1.0×10^6 conidia/ml), which draws attention to the issue of doses that are recommended for field applications. In individual experiments, we have verified this same effect in tomato, potato, and garlic crops, which has made it clear that observed effect is not always greater at a higher conidia dose. Several investigations have pointed in same direction (Bal and Altintas 2006), which is why it is important to work on control experiments on dose adjustment of a bioproduct.

Vitavax® is a commercial blend of carboxin and captan, used to protect seeds at planting and to facilitate plant emergence. This mixture presented the greatest length in the length of the aerial part. This is a differential effect of treatments, while *T. asperellum* allowed plant to concentrate its energy on root growth; conventional treatment gives a better appearance in initial growth stages. Analysis of root dry weights and aerial part showed again that *T. asperellum* at 1.0×10^6 conidia/ml promoted the growth. In this way it is important to note that fungus stimulated increase of dry biomass, even in aerial part where it did not have the greatest effect on size.

In other experiments, special attention was paid to results of *Trichoderma* filtrates, since it was important to see if effects were related more to conidia germination and colonization process or to effect of metabolites secreted by the fungus. The filtrate diluted to 50% had the best effects on the four variables observed, that is, length of main root, aerial part length, root dry weight, and aerial part. Although in some cases it shares significance with other treatments, especially with the use of *T. asperellum* at 1.0×10^6 conidia/ml, 100% filtration treatment was not as effective, except in aerial length. Conventional treatment, except for aerial part length, was the treatment with the least effects in other variables; in some cases, it had a lower result than control. Coherence of results between cycles and even between the 2 years of observation made it considered valid to try to evaluate effects of these treatments on performance in trials in commercial greenhouses.

The percentage of root colonization obtained after 30 days from seed inoculation and at evaluation time of other variables was evaluated. In the first year *T. asperellum* at 1.0×10^6 conidia/ml had an average of 68%, and *T. asperellum* at 1.0×10^7 conidia/ml had 73%. For the second year, *T. asperellum* at 1.0×10^6 conidia/ml obtained a percentage of 57, and *T. asperellum* at 1.0×10^7 conidia/ml had 64%. For trials conducted on greenhouse substrate, colonization was high.

These tests were conducted during 2 consecutive years evaluating 2 months of production each year. Four treatments were used the first year and six the second. The treatments were same as in the previous experiments, except that doses were adjusted to a commercial dose of ml/ha. The greenhouses were in Cartago in Costa Rica, at an altitude of 1450 m with an average temperature range between 16 and 26 °C, with an average annual rainfall of 1561 mm. The yield in grams and quality of fruits were evaluated (classified in three categories according to their length) under a planting density of 2 plants/m^2 and a conventional management for control of diseases and pests (agrochemicals) and with drip fertilization in lines of 30 plants. However, after analyzing the results, no significant differences were observed between the treatments in any of the 2 years of follow-up. Quality of fruits did not show any effect due to different treatments.

It is possible that effect of treatments based on conidia does not establish a difference with respect to others because of its low root colonization, determined when extracting the plants roots at the end of experiment (Table 15.4). Colonization percentages of 8% and 9% are low, especially when compared to those found in experiments with seedlings. It seems that this difference between colonization is determined by the method of inoculation used in seeds, in seedbeds, or in plants at the time of sowing.

Table 15.4 Yield and root colonization in greenhouse experiment with *T. asperellum* in two concentrations and its filtrates in two concentrations

Treatment	Yield per 15 m^2 (line in the greenhouse)	Calculate yield per hectare (kg)	% Colonized roots
Vitavax® (carboxin + captan)	*18.0ab*	12,000	–
T. asperellum (1.0×10^6 conidia/ml)	*21.38a*	14,253.1	8
T. asperellum (1.0×10^7 conidia/ml)	*20.55a*	13,699.8	9
T. asperellum filtrate 50%	*21.61a*	14,406.5	–
T. asperellum filtrate 100%	*18.63ab*	12,419.8	–
Control (water)	*19.63a*	13,086.5	–

Values followed by different letters are significantly different according to the Fisher test ($P < 0.05$)

15.7 Comparing the Effects of *Trichoderma* in Two Plant Models

The work carried out in field experiments, greenhouse, in vitro tests, and molecular tests has allowed us to evaluate the most convenient ways to integrate *Trichoderma* in rhizosphere of two crops with different characteristics. One monocot species (onion) and one dicotyledon species (sweet pepper), in addition to differences in plant size and cultivation system, but in a similar tropical climate and over several years of observation.

The application of different *Trichoderma* species was supported by efficacy tests that demonstrated the biocontrol mechanisms that mediate established interactions, considering the other populations of microorganisms' community and their association with a specific plant species, with its phenology and soil management during interaction. The knowledge generated in research laboratories should accompany the development of bioproducts and strategies for establishment of fungus in the soil.

All *Trichoderma* species tested in onion have been shown to have a positive effect by increasing yields and reducing incidence of infected plants; and this increase is closely related to high percentages of root colonization that were achieved. On the contrary, in sweet pepper, despite size of actual effect on production, statistical analysis of treatments does not show an influence on yield, and this could be associated with the low root colonization that was achieved with *T. asperellum.*

Achieving that *Trichoderma* exhibits its characteristics of rhizosphere and endophytic organisms is one of main challenges when establishing appropriate inoculation forms. In our experience, we have tested four main ways to inoculate crops. The first is immersion of seeds (or bulbs) in a fungus solution and then letting them dry in room temperature; once dried these can be sown. Another way is to flood the substrates with microorganism solutions and sow the seeds without inoculating them. The third way is to carry out the normal sowing in substrate, and once the seeds are covered, a conidia solution is sprinkled. However, the best result was obtained with the inoculation of the seeds and drying them overnight, and once the sowing was done, apply a solution of conidia on the substrate that covered the seeds. With this technique, percentages of colonization have been more than 50% in every single experiment.

Effect on yield in onion and garlic and the defense induction in onion, whether in greenhouse tests and in field production, has been evident for the species *T. asperellum*, although its effect as mycoparasite does not seem to be so strong. *T. virens* and *T. harzianum* exercise better control over diseases in the cultivation of onions. However, we must be careful when talking about potential of a species because we must consider differences that can occur between different isolates even in same species. Tropics seem particularly promising to find isolates with very distinctive capabilities in relatively small geographic areas (Almeida et al. 2018; Cummings et al. 2016), so bioprospecting efforts should point to potential of microclimate sites or the rhizosphere of native plants integrated into agroecosystems.

15.8 Next Steps in the Study of *Trichoderma* in Tropical Soils in Central America

Advance in understanding of biological control mechanisms deployed by *Trichoderma* in its interactions with plants and plant pathogens has made clear the role that this fungus plays in the rhizosphere, but it is necessary to deepen its role as an integrating or regulating organism, a key organism for the communities developed around the root. It is also necessary to work on inoculation strategies to ensure adequate root colonization.

Biocontrol Lab of Costa Rica Technological Institute focuses now on describing the structure of rhizospheric communities in onion and various tropical fruits, in various types of soil and under different cropping systems in tropical microclimates of Central America. For this, we use some microorganisms' indicators of health soils and pathogenicity that can help us to understand the functionality of established system. This relationship between structure and function also allows to evaluate most appropriate ways to incorporate *Trichoderma* inoculum (different species or combinations) and achieve successful colonization.

Currently, all regions are experimenting the consequences of global climatic change, but in intertropical countries modifications in weather patterns have been particularly severe, and in this situation the study of how soil microorganisms' communities are modified, their resilience, and the ability to adapt them is fundamental. Likewise, the idea of reengineering or redesigning the rhizosphere through the introduction of microorganisms should be considered an area of interest for research in agricultural microbiology.

The SynComs could be a solution for loss of soil fertility and drought problems (Johns et al. 2016; Hamilton et al. 2016). Development of microorganisms' consortiums can be used for recovery of soils affected by agrochemical residues and fertilizers, changes in salinity and in aggregation of soil particles, low fertility, and a poor percentage of organic matter. There are scientific groups worldwide applying consortiums in different kind of soils and crops with interesting results to analyze (Jain et al. 2013).

Further it is necessary to deepen the study of communities' composition using tools of metagenomics, proteomics, and secretomes analysis of many species linked with crops but also with the native plants. This, together with new sequencing tools, will allow a greater analysis of different species found throughout the world, their evolution, and the relationship of their genes.

The technology transfer to farmers and agricultural technicians plays a key role in a successful soil adaptation to new climatic conditions. Using microorganisms to help in nutrition and plant protection is a sustainable way to produce food that must be supported not only by scientist but also by politicians, traders, and consumers. Nevertheless, in the base line, farmers must receive training for an adequate selection and application of bioproducts in his crops. In countries that are qualified in ways of development, the companies that produce microorganisms like *Trichoderma* need the collaboration of research centers or the universities and supervision of government agencies. And it is vital to remember that knowledge generated should be reflected in sustainability of agroecosystems, particularly in the most fragile systems of planet, as is the case in tropics. Besides, the farmers should be seen as important actors, for which it is necessary that they have access to adequate technologies for management of water, soil, and inputs needed in production.

References

Afsharmanesh H, Ahmadzadeh M, Javan-Nikkhah M, Behboudi K (2010) Characterization of the antagonistic activity of a new indigenous strain of *Pseudomonas fluorescens* isolated from onion rhizosphere. J Plant Pathol 92:187–194

Almeida KA, Armesto C, Monteiro FP, de Souza JT (2018) Diversity of *Trichoderma* species isolated from dead branches and sapwood of *Theobroma cacao* trees. Trop Plant Pathol 43 (1):90–94

Alvarado-Marchena L, Rivera-Méndez W (2016) Molecular identification of *Trichoderma spp.* in garlic and onion fields and *in vitro* antagonism trials on *Sclerotium cepivorum*. Rev Bras Ciênc Solo 40

Atanasova L, Le Crom S, Gruber S, Coulpier F, Seidl-Seiboth V, Kubicek CP, Druzhinina IS (2013) Comparative transcriptomics reveals different strategies of En la comprensión *Trichoderma* mycoparasitism. BMC Genomics 14(1):121

Bailey BA, Bae H, Strem MD, Crozier J, Thomas SE, Samuels GJ et al (2008) Antibiosis, mycoparasitism, and colonization success for endophytic *Trichoderma* isolates with biological control potential in *Theobroma cacao*. Biol Control 46(1):24–35

Bal U, Altintas S (2006) Application of the antagonistic fungus *Trichoderma harzianum* (TrichoFlow WP™) to root zone increases yield of bell peppers grown in soil. Biol Agric Hortic 24(2):149–163

Barlow J, França F, Gardner TA, Hicks CC, Lennox GD, Berenguer E et al (2018) The future of hyperdiverse tropical ecosystems. Nature 559(7715):517

Berg G, Rybakova D, Grube M, Köberl M (2015) The plant microbiome explored: implications for experimental botany. J Exp Bot 67(4):995–1002

Bettiol W, Rivera MC, Mondino P, Montealegre JR, Colmenarez Y (2014). Control biológico de enfermedades de plantas en América Latina y el Caribe. Embrapa Meio Ambiente-Livro científico (ALICE)

Brenes-Madriz J, Zúñiga-Vega C, Villalobos-Araya M, Zúñiga-Poveda C, Rivera-Méndez W (2019) Efectos de *Trichoderma asperellum* en la estimulación del crecimiento en chile dulce (*Capsicum annum*) variedad Nathalie en ambientes protegidos. Rev Tecnol Marcha. https://doi.org/10.18845/tm.v32i3.4481

Bulgarelli D, Schlaeppi K, Spaepen S, van Themaat EVL, Schulze-Lefert P (2013) Structure and functions of the bacterial microbiota of plants. Annu Rev Plant Biol 64:807–838

Coşkuntuna A, Özer N (2008) Biological control of onion basal rot disease using *Trichoderma harzianum* and induction of antifungal compounds in onion set following seed treatment. Crop Prot 27(3–5):330–336

Cummings NJ, Ambrose A, Braithwaite M, Bissett J, Roslan HA, Abdullah J et al (2016) Diversity of root-endophytic *Trichoderma* from Malaysian Borneo. Mycol Prog 15(5):50

De Vries FT, Thébault E, Liiri M, Birkhofer K, Tsiafouli MA, Bjørnlund L et al (2013) Soil food web properties explain ecosystem services across European land use systems. Proc Natl Acad Sci 110(35):14296–14301

Fierer N (2017) Embracing the unknown: disentangling the complexities of the soil microbiome. Nat Rev Microbiol 15(10):579

Fierer N, Jackson RB (2006) The diversity and biogeography of soil bacterial communities. Proc Natl Acad Sci 103(3):626–631

Gardner T, Acosta-Martinez V, Senwo Z, Dowd SE (2011) Soil rhizosphere microbial communities and enzyme activities under organic farming in Alabama. Diversity 3(3):308–328

Geerts B, Linacre E (2002) Climates and weather explained. Routledge, London

Ghini R, Bettiol W, Hamada E (2011) Diseases in tropical and plantation crops as affected by climate changes: current knowledge and perspectives. Plant Pathol 60(1):122–132

Gomes NCM, Fagbola O, Costa R, Rumjanek NG, Buchner A, Mendona-Hagler L, Smalla K (2003) Dynamics of fungal communities in bulk and maize rhizosphere soil in the tropics. Appl Environ Microbiol 69(7):3758–3766

Green JL et al (2004) Spatial scaling of microbial eukaryote diversity. Nature 432(7018):747

Hamilton CE, Bever JD, Labbé J, Yang X, Yin H (2016) Mitigating climate change through managing constructed-microbial communities in agriculture. Agric Ecosyst Environ 216:304–308

Jacobs JL, Fasi AC, Ramette A, Smith JJ, Hammerschmidt R, Sundin GW (2008) Identification and onion pathogenicity of *Burkholderia cepacia* complex isolates from the onion rhizosphere and onion field soil. Appl Environ Microbiol 74(10):3121–3129

Jain A, Singh A, Singh BN, Singh S, Upadhyay RS, Sarma BK, Singh HB (2013) Biotic stress management in agricultural crops using microbial consortium. In: Bacteria in agrobiology: disease management. Springer, Berlin, pp 427–448

Johns NI, Blazejewski T, Gomes AL, Wang HH (2016) Principles for designing synthetic microbial communities. Curr Opin Microbiol 31:146–153

Jung BK, Sang-Dal K, Khan AR, Jong-Hui L, Kim YH, Song JH et al (2015) Rhizobacterial communities and red pepper (*Capsicum annum*) yield under different cropping systems. Int J Agric Biol 17(4):734

Kashyap PL, Rai P, Srivastava AK, Kumar S (2017) *Trichoderma* for climate resilient agriculture. World J Microbiol Biotechnol 33(8):155

Kim YT, Cho M, Jeong JY, Lee HB, Kim SB (2010) Application of terminal restriction fragment length polymorphism (T-RFLP) analysis to monitor effect of biocontrol agents on rhizosphere microbial community of hot pepper (*Capsicum annuum* L.). J Microbiol 48(5):566–572

Kolton M, Harel YM, Pasternak Z, Graber ER, Elad Y, Cytryn E (2011) Impact of biochar application to soil on the root-associated bacterial community structure of fully developed greenhouse pepper plants. Appl Environ Microbiol 7(14):4924–4930

Lamdan NL, Shalaby S, Ziv T, Kenerley CM, Horwitz BA (2015) Secretome of *Trichoderma* interacting with maize roots: role in induced systemic resistance. Mol Cell Proteomics 14 (4):1054–1063

López-Bucio J, Pelagio-Flores R, Herrera-Estrella A (2015) *Trichoderma* as biostimulant: exploiting the multilevel properties of a plant beneficial fungus. Sci Hortic 196:109–123

Martínez-Medina A, Fernández I, Sánchez-Guzmán MJ, Jung SC, Pascual JA, Pozo MJ (2013) Deciphering the hormonal signalling network behind the systemic resistance induced by *Trichoderma harzianum* in tomato. Front Plant Sci 4:206

Mayfield MM, Bonser SP, Morgan JW, Aubin I, McNamara S, Vesk PA (2010) What does species richness tell us about functional trait diversity? Predictions and evidence for responses of species and functional trait diversity to land-use change. Glob Ecol Biogeogr 19(4):423–431

Mendes LW, Tsai SM, Navarrete AA, De Hollander M, van Veen JA, Kuramae EE (2015) Soil-borne microbiome: linking diversity to function. Microb Ecol 70(1):255–265

Mendoza-Mendoza A, Zaid R, Lawry R, Hermosa R, Monte E, Horwitz BA, Mukherjee PK (2018) Molecular dialogues between *Trichoderma* and roots: role of the fungal secretome. Fungal Biol Rev 32(2):62–85

Metcalf DA, Dennis JJC, Wilson CR (2004) Effect of inoculum density of *Sclerotium cepivorum* on the ability of *Trichoderma koningii* to suppress white rot of onion. Plant Dis 88(3):287–291

Miranda MA, Estrella AH, Cabriales JP (2006) Colonization of the rhizosphere, rhizoplane and endorhiza of garlic (*Allium sativum L.*) by strains of *Trichoderma harzianum* and their capacity to control allium white-rot under field conditions. Soil Biol Biochem 38(7):1823–1830

Mukherjee M, Mukherjee PK, Horwitz BA, Zachow C, Berg G, Zeilinger S (2012) *Trichoderma*–plant–pathogen interactions: advances in genetics of biological control. Indian J Microbiol 52 (4):522–529

Mukherjee PK, Horwitz BA, Herrera-Estrella A, Schmoll M, Kenerley CM (2013) *Trichoderma* research in the genome era. Annu Rev Phytopathol 51:105–129

Nuccio EE, Anderson-Furgeson J, Estera KY, Pett-Ridge J, de Valpine P, Brodie EL, Firestone MK (2016) Climate and edaphic controllers influence rhizosphere community assembly for a wild annual grass. Ecology 97(5):1307–1318

Okoth SA, Okoth P, Muya E (2009) Influence of soil chemical and physical properties on occurrence of Trichoderma spp. in Embu, Kenya. Trop Subtrop Agroecosyst 11(2):303–312

Özer N, Köycü ND (2004) Seed-borne fungal diseases of onion, and their control. In: Fruit and vegetable diseases. Springer, Dordrecht, pp 281–306

Rivera-Méndez W (2016) Microbiological control as experience of local sustainability in Central American agriculture. Tecnol Marcha 30(4):31–40

Rivera-Méndez W, Brenes-Madriz J, Zúñiga-Vega O (2018a) Efectos de la aplicación de *Trichoderma asperellum* y su filtrado en el crecimiento de almácigos de cebolla (*Allium cepa*). Rev Tecnol Marcha 31(2):98–105

Rivera-Méndez W, Fuentes-Alfaro R, Courrau-López K, Aguilar-Ulloa W, Zúñiga-Vega C, Brenes-Madriz J (2018b) Biological control of *Setophoma terrestris* isolated from onion rhizosphere in Costa Rica. Arch Phytopathol Plant Protect:1–12. https://doi.org/10.1080/03235408.2018.1548258

Rivera-Méndez W, Obregón M, Morán-Diez ME, Hermosa R, Monte E (2020) *Trichoderma asperellum* biocontrol activity and induction of systemic defenses against *Sclerotium cepivorum* in onion plants under tropical climate conditions. Biol Control 141:104145

Sanchez PA (2019) Properties and management of soils in the tropics. Cambridge University Press, Cambridge

Serna-Chavez HM, Fierer N, Van Bodegom PM (2013) Global drivers and patterns of microbial abundance in soil. Glob Ecol Biogeogr 22(10):1162–1172

Shade A, Handelsman J (2011) Beyond the Venn diagram: the hunt for a core microbiomeemi_2585

Singh V, Ray S, Bisen K, Keswani C, Upadhyay RS, Sarma BK, Singh HB (2017) Unravelling the dual applications of *Trichoderma spp.* as biopesticide and biofertilizer. Adv PGPR Res 364

Singh P, Hussain T, Patel S, Akhtar N (2018) Impact of climate change on root–pathogen interactions. In: Root biology. Springer, Cham, pp 409–427

Uzo JO, Currah L (2018) Cultural systems and agronomic practices in tropical climates. In: Onions and allied crops: Volume II: Agronomy biotic interactions. CRC Press, Boca Raton, FL

Viterbo A, Ramot O, Chernin L, Chet I (2002) Significance of lytic enzymes from *Trichoderma spp.* in the biocontrol of fungal plant pathogens. Antonie Van Leeuwenhoek 81(1–4):549–556

Wang P, Marsh EL, Ainsworth EA, Leakey AD, Sheflin AM, Schachtman DP (2017) Shifts in microbial communities in soil, rhizosphere and roots of two major crop systems under elevated CO_2 and O_3. Sci Rep 7(1):15019

Woo SL, Ruocco M, Vinale F, Nigro M, Marra R, Lombardi N et al (2014) *Trichoderma*-based products and their widespread use in agriculture. Open Mycol J 8(1):71

Yeoh YK, Dennis PG, Paungfoo-Lonhienne C, Weber L, Brackin R, Ragan MA et al (2017) Evolutionary conservation of a core root microbiome across plant phyla along a tropical soil chronosequence. Nat Commun 8(1):215

Zachow C, Berg C, Müller H, Monk J, Berg G (2016) Endemic plants harbour specific *Trichoderma* communities with an exceptional potential for biocontrol of phytopathogens. J Biotechnol 235:162–170

Index

A. K. Sharma, P. Sharma (eds.), *Trichoderma*, Rhizosphere Biology,
https://doi.org/10.1007/978-981-15-3321-1

The manufacturer's authorised representative in the EU is Springer Nature Customer Service Centre GmbH, Europaplatz 3, 69115 Heidelberg, Germany. If you have any concerns regarding our products, please contact ProductSafety@springernature.com

Printed and bound by CPI Group (UK) Ltd, Croydon, CR0 4YY
30/09/2026
02225557-0008